Student's Solutions Manual

YOUNG • FREEDMAN
UNIVERSITY PHYSICS

ninth edition
Volume One

A. Lewis Ford
Texas A&M University

Addison-Wesley Publishing Company

Reading, Massachusetts • Menlo Park, California • New York • Don Mills, Ontario
Harlow, United Kingdom • Amsterdam • Bonn • Sydney • Singapore • Tokyo
Madrid • San Juan • Milan • Paris

Copyright © 1996 by Addison-Wesley Publishing Company, Inc.

All rights reserved. No part of this publication may be reproduced, stored in a retrieval system, or transmitted, in any form or by any means, electronic, mechanical, photocopying, recording, or otherwise, without the prior written permission of the publisher. Printed in the United States of America.

ISBN 0-201-64059-7

3 4 5 6 7 8 9 10—CRS—00999897

PREFACE

The Student's Solution Manual contains detailed solutions for approximately one-third of the Exercises and Problems in the 9th edition of University Physics. The Exercises and Problems included here were selected solely from the odd-numbered Exercises and Problems in the text, for which the answers are tabulated in the back of the textbook. The Exercises and Problems to be included were not selected at random, but rather have been carefully selected so as to include at least one representative example of each problem type. This solution manual greatly expands the set of worked-out examples that go along with the presentation of the physics laws and concepts in the text. The remaining Exercises and Problems, for which solutions are not given, constitute an ample set of problems for the students to tackle on their own. In addition, there are Challenge Problems for which no solutions are given here.

This solution manual is written for student use. A primary function of the manual is to provide the student with models to follow in working physics problems. The problems are worked out in the manual in the manner and style in which the students should carry out their own problem solutions.

The author will gratefully receive comments as to style, points of physics, errors, or anything else relating to the manual.

A.L.F.

Physics Department
Texas A and M University
College Station, TX 77843
January, 1996

CONTENTS

Chapter			Page
1	Exercises	3, 5, 7, 13, 15, 19, 21, 31, 33, 35, 37, 39, 41, 43, 47	1
	Problems	49, 53, 55, 57, 59, 63, 65, 69	
2	Exercises	3, 5, 7, 11, 15, 19, 23, 27, 29, 35, 39, 41, 43, 47	12
	Problems	51, 53, 57, 61, 63, 65, 69, 73, 75, 77	
3	Exercises	1, 3, 5, 7, 11, 15, 19, 21, 25, 29, 33, 35	28
	Problems	37, 43, 45, 49, 53, 55, 57, 59, 61, 63, 67	
4	Exercises	5, 11, 13, 15, 17, 21, 23, 25, 27, 29	48
	Problems	33, 35, 37, 41, 43, 45, 49	
5	Exercises	3, 5, 7, 11, 13, 15, 17, 19, 21, 25, 27, 31, 35, 37, 39, 45, 47, 53, 55	57
	Problems	63, 65, 67, 69, 73, 75, 77, 79, 83, 87, 93, 95, 97, 99, 101	
6	Exercises	1, 3, 7, 11, 15, 17, 21, 23, 25, 27, 29, 31, 37, 41, 45	80
	Problems	47, 49, 55, 57, 59, 61, 65, 71, 73, 77, 79	
7	Exercises	3, 7, 11, 15, 17, 19, 21, 23, 27, 31, 33, 35, 37	92
	Problems	41, 45, 47, 49, 51, 53, 55, 61, 63, 65, 69	
8	Exercises	3, 7, 11, 13, 15, 21, 23, 25, 27, 31, 33, 35, 39, 41, 43, 45, 47, 53, 57	107
	Problems	59, 65, 69, 71, 73, 77, 81, 85, 87, 89, 93	
9	Exercises	5, 7, 9, 11, 17, 19, 21, 25, 27, 31, 33, 35, 37, 43, 45, 47, 49, 51, 55	126
	Problems	59, 61, 63, 65, 67, 71, 73, 75, 83	
10	Exercises	1, 3, 5, 9, 13, 17, 19, 23, 25, 29, 31, 33, 37, 39, 43	137
	Problems	49, 51, 55, 57, 59, 61, 63, 67, 69, 73, 75, 79, 81	
11	Exercises	1, 3, 5, 9, 13, 15, 19, 23, 27, 29, 33, 35, 37	154
	Problems	41, 45, 49, 51, 61, 63, 65, 67, 69, 71, 73, 75, 77, 79	
12	Exercises	1, 3, 7, 9, 13, 17, 21, 23, 25, 29, 33, 35, 37	169
	Problems	41, 43, 45, 51, 53, 57, 59, 63, 65, 69, 71	
13	Exercises	3, 7, 9, 11, 13, 17, 19, 21, 23, 29, 31, 35, 37, 43, 45, 49	180
	Problems	53, 55, 59, 61, 63, 65, 67, 75, 77, 81, 83	
14	Exercises	3, 5, 11, 13, 17, 19, 21, 23, 27, 29, 31, 33, 35, 41, 43, 45	194
	Problems	51, 53, 55, 59, 61, 69, 71, 73, 77, 79, 87, 89, 91, 95, 97	

Chapter			Page
15	Exercises	5, 11, 17, 19, 23, 25, 27, 31, 33, 35, 37, 39, 41, 45, 47, 49, 53, 57, 59, 61, 63, 67	207
	Problems	71, 73, 75, 79, 81, 83, 87, 89, 93, 97, 99, 101, 103	
16	Exercises	3, 7, 11, 13, 15, 19, 21, 23, 27, 29, 31, 33, 39	219
	Problems	41, 45, 47, 49, 51, 55, 57, 59, 63, 67, 71	
17	Exercises	3, 5, 7, 11, 15, 17, 21, 23, 25	229
	Problems	29, 31, 35, 37, 41, 43, 45	
18	Exercises	5, 7, 9, 11, 15, 17, 21, 23, 25, 29, 35, 37	236
	Problems	39, 41, 43, 45, 53	
19	Exercises	3, 5, 9, 13, 17, 19, 21, 23, 25, 29, 31	247
	Problems	33, 35, 43	
20	Exercises	5, 7, 11, 13, 17, 19, 25, 27	253
	Problems	29, 31, 33, 35, 39, 41	
21	Exercises	1, 7, 11, 13, 15, 17, 21	260
	Problems	25, 27, 29, 31, 37	

CHAPTER 1

Exercises 3, 5, 7, 13, 15, 19, 21, 31, 33, 35, 37, 39, 41, 43, 47

Problems 49, 53, 55, 57, 59, 63, 65, 69

Exercises.

1-3

1.00 in. = 2.54 cm

$$1.00 \text{ Km} = 1.00 \text{ Km} \left(\frac{1000 \text{ m}}{1 \text{ Km}}\right)\left(\frac{100 \text{ cm}}{1 \text{ m}}\right)\left(\frac{1.00 \text{ in.}}{2.54 \text{ cm}}\right)\left(\frac{1 \text{ ft}}{12 \text{ in}}\right)\left(\frac{1 \text{ mi}}{5280 \text{ ft}}\right) = \underline{0.621 \text{ mi}}$$

1-5

a) $1450 \text{ mi/h} = 1450 \text{ mi/h} \left(\frac{1 \text{ h}}{60 \text{ min}}\right)\left(\frac{1 \text{ min}}{60 \text{ s}}\right) = \underline{0.403 \text{ mi/s}}$

b) $0.403 \text{ mi/s} = 0.403 \text{ mi/s} \left(\frac{1.609 \text{ Km}}{1 \text{ mi}}\right)\left(\frac{1000 \text{ m}}{1 \text{ Km}}\right) = \underline{648 \text{ m/s}}$

1-7

$$12.0 \text{ km/L} = (12.0 \text{ km/L})\left(\frac{0.6214 \text{ mi}}{1 \text{ Km}}\right)\left(\frac{3.788 \text{ L}}{1 \text{ gal}}\right) = \underline{28.2 \text{ mi/gal}}$$

1-13

The volume of a disk of diameter d and thickness t is $V = \pi \left(\frac{d}{2}\right)^2 t$.

The average volume is $V = \pi \left(\frac{8.50 \text{ cm}}{2}\right)^2 (0.050 \text{ cm}) = 2.837 \text{ cm}^3$. But t is given to only two significant figures, so the answer should be expressed to two significant figures: $V = \underline{2.8 \text{ cm}^3}$.

We can find the uncertainty in the volume as follows. The volume could be as large as $V = \pi \left(\frac{8.52 \text{ cm}}{2}\right)^2 (0.055 \text{ cm}) = 3.1 \text{ cm}^3$, which is 0.3 cm^3 larger than the average value. The volume could be as small as $V = \pi \left(\frac{8.48 \text{ cm}}{2}\right)^2 (0.045 \text{ cm}) = 2.5 \text{ cm}^3$, which is 0.3 cm^3 smaller than the average value. The uncertainty is $\pm 0.3 \text{ cm}^3$; $V = \underline{2.8 \pm 0.3 \text{ cm}^3}$.

1-15

a) average density $= \dfrac{m}{V} = \dfrac{m}{\frac{4}{3}\pi r^3} = \dfrac{5.69 \times 10^{26} \text{ kg}}{\frac{4}{3}\pi (6.03 \times 10^7 \text{ m})^3} = \underline{6.20 \times 10^2 \text{ kg/m}^3}$

b) $6.20 \times 10^2 \text{ kg/m}^3 = (6.20 \times 10^2 \text{ kg/m}^3)\left(\frac{1 \text{ m}}{10^2 \text{ cm}}\right)^3 \left(\frac{10^3 \text{ g}}{1 \text{ kg}}\right) = 0.620 \text{ g/cm}^3$

The average density of Saturn is less than the density of water, so <u>yes</u> it will float.

1-19

I _estimate_ that my scalp's area is about that of a 10 in. diameter circle: $d = 10 \text{ in.} \left(\frac{2.54 \text{ cm}}{1 \text{ in.}}\right)\left(\frac{10 \text{ mm}}{1 \text{ cm}}\right) = 250 \text{ mm}$

The estimated area is thus $A = \pi r^2$ with $r = \frac{d}{2} = 125 \text{ mm}$:
$A = \pi (125 \text{ mm})^2 = 5 \times 10^4 \text{ mm}^2$.

I further _estimate_ that on my scalp there are 5 hairs per mm². The number of hairs on my head is thus estimated to be about
$(5 \text{ hairs/mm}^2)(5 \times 10^4 \text{ mm}^2) = \underline{2 \times 10^5 \text{ hairs}}$.

1-21

Estimate that the pile is 18 in. × 18 in. × 5 ft 8 in., so the volume of gold in the pile is $V = 18 \text{ in.} \times 18 \text{ in.} \times 68 \text{ in.} = 22,000 \text{ in.}^3$

Convert to cm³: $V = 22,000 \text{ in.}^3 \left(\frac{1000 \text{ cm}^3}{61.02 \text{ in.}^3}\right) = 3.6 \times 10^5 \text{ cm}^3$.

From Example 1-4 the density of gold is 19.3 g/cm³, so the mass of this volume of gold is $m = (19.3 \text{ g/cm}^3)(3.6 \times 10^5 \text{ cm}^3) = 7 \times 10^6 \text{ g}$.

Convert to ounces: $m = 7 \times 10^6 \text{ g} \left(\frac{1 \text{ ounce}}{30 \text{ g}}\right) = 2 \times 10^5 \text{ ounces}$.

Also from Example 1-4, the monetary value of one ounce is $400, so the gold has a value of $(\$400/\text{ounce})(2 \times 10^5 \text{ ounces}) = \$8 \times 10^7$, or about $\$100 \times 10^6$ (one hundred million dollars).

1-31

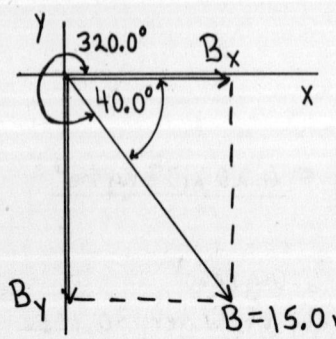

$A_x = A \cos 53.0° = (12.0 \text{ m}) \cos 53.0° = \underline{+7.22 \text{ m}}$

$A_y = A \sin 53.0° = (12.0 \text{ m}) \sin 53.0° = \underline{+9.58 \text{ m}}$

(The sketch shows that A_x and A_y should both be positive.)

$B_x = B \cos 320.0° = (15.0 \text{ m}) \cos 320.0° = \underline{+11.5 \text{ m}}$

$B_y = B \sin 320.0° = (15.0 \text{ m}) \sin 320.0° = \underline{-9.64 \text{ m}}$

(The sketch shows that B_x is positive and B_y is negative.)

1-31 (cont)

$C_x = C \cos 240.0° = (6.0 \text{ m}) \cos 240.0° = \underline{-3.0 \text{ m}}$

$C_y = C \sin 240.0° = (6.0 \text{ m}) \sin 240.0° = \underline{-5.2 \text{ m}}$

(The sketch shows that both C_x and C_y are negative.)

1-33

$A_x = -12.0 \text{ m}, \quad A_y = 0$

$B_x = B \cos 37° = (18.0 \text{ m}) \cos 37° = 14.38 \text{ m}$
$B_y = B \sin 37° = (18.0 \text{ m}) \sin 37° = 10.83 \text{ m}$

Note that both B_x and B_y are positive. A_x is negative because $\vec{A}$ is in the negative x-direction.

a) Let $\vec{R} = \vec{A} + \vec{B}$
$R_x = A_x + B_x = -12.0 \text{ m} + 14.38 \text{ m} = +2.38 \text{ m}$
$R_y = A_y + B_y = 0 + 10.83 \text{ m} = +10.83 \text{ m}$

$R = \sqrt{R_x^2 + R_y^2} = \sqrt{(2.38 \text{ m})^2 + (10.83 \text{ m})^2} = \underline{11.1 \text{ m}}$

$\tan \theta = \dfrac{R_y}{R_x} = \dfrac{10.8 \text{ m}}{2.38 \text{ m}} = 4.538 \Rightarrow \theta = \underline{77.6°}$,

measured counterclockwise from +x-axis.

b) Now let $\vec{R} = \vec{A} - \vec{B}$
$R_x = A_x - B_x = -12.0 \text{ m} - 14.38 \text{ m} = -26.38 \text{ m}$
$R_y = A_y - B_y = 0 - 10.83 \text{ m} = -10.83 \text{ m}$

Now both R_x and R_y are negative.

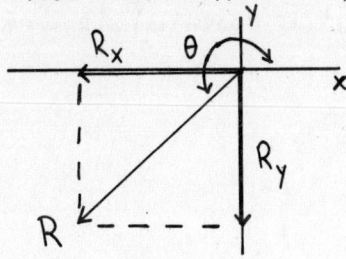

$R = \sqrt{R_x^2 + R_y^2} = \sqrt{(-26.38 \text{ m})^2 + (-10.83 \text{ m})^2} = \underline{28.5 \text{ m}}$

$\tan \theta = \dfrac{R_y}{R_x} = \dfrac{-10.83 \text{ m}}{-26.38 \text{ m}} = +0.4105 \Rightarrow \underline{\theta = 202°}$,

measured counterclockwise from +x-axis.

Note: My calculator gives $\arctan(+0.4105) = 22.3°$. But the sketch shows

1-33 (cont)
that $\vec{R}$ is in the third quadrant. The angle $22.3° + 180° = 202°$ also has a tangent of $+0.4105$ and the sketch shows that this is the correct answer.

c) $\vec{R} = \vec{B} - \vec{A}$
Then $R_x = B_x - A_x = -(A_x - B_x) = +26.38$ m and $R_y = B_y - A_y = -(A_y - B_y) = +10.83$ m, using the results of part (b). R_x and R_y now have the same magnitudes but opposite signs as in part (b), so $\vec{R}$ has the same magnitude and opposite direction as in part (b).

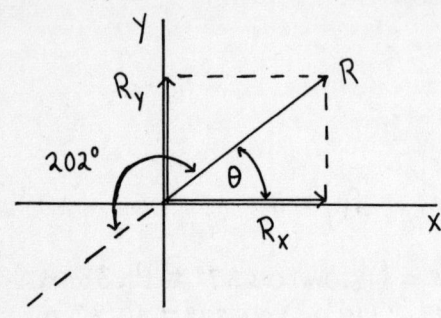

$R = 28.5$ m

$\tan \theta = \dfrac{R_y}{R_x} = +0.4105 \Rightarrow \theta = 22.3°$

1-35

$A = 4.25$ km

$B = 2.75$ km

$C = 1.50$ km

Select a coordinate system where $+x$ is east and $+y$ is north. Let $\vec{A}$, $\vec{B}$, and $\vec{C}$ be the three displacements of the professor. Then the resultant displacement $\vec{R}$ is given by $\vec{R} = \vec{A} + \vec{B} + \vec{C}$. By the method of components, $R_x = A_x + B_x + C_x$ and $R_y = A_y + B_y + C_y$. Find the x and y components of each vector; add them to find the components of the resultant. Then the magnitude and direction of the resultant can be found from its x and y components that we have calculated. As always, it is essential to draw a sketch.

$A_x = 0, \ A_y = -4.25$ km
$B_x = -2.75$ km, $B_y = 0$
$C_x = 0, \ C_y = +1.50$ km

$R_x = A_x + B_x + C_x = 0 - 2.75 \text{ km} + 0 = -2.75$ km
$R_y = A_y + B_y + C_y = -4.25 \text{ km} + 1.50 \text{ km} = -2.75$ km

1-35 (cont)

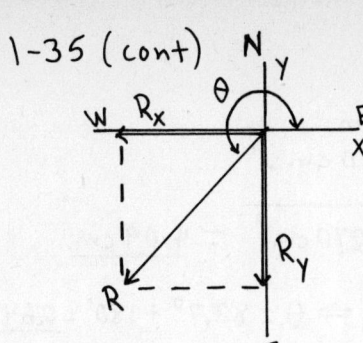

$R = \sqrt{R_x^2 + R_y^2} = \sqrt{(-2.75\,\text{km})^2 + (-2.75\,\text{km})^2} = \underline{3.89\,\text{km}}$

$\tan\theta = \dfrac{R_y}{R_x} = \dfrac{-2.75\,\text{km}}{-2.75\,\text{km}} = +1.00 \Rightarrow \theta = 225°$

The angle θ is measured counterclockwise from the $+x$-axis. In terms of compass directions, the resultant displacement is in the direction 45.0° W of S.

1-37

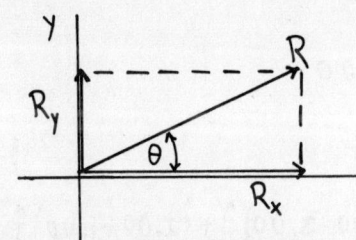

$A_x = A\cos 60.0° = (2.80\,\text{cm})\cos 60.0° = +1.40\,\text{cm}$

$A_y = A\sin 60.0° = (2.80\,\text{cm})\sin 60.0° = +2.425\,\text{cm}$

$B_x = B\cos(-60.0°) = (1.90\,\text{cm})\cos(-60.0°) = +0.95\,\text{cm}$

$B_y = B\sin(-60.0°) = (1.90\,\text{cm})\sin(-60.0°) = -1.645\,\text{cm}$

Note that the signs of the components correspond to the directions of the component vectors.

a) Let $\vec{R} = \vec{A} + \vec{B}$.
Then $R_x = A_x + B_x = +1.40\,\text{cm} + 0.95\,\text{cm} = +2.35\,\text{cm}$
$R_y = A_y + B_y = +2.425\,\text{cm} - 1.645\,\text{cm} = +0.779\,\text{cm}$

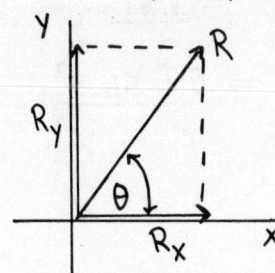

$R = \sqrt{R_x^2 + R_y^2} = \sqrt{(2.35\,\text{cm})^2 + (0.779\,\text{cm})^2} = \underline{2.48\,\text{cm}}$

$\tan\theta = \dfrac{R_y}{R_x} = \dfrac{+0.779\,\text{cm}}{+2.35\,\text{cm}} = +0.3315 \Rightarrow \underline{\theta = 18.3°}$

b) Let $\vec{R} = \vec{A} - \vec{B}$.
Then $R_x = A_x - B_x = 1.40\,\text{cm} - 0.95\,\text{cm} = 0.45\,\text{cm}$
$R_y = A_y - B_y = 2.425\,\text{cm} + 1.645\,\text{cm} = 4.070\,\text{cm}.$

$R = \sqrt{R_x^2 + R_y^2} = \sqrt{(0.45\,\text{cm})^2 + (4.070\,\text{cm})^2} = \underline{4.09\,\text{cm}}$

$\tan\theta = \dfrac{R_y}{R_x} = \dfrac{4.070\,\text{cm}}{0.45\,\text{cm}} = 9.044 \Rightarrow \underline{\theta = 83.7°}$

1-37 (cont)

c) Let $\vec{R} = \vec{B} - \vec{A}$.
Then $R_x = B_x - A_x = +0.95\,cm - 1.40\,cm = -0.45\,cm$
$R_y = B_y - A_y = -1.645\,cm - 2.425\,cm = -4.070\,cm.$

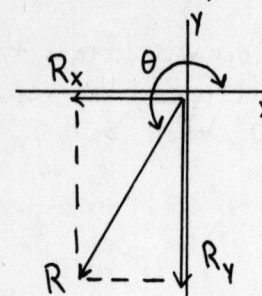

$R = \sqrt{R_x^2 + R_y^2} = \sqrt{(-0.45\,cm)^2 + (-4.070\,cm)^2} = \underline{4.09\,cm}$

$\tan\theta = \dfrac{R_y}{R_x} = \dfrac{-4.070\,cm}{-0.45\,cm} = 9.044 \Rightarrow \theta = 83.7° + 180° = \underline{264°}$

Note: $\vec{B} - \vec{A} = -(\vec{A} - \vec{B})$; $\vec{B} - \vec{A}$ and $\vec{A} - \vec{B}$ are equal in magnitude and opposite in direction.

1-39

We can use the x- and y-components of each vector as calculated in Exercise 1-31.

$\vec{A} = (+7.22\,m)\,\hat{\imath} + (9.58\,m)\,\hat{\jmath}$
$\vec{B} = (+11.5\,m)\,\hat{\imath} + (-9.64\,m)\,\hat{\jmath}$
$\vec{C} = (-3.0\,m)\,\hat{\imath} + (-5.2\,m)\,\hat{\jmath}$

1-41

a) $\vec{A} = 5.00\,\hat{\imath} + 2.00\,\hat{\jmath} \Rightarrow A_x = +5.00,\ A_y = +2.00$
$A = \sqrt{A_x^2 + A_y^2} = \sqrt{(5.00)^2 + (2.00)^2} = \underline{5.39}$

$\vec{B} = 3.00\,\hat{\imath} - 1.00\,\hat{\jmath} \Rightarrow B_x = 3.00,\ B_y = -1.00$
$B = \sqrt{B_x^2 + B_y^2} = \sqrt{(3.00)^2 + (-1.00)^2} = \underline{3.16}$

b) $\vec{A} - \vec{B} = 5.00\,\hat{\imath} + 2.00\,\hat{\jmath} - (3.00\,\hat{\imath} - 1.00\,\hat{\jmath}) = (5.00 - 3.00)\,\hat{\imath} + (2.00 + 1.00)\,\hat{\jmath}$
$\underline{\vec{A} - \vec{B} = 2.00\,\hat{\imath} + 3.00\,\hat{\jmath}}$

c) Let $\vec{R} = \vec{A} - \vec{B} = 2.00\,\hat{\imath} + 3.00\,\hat{\jmath}$. Then $R_x = 2.00,\ R_y = 3.00$.

$R = \sqrt{R_x^2 + R_y^2} = \sqrt{(2.00)^2 + (3.00)^2} = \underline{3.61}$

$\tan\theta = \dfrac{R_y}{R_x} = \dfrac{+3.00}{+2.00} = +1.50 \Rightarrow \underline{\theta = 56.3°}$

1-43

$\vec{A} = 5.00\hat{i} + 2.00\hat{j}$, $\vec{B} = 3.00\hat{i} - 1.00\hat{j}$

$\vec{A} \cdot \vec{B} = (5.00\hat{i} + 2.00\hat{j}) \cdot (3.00\hat{i} - 1.00\hat{j}) = (5.00)(3.00) + (2.00)(-1.00) = 15.0 - 2.0 = \underline{+13.0}$

1-47

$\vec{A} = 5.00\hat{i} + 2.00\hat{j}$, $\vec{B} = 3.00\hat{i} - 1.00\hat{j}$

$\vec{A} \times \vec{B} = (5.00\hat{i} + 2.00\hat{j}) \times (3.00\hat{i} - 1.00\hat{j}) = 15.0\,\hat{i}\times\hat{i} - 5.00\,\hat{i}\times\hat{j} + 6.00\,\hat{j}\times\hat{i} - 2.00\,\hat{j}\times\hat{j}$

But $\hat{i}\times\hat{i} = \hat{j}\times\hat{j} = 0$ and $\hat{i}\times\hat{j} = \hat{k}$, $\hat{j}\times\hat{i} = -\hat{k}$ so

$\vec{A} \times \vec{B} = -5.00\hat{k} + 6.00(-\hat{k}) = -11.0\hat{k}$

The magnitude of $\vec{A} \times \vec{B}$ is $|\vec{A} \times \vec{B}| = \underline{11.0}$.

Note: Sketch the vectors $\vec{A}$ and $\vec{B}$ in a coordinate system where the xy-plane is in the plane of the paper and the z-axis is directed out toward you.

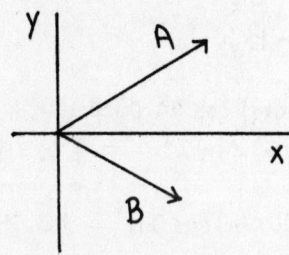

By the right-hand rule $\vec{A} \times \vec{B}$ is directed into the plane of the paper, in the $-z$-direction. This agrees with the above calculation that used unit vectors.

Problems

1-49

a) $f = 1.420 \times 10^9$ cycles/s $\Rightarrow \frac{1}{1.420 \times 10^9}$ s $= \underline{7.04 \times 10^{-10}\text{s}}$ for one cycle

b) $\frac{3600 \text{ s/h}}{7.04 \times 10^{-10} \text{s/cycle}} = \underline{5.11 \times 10^{12} \text{ cycles/h}}$

c) Calculate the number of seconds in 4600 million years $= 4.60 \times 10^9$ y and divide by the time for 1 cycle:

$\frac{(4.60 \times 10^9 \text{y})(3.156 \times 10^7 \text{s/y})}{7.04 \times 10^{-10} \text{s/cycle}} = \underline{2.06 \times 10^{26} \text{ cycles}}$

d) The clock is off by 1 s in $100{,}000 \text{ y} = 1 \times 10^5$ y, so in 4.60×10^9 y it is off by $1 \text{s} \left(\frac{4.60 \times 10^9}{1 \times 10^5} \right) = \underline{4.6 \times 10^4 \text{s}}$ (about 13 h)

1-53

a) The volume of the condo is $(75.0 \text{ m}^2)(3.10 \text{ m}) = 232.5 \text{ m}^3$.

The cost of one cubic meter is $\frac{\$380,000}{232.5 \text{ m}^3} = \underline{\$1630/\text{m}^3}$.

b) Find how much the worker makes per hour.
In one year the worker works $(40 \text{ h/wk})(50 \text{ wk/yr}) = 2000 \text{ h/yr}$.
He earns $\$60,000/\text{yr}$, so his hourly wage is $(\$60,000/\text{yr})\left(\frac{1}{2000 \text{ h/yr}}\right) = \$30/\text{h}$.

One cubic meter costs $\$1630$, so the hours he must work for one cubic meter is $\frac{\$1630/\text{m}^3}{\$30/\text{h}} = \underline{54 \text{ h/m}^3}$.

1-55

a)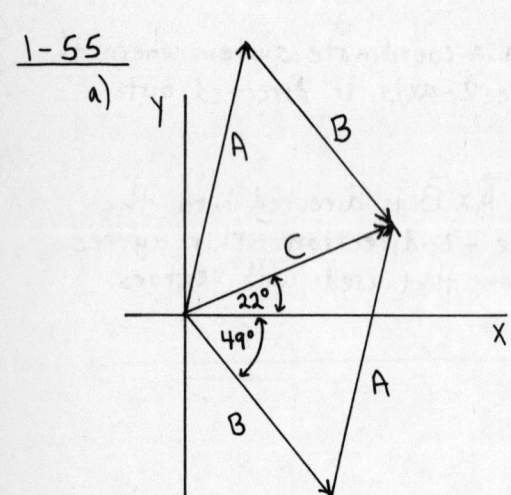

$\vec{A} + \vec{B} = \vec{C}$ (or $\vec{B} + \vec{A} = \vec{C}$)

$C_x = A_x + B_x \Rightarrow A_x = C_x - B_x$
$C_y = A_y + B_y \Rightarrow A_y = C_y - B_y$

$C_x = C \cos 22.0° = (4.80 \text{ cm}) \cos 22.0° = +4.450 \text{ cm}$
$C_y = C \sin 22.0° = (4.80 \text{ cm}) \sin 22.0° = +1.798 \text{ cm}$

$B_x = B \cos(360° - 49°) = (4.80 \text{ cm}) \cos 311° = +3.149 \text{ cm}$
$B_y = B \sin 311° = (4.80 \text{ cm}) \sin 311° = -3.623 \text{ cm}$

b) $A_x = C_x - B_x = +4.450 \text{ cm} - 3.149 \text{ cm} = \underline{+1.30 \text{ cm}}$
$A_y = C_y - B_y = +1.798 \text{ cm} - (-3.623 \text{ cm}) = \underline{+5.42 \text{ cm}}$

c)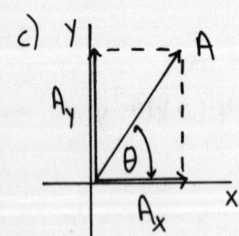

$A = \sqrt{A_x^2 + A_y^2} = \sqrt{(1.30 \text{ cm})^2 + (5.42 \text{ cm})^2} = \underline{5.57 \text{ cm}}$

$\tan \theta = \frac{A_y}{A_x} = \frac{5.42 \text{ cm}}{1.30 \text{ cm}} = 4.17$, so $\theta = \underline{76.5°}$

1-57

Use a coordinate system where east is in the +x-direction and north is in the +y-direction.
Let $\vec{A}, \vec{B}$, and $\vec{C}$ be the three displacements that are given and let $\vec{D}$ be the fourth unmeasured displacement. Then the resultant displacement is $\vec{R} = \vec{A} + \vec{B} + \vec{C} + \vec{D}$. And since he ends up back where he started, $\vec{R} = 0$.

1-57 (cont)

$0 = \vec{A} + \vec{B} + \vec{C} + \vec{D} \Rightarrow \vec{D} = -(\vec{A} + \vec{B} + \vec{C})$

$D_x = -(A_x + B_x + C_x)$ and $D_y = -(A_y + B_y + C_y)$

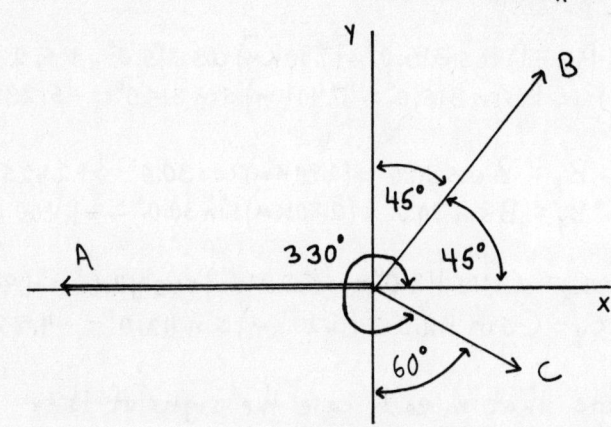

$A_x = -210 \text{ m}, \quad A_y = 0$

$B_x = B\cos 45° = (180 \text{ m})\cos 45° = +127.3 \text{ m}$
$B_y = B\sin 45° = (180 \text{ m})\sin 45° = +127.3 \text{ m}$

$C_x = C\cos 330° = (110 \text{ m})\cos 330° = +95.3 \text{ m}$
$C_y = C\sin 330° = (110 \text{ m})\sin 330° = -55.0 \text{ m}$

$D_x = -(A_x + B_x + C_x) = -(-210\text{ m} + 127.3\text{ m} + 95.3\text{ m}) = -12.6 \text{ m}$
$D_y = -(A_y + B_y + C_y) = -(0 + 127.3\text{ m} - 55.0\text{ m}) = -72.3 \text{ m}$

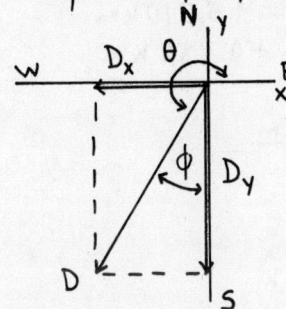

$D = \sqrt{D_x^2 + D_y^2} = \sqrt{(-12.6\text{ m})^2 + (-72.3\text{ m})^2} = \underline{73.4 \text{ m}}$

$\tan\theta = \dfrac{D_y}{D_x} = \dfrac{-72.3\text{ m}}{-12.6\text{ m}} = +5.74$

$\Rightarrow \theta = 180° + 80.1° = 260.1°$

($\vec{D}$ is in the third quadrant since both D_x and D_y are negative.)

The direction of $\vec{D}$ can also be specified in terms of $\phi = 270° - \theta = 9.9°$; $\vec{D}$ is $\underline{9.9° \text{ W of S}}$.

1-59
a)

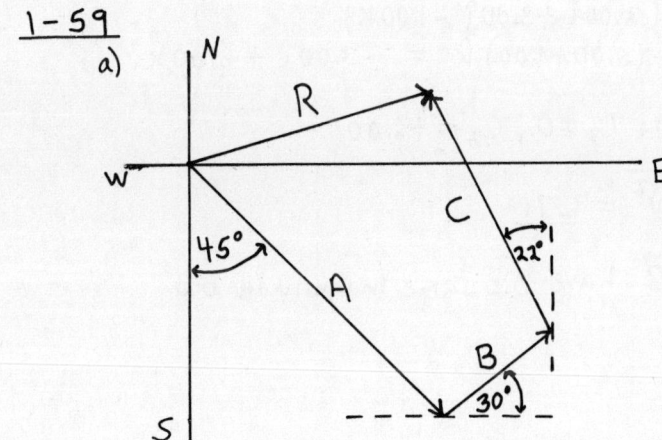

Let the three displacements that are given in the problem be called $\vec{A}$, $\vec{B}$, and $\vec{C}$.

Let the resultant displacement be $\vec{R}$; $\vec{R} = \vec{A} + \vec{B} + \vec{C}$.

Part (b) of the problem is asking for the magnitude of $\vec{R}$.

1-59 (cont)

b) Use a coordinate system where the x-direction is east and the y-direction is north.

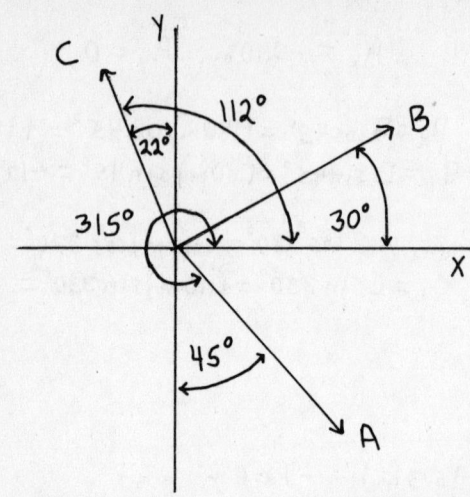

$A_x = A \cos 315.0° = (7.40 \text{ km}) \cos 315.0° = +5.233 \text{ km}$
$A_y = A \sin 315.0° = (7.40 \text{ km}) \sin 315.0° = -5.233 \text{ km}$

$B_x = B \cos 30.0° = (2.80 \text{ km}) \cos 30.0° = +2.425 \text{ km}$
$B_y = B \sin 30.0° = (2.80 \text{ km}) \sin 30.0° = +1.400 \text{ km}$

$C_x = C \cos 112.0° = (5.20 \text{ km}) \cos 112.0° = -1.948 \text{ km}$
$C_y = C \sin 112.0° = (5.20 \text{ km}) \sin 112.0° = +4.821 \text{ km}$

Note that in each case the signs of the components correspond to the directions of the component vectors.

$R_x = A_x + B_x + C_x = +5.233 \text{ km} + 2.425 \text{ km} - 1.948 \text{ km} = +5.710 \text{ km}$
$R_y = A_y + B_y + C_y = -5.233 \text{ km} + 1.400 \text{ km} + 4.821 \text{ km} = +0.988 \text{ km}$

$R = \sqrt{R_x^2 + R_y^2} = \sqrt{(+5.710 \text{ km})^2 + (0.988 \text{ km})^2} = \underline{5.79 \text{ km}}$

1-63

$\vec{A} = -1.00\hat{i} + 3.00\hat{j} + 5.00\hat{k}, \quad \vec{B} = 2.00\hat{i} + 3.00\hat{j} - 1.00\hat{k}$

a) $A = \sqrt{A_x^2 + A_y^2 + A_z^2} = \sqrt{(-1.00)^2 + (3.00)^2 + (5.00)^2} = \sqrt{35} = \underline{5.92}$

$B = \sqrt{B_x^2 + B_y^2 + B_z^2} = \sqrt{(2.00)^2 + (3.00)^2 + (-1.00)^2} = \sqrt{14} = \underline{3.74}$

b) $\vec{A} - \vec{B} = (-1.00\hat{i} + 3.00\hat{j} + 5.00\hat{k}) - (2.00\hat{i} + 3.00\hat{j} - 1.00\hat{k})$
$\vec{A} - \vec{B} = (-1.00 - 2.00)\hat{i} + (3.00 - 3.00)\hat{j} + (5.00 + 1.00)\hat{k} = -3.00\hat{i} + 6.00\hat{k}$

c) Let $\vec{C} = \vec{A} - \vec{B}$, So $C_x = -3.00$, $C_y = 0$, $C_z = +6.00$

$C = \sqrt{C_x^2 + C_y^2 + C_z^2} = \sqrt{(-3.00)^2 + (6.00)^2} = \underline{6.71}$

$\vec{B} - \vec{A} = -(\vec{A} - \vec{B})$, so $\vec{B} - \vec{A}$ and $\vec{A} - \vec{B}$ have the same magnitude but opposite directions.

1-65

Use $\cos\phi = \dfrac{\vec{A} \cdot \vec{B}}{AB}$

a) $\vec{A} = \hat{k}$ (along line ab)
$\vec{B} = \hat{i} + \hat{j} + \hat{k}$ (along line ad)

1-65 (cont)

$A = 1$, $B = \sqrt{1^2+1^2+1^2} = \sqrt{3}$

$\vec{A} \cdot \vec{B} = \hat{k} \cdot (\hat{i} + \hat{j} + \hat{k}) = 1$

So $\cos\phi = \dfrac{\vec{A} \cdot \vec{B}}{AB} = \dfrac{1}{\sqrt{3}} \Rightarrow \phi = \underline{54.7°}$

b) $\vec{A} = \hat{i} + \hat{j} + \hat{k}$ (along line ad)
$\vec{B} = \hat{j} + \hat{k}$ (along line ac)

$A = \sqrt{1^2+1^2+1^2} = \sqrt{3}$; $B = \sqrt{1^2+1^2} = \sqrt{2}$

$\vec{A} \cdot \vec{B} = (\hat{i} + \hat{j} + \hat{k}) \cdot (\hat{j} + \hat{k}) = 1 + 1 = 2$

So $\cos\phi = \dfrac{\vec{A} \cdot \vec{B}}{AB} = \dfrac{2}{\sqrt{3}\sqrt{2}} = \dfrac{2}{\sqrt{6}} \Rightarrow \phi = \underline{35.3°}$

1-69

a) $\vec{A} \cdot (\vec{B} \times \vec{C}) = A_x (\vec{B} \times \vec{C})_x + A_y (\vec{B} \times \vec{C})_y + A_z (\vec{B} \times \vec{C})_z$

$\vec{A} \cdot (\vec{B} \times \vec{C}) = A_x (B_y C_z - B_z C_y) + A_y (B_z C_x - B_x C_z) + A_z (B_x C_y - B_y C_x)$

$(\vec{A} \times \vec{B}) \cdot \vec{C} = (\vec{A} \times \vec{B})_x C_x + (\vec{A} \times \vec{B})_y C_y + (\vec{A} \times \vec{B})_z C_z$

$(\vec{A} \times \vec{B}) \cdot \vec{C} = (A_y B_z - A_z B_y) C_x + (A_z B_x - A_x B_z) C_y + (A_x B_y - A_y B_x) C_z$

Comparison of the expressions for $\vec{A} \cdot (\vec{B} \times \vec{C})$ and $(\vec{A} \times \vec{B}) \cdot \vec{C}$ show they contain the same terms, so $\vec{A} \cdot (\vec{B} \times \vec{C}) = (\vec{A} \times \vec{B}) \cdot \vec{C}$.

b) $A_x = A\cos\theta_A = (6.00)\cos 64.0° = 2.630$
$A_y = A\sin\theta_A = (6.00)\sin 64.0° = 5.393$
$A_z = 0$

$B_x = B\cos\theta_B = (4.00)\cos 28.0° = 3.532$
$B_y = B\sin\theta_B = (4.00)\sin 28.0° = 1.878$
$B_z = 0$

$C_x = 0$, $C_y = 0$, $C_z = 5.00$

$(\vec{A} \times \vec{B}) \cdot \vec{C} = (\vec{A} \times \vec{B})_z C_z = (A_x B_y - A_y B_x) C_z = ((2.630)(1.878) - (5.393)(3.532))(5.00)$

$(\vec{A} \times \vec{B}) \cdot \vec{C} = (-14.11)(5.00) = \underline{-70.5}$

CHAPTER 2

Exercises 3, 5, 7, 11, 15, 19, 23, 27, 29, 35, 39, 41, 43, 47

Problems 51, 53, 57, 61, 63, 65, 69, 73, 75, 77

Exercises

2-3

$$V_{av} = \frac{\Delta X}{\Delta t} \Rightarrow \Delta X = V_{av} \Delta t \text{ and } \Delta t = \frac{\Delta X}{V_{av}}$$

Use the information given for normal driving conditions to calculate the distance between the two cities:

$$\Delta X = V_{av} \Delta t = (96 \text{ km/h})\left(\frac{1 \text{ h}}{60 \text{ min}}\right)(130 \text{ min}) = 208 \text{ km}$$

Now use V_{av} for a rainy day to calculate Δt; ΔX is the same as before:

$$\Delta t = \frac{\Delta X}{V_{av}} = \frac{208 \text{ km}}{80 \text{ km/h}} = 2.60 \text{ h} = 2 \text{ h and } 36 \text{ min}$$

The trip takes an additional 26 min.

2-5

a) $V_{av} = \frac{\Delta X}{\Delta t}$, where ΔX is the total displacement for the entire 120 s

$$\Delta X = (8.0 \text{ m/s})(60 \text{ s}) + (24.0 \text{ m/s})(60 \text{ s}) = 1920 \text{ m}$$

$$V_{av} = \frac{1920 \text{ m}}{120 \text{ s}} = \underline{16.0 \text{ m/s}}$$

b) $V_{av} = \frac{\Delta X}{\Delta t}$

Now we know that the total displacement is $480 \text{ m} + 480 \text{ m} = 960 \text{ m}$, but we must calculate the total elapsed time Δt.

$$\Delta t = \frac{\Delta X}{V_{av}}, \text{ so } \Delta t = \frac{480 \text{ m}}{8.0 \text{ m/s}} + \frac{480 \text{ m}}{24.0 \text{ m/s}} = 60.0 \text{ s} + 20.0 \text{ s} = 80.0 \text{ s}$$

Then $V_{av} = \frac{960 \text{ m}}{80.0 \text{ s}} = \underline{12.0 \text{ m/s}}$.

c) In part (a) the numerical average of the two speeds is $\frac{8.0 \text{ m/s} + 24.0 \text{ m/s}}{2} = 16.0 \text{ m/s}$, which does equal V_{av}.

In part (b) the numerical average of the two speeds is 16.0 m/s, but this does not equal V_{av}.

The numerical average of the two speeds equals V_{av} in part (a) since there the two speeds are maintained for equal amounts of time; this is not the case in part (b).

2-7

$x = \alpha t^2 + \beta t^3$, with $\alpha = 1.50 \text{ m/s}^2$, $\beta = 0.250 \text{ m/s}^3$

$$v_{av} = \frac{\Delta x}{\Delta t} = \frac{x_2 - x_1}{t_2 - t_1}$$

We can use the equation for x as a function of t to calculate Δx.

a) $t_1 = 0 \Rightarrow x_1 = 0$
$t_2 = 2.00 \text{ s} \Rightarrow x_2 = (1.50 \text{ m/s}^2)(2.00 \text{ s})^2 + (0.250 \text{ m/s}^3)(2.00 \text{ s})^3 = 8.00 \text{ m}$

Then $v_{av} = \frac{x_2 - x_1}{t_2 - t_1} = \frac{8.00 \text{ m} - 0}{2.00 \text{ s} - 0} = \underline{4.00 \text{ m/s}}$.

b) $t_1 = 0 \Rightarrow x_1 = 0$
$t_2 = 4.00 \text{ s} \Rightarrow x_2 = (1.50 \text{ m/s}^2)(4.00 \text{ s})^2 + (0.250 \text{ m/s}^3)(4.00 \text{ s})^3 = 40.0 \text{ m}$

$v_{av} = \frac{x_2 - x_1}{t_2 - t_1} = \frac{40.0 \text{ m} - 0}{4.00 \text{ s} - 0} = \underline{10.0 \text{ m/s}}$

c) $t_1 = 2.00 \text{ s} \Rightarrow x_1 = 8.00 \text{ m}$
$t_2 = 4.00 \text{ s} \Rightarrow x_2 = 40.0 \text{ m}$

$v_{av} = \frac{x_2 - x_1}{t_2 - t_1} = \frac{40.0 \text{ m} - 8.00 \text{ m}}{4.00 \text{ s} - 2.00 \text{ s}} = \frac{32.0 \text{ m}}{2.00 \text{ s}} = \underline{16.0 \text{ m/s}}$

Note that v_{av} depends on the Δt interval.

2-11

a) The acceleration a is the slope of the v versus t curve. a has its most positive value when the curve has its largest positive slope; this occurs between approximately 4s and 7s.

b) a has its most negative value when the curve has its most negative slope; this occurs between approximately 30s to 40s.

c) At $t = 20 \text{ s}$ the v versus t curve is a horizontal straight line with zero slope, so $a = 0$.

d) Between 30s and 40s the v versus t curve is a straight line with slope $\frac{0 - 60 \text{ km/h}}{40 \text{ s} - 30 \text{ s}} = \frac{-60 \text{ km/h} \left(\frac{10^3 \text{ m}}{1 \text{ km}}\right)\left(\frac{1 \text{ h}}{3600 \text{ s}}\right)}{10 \text{ s}} = \frac{-16.67 \text{ m/s}}{10 \text{ s}} = -1.67 \text{ m/s}^2$

The acceleration is constant with this value in this time interval, so at $t = 35 \text{ s}$ it is $\underline{a = -1.67 \text{ m/s}^2}$.

e) $\underline{t = 5s}$
The average velocity for $t = 0$ to $t = 5 \text{ s}$ is approximately 15 km/h or 4.1 m/s, so $\Delta x = v_{av} \Delta t = 20 \text{ m}$; if the car is at $x_0 = 0$ at $t = 0$ then it is at $x = 20 \text{ m}$ at $t = 5 \text{ s}$.

2-11 (cont)

The velocity at $t=5s$ is 30 km/h $= 8$ m/s.

The acceleration at $t=5s$ is approximately $\frac{60 \text{km/h} - 30 \text{km/h}}{10s - 5s} \left(\frac{10^3 m}{1 km}\right)\left(\frac{1h}{3600s}\right) = +1.7 \text{m/s}^2$

$t = 15s$

In the first 10s the car has $\Delta x \approx (8 \text{m/s})(10s) = 80$ m. From 10s to 15s the car has constant speed 60 km/h = 17 m/s, so $\Delta x = (17 \text{m/s})(5s) = 85$ m. At $t=15s$ the car is at $x = 80m + 85m = 165$ m. The velocity at $t=15s$ is 17 m/s. The acceleration is zero.

$t = 25s$

For the time interval 10s to 25s the car has constant speed 17 m/s, so $\Delta x = (17 \text{m/s})(15s) = 255$ m. At $t=25s$ the car is at $x = 80m + 255m = 335$ m. The velocity is 17 m/s. The acceleration is zero.

$t = 35s$

For the time interval 10s to 30s, $\Delta x = (17 \text{m/s})(20s) = 340$ m. From $t=30s$ to 35s the average velocity is 45 km/h = 12 m/s, so $\Delta x = (12 \text{m/s})(5s) = 60$ m. At $t=35s$ the car is at $x = 80m + 340m + 60m = 480$ m. $v = 30$ km/h $= 8$ m/s. From part (d), $a = -1.7 \text{m/s}^2$.

These results allow construction of the motion diagrams.

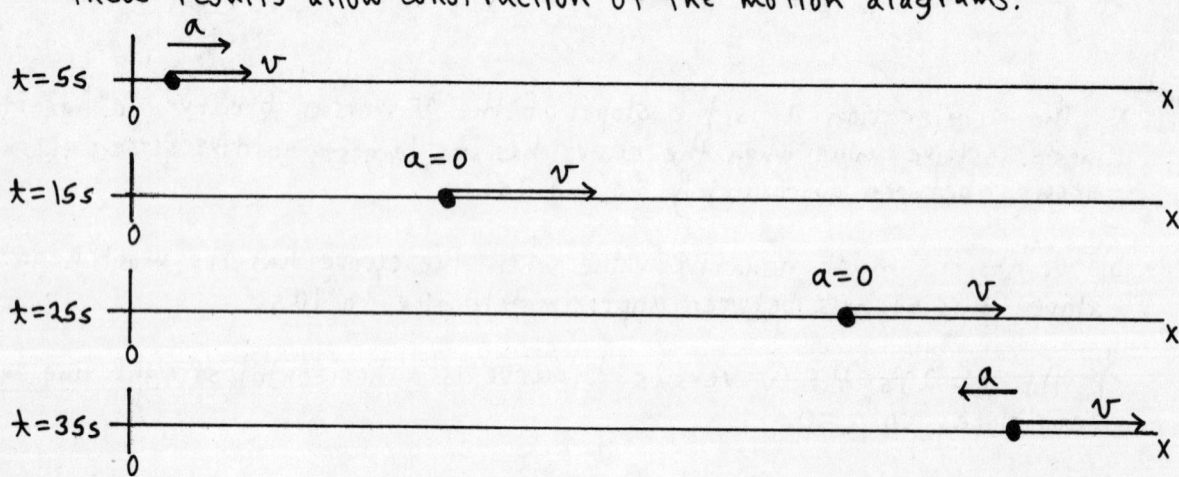

2-15

$t = 0$ to $5s$
x versus t is a parabola so a is constant. The curvature is positive so a is positive. v versus t is a straight line with positive slope. $v_0 = 0$.

$t = 5s$ to $15s$
x versus t is a straight line so v is constant and $a = 0$. The slope of x versus t is positive so v is positive.

2-15 (cont)

t = 15s to 25s   X versus t is a parabola with negative curvature, so a is constant and negative. v versus t is a straight line with negative slope. The velocity is zero at 20s, positive for 15s to 20s, and negative for 20s to 25s.

t = 25s to 35s   X versus t is a straight line with negative slope, so v is constant and negative, and a=0.

t = 35s to 40s   X versus t is a parabola with positive curvature, so a is constant and positive. v versus t is a straight line with positive slope. The velocity reaches zero at t = 40s.

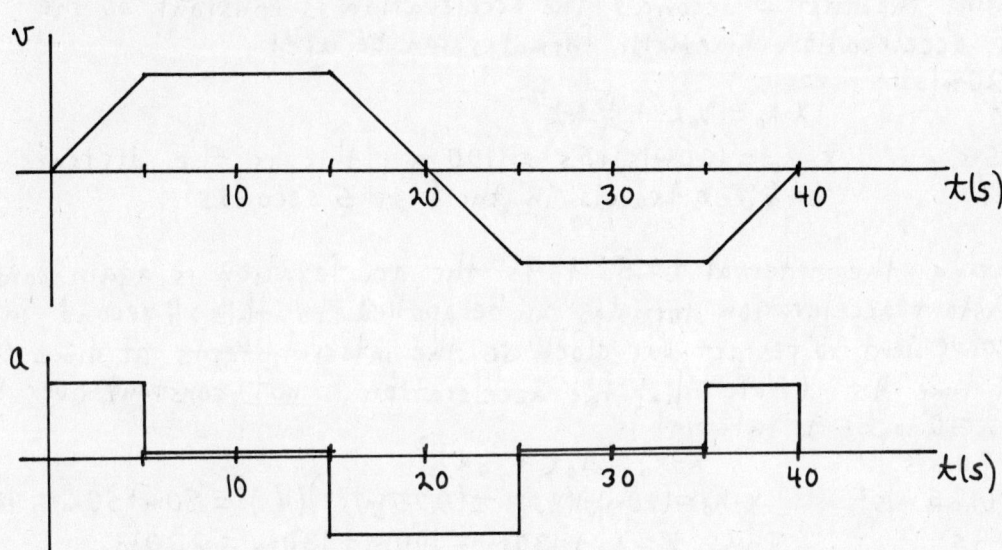

2-19

a)

$x_0 = 0$, $t = 0$
$x = 80$ m, $t = 7.00$ s
$v = 15.0$ m/s

$x - x_0 = 80$ m
$t = 7.00$ s
$v = 15.0$ m/s
$v_0 = ?$

Use $x - x_0 = \left(\frac{v_0 + v}{2}\right) t \Rightarrow v_0 = \frac{2(x-x_0)}{t} - v = \frac{2(80\text{m})}{7.00\text{s}} - 15.0 \text{ m/s} = \underline{7.86 \text{ m/s}}$

b) Use $v = v_0 + at \Rightarrow a = \frac{v - v_0}{t} = \frac{15.0 \text{ m/s} - 7.86 \text{ m/s}}{7.00 \text{ s}} = \underline{1.02 \text{ m/s}^2}$

2-23

a) The acceleration a at time t is the slope of the tangent to the v versus t curve at time t.

At t = 3s, the v versus t curve is a horizontal straight line, with zero slope. Thus $\underline{a = 0}$.

2-23 (cont)

At $t = 7s$, the v versus t curve is a straight-line segment with slope $\frac{45 m/s - 20 m/s}{9s - 5s} = 6.25 m/s^2$. Thus $\underline{a = 6.25 m/s^2}$.

At $t = 11s$ the curve is again a straight-line segment, now with slope $\frac{0 - 45 m/s}{13s - 9s} = -11.2 m/s^2$. Thus $\underline{a = -11.2 m/s^2}$.

b) For the time interval $t = 0$ to $t = 5s$ the acceleration is constant and equal to zero. For the time interval $t = 5s$ to $t = 9s$ the acceleration is constant and equal to $6.25 m/s^2$. For the interval $t = 9s$ to $t = 13s$ the acceleration is constant and equal to $-11.2 m/s^2$.

During the first 5 seconds the acceleration is constant, so the constant acceleration kinematic formulas can be used.

$v_0 = 20 m/s$
$a = 0$
$t = 5s$
$x - x_0 = ?$

$x - x_0 = v_0 t + \frac{1}{2} a t^2$
$x - x_0 = (20 m/s)(5s) = \underline{100 m}$; this is the distance the officer travels in the first 5 seconds

During the interval $t = 5s$ to $9s$ the acceleration is again constant. The constant acceleration formulas can be applied to this 4 second interval. It is convenient to restart our clock so the interval starts at time 0 and ends at time 4s. (Note that the acceleration is <u>not</u> constant over the entire $t = 0$ to $t = 9s$ interval.)

$v_0 = 20 m/s$
$a = 6.25 m/s^2$
$t = 4s$
$x_0 = 100 m$
$x - x_0 = ?$

$x - x_0 = v_0 t + \frac{1}{2} a t^2$
$x - x_0 = (20 m/s)(4s) + \frac{1}{2}(6.25 m/s^2)(4s)^2 = 80m + 50m = 130 m$
Thus $x = x_0 + 130 m = 100 m + 130 m = 230 m$.
At $t = 9s$ the officer is at $x = 230 m$, so she has traveled $\underline{230 m}$ in the first 9 seconds.

During the interval $t = 9s$ to $t = 13s$ the acceleration is again constant. The constant acceleration formulas can be applied for this 4 second interval but <u>not</u> for the whole $t = 0$ to $t = 13s$ interval. To use the equations restart our clock so this interval begins at time 0 and ends at time 4s.

$v_0 = 45 m/s$ (at the start of the time interval)
$a = -11.2 m/s^2$
$t = 4s$
$x_0 = 230 m$
$x - x_0 = ?$

$x - x_0 = v_0 t + \frac{1}{2} a t^2$
$x - x_0 = (45 m/s)(4s) + \frac{1}{2}(-11.2 m/s^2)(4s)^2$
$x - x_0 = 180 m - 89.6 m = 90.4 m$
Thus $x = x_0 + 90.4 m = 230 m + 90.4 m = 320 m$.

At $t = 13s$ the officer is at $x = 320 m$, so she traveled $\underline{320 m}$ in the first 13 seconds.

2-27

a) The maximum speed occurs at the end of the initial acceleration period.
$a = 20.0 \text{ m/s}^2$
$t = 10.0 \text{ min} = 600 \text{ s}$
$v_0 = 0$
$v = ?$

$v = v_0 + at$
$v = 0 + (20.0 \text{ m/s}^2)(600 \text{ s}) = 12.0 \times 10^3 \text{ m/s}$

b) The motion consists of three constant acceleration intervals. In the middle segment of the trip $a=0$ and $v = 12.0 \times 10^3 \text{ m/s}$, but we can't directly find the distance traveled during this part of the trip because we don't know the time. Instead, find the distance traveled in the first ($a = +20.0 \text{ m/s}^2$) part of the trip and in the last ($a = -20.0 \text{ m/s}^2$) part of the trip. Subtract these two distances from the total distance of 4.0×10^5 km to find the distance traveled in the middle ($a=0$) part of the trip.

first segment
$x - x_0 = ?$
$t = 10.0 \text{ min} = 600 \text{ s}$
$a = +20.0 \text{ m/s}^2$
$v_0 = 0$

$x - x_0 = v_0 t + \frac{1}{2} a t^2$
$x - x_0 = \frac{1}{2}(20.0 \text{ m/s}^2)(600 \text{ s})^2$
$x - x_0 = 3.60 \times 10^6 \text{ m} = 3.60 \times 10^3 \text{ km}$

third segment
$x - x_0 = ?$
$t = 10.0 \text{ min} = 600 \text{ s}$
$a = -20.0 \text{ m/s}^2$
$v_0 = 12.0 \times 10^3 \text{ m/s}$

$x - x_0 = v_0 t + \frac{1}{2} a t^2$
$x - x_0 = (12.0 \times 10^3 \text{ m/s})(600 \text{ s}) + \frac{1}{2}(-20.0 \text{ m/s}^2)(600 \text{ s})^2$
$x - x_0 = 7.20 \times 10^6 \text{ m} - 3.60 \times 10^6 \text{ m} = 3.60 \times 10^6 \text{ m}$
$x - x_0 = 3.60 \times 10^3 \text{ km}$
(The same distance as in the first segment.)

Therefore the distance traveled at constant speed is
$4.00 \times 10^5 \text{ km} - 3.60 \times 10^3 \text{ km} - 3.60 \times 10^3 \text{ km} = 3.928 \times 10^5 \text{ km} = 3.928 \times 10^8 \text{ m}$

The fraction this is of the total distance is $\frac{3.928 \times 10^5 \text{ km}}{4.00 \times 10^5 \text{ km}} = \underline{0.982}$

c) Find the time for the constant speed segment:
$x - x_0 = 3.928 \times 10^8 \text{ m}$
$v = 12.0 \times 10^3 \text{ m/s}$
$a = 0$
$t = ?$

$x - x_0 = v_0 t + \frac{1}{2} a t^2$
$t = \frac{x - x_0}{v_0} = \frac{3.928 \times 10^8 \text{ m}}{12.0 \times 10^3 \text{ m/s}} = 3.273 \times 10^4 \text{ s}$

The total time for the whole trip is thus
$600 \text{ s} + 3.273 \times 10^4 \text{ s} + 600 \text{ s} = (3.393 \times 10^4 \text{ s})\left(\frac{1 \text{ h}}{3600 \text{ s}}\right) = \underline{9.43 \text{ h}}$

2-29

a) For the car v is constant and $a=0$, so x versus t is a straight line with positive slope. For the motorcycle $v_0 = 0$ and a is constant and positive, so x versus t is a parabola with positive curvature and zero slope at $t=0$. Let both vehicles be at $x_0 = 0$ at $t = 0$.

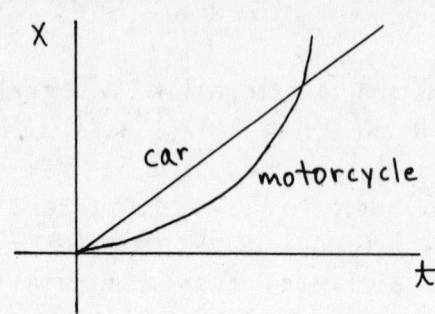

Let T be the time t when the motorcycle overtakes the car and let d be the displacement of the car (and motorcycle) when this occurs.

car

$v_0 = v_c$
$a = 0$
$t = T$
$x - x_0 = d$

$x - x_0 = v_0 t + \tfrac{1}{2} a t^2$

$\boxed{d = v_c T}$

motorcycle

$v_0 = 0$
$a = a_M$
$t = T$
$x - x_0 = d$

$x - x_0 = v_0 t + \tfrac{1}{2} a t^2$

$\boxed{d = \tfrac{1}{2} a_M T^2}$

Combine these two equations to eliminate d: $v_c T = \tfrac{1}{2} a_M T^2$

$$T = \frac{2 v_c}{a_M}$$

Then $v = v_0 + at$ for the motorcycle gives $v_M = 0 + a_M \left(\frac{2 v_c}{a_M} \right) = 2 v_c$

when the motorcycle has overtaken the car.

b) From part (a), $d = v_c T = v_c \left(\frac{2 v_c}{a_M} \right) = \frac{2 v_c^2}{a_M}$ and $v_c^2 = \frac{a_M d}{2}$.

Apply $v^2 = v_0^2 + 2a(x - x_0)$ to the motorcycle $\Rightarrow v_M^2 = 2 a_M (x - x_0)$.
Set $v_M = v_c$; then $x - x_0$ is the distance the motorcycle has traveled when its velocity equals the velocity of the car.

$$x - x_0 = \frac{v_M^2}{2 a_M} = \frac{v_c^2}{2 a_M} = \frac{1}{2 a_M} \left(\frac{a_M d}{2} \right) = \frac{d}{4}.$$

The motorcycle has traveled a distance $\frac{d}{4}$ when its velocity equals v_c.

2-35

Take the origin at the ground and the positive direction to be upward.

a) At the maximum height $v = 0$.

$v = 0$
$y - y_0 = 0.520 \text{ m}$
$a = -9.80 \text{ m/s}^2$
$v_0 = ?$

$v^2 = v_0^2 + 2a(y - y_0)$

$v_0 = \sqrt{-2a(y - y_0)} = \sqrt{-2(-9.80 \text{ m/s}^2)(0.520 \text{ m})}$

$v_0 = 3.19 \text{ m/s}$

2-35 (cont)
b) When the flea has returned to the ground $y-y_0 = 0$.

$y-y_0 = 0$
$v_0 = +3.19 \text{ m/s}$
$a = -9.80 \text{ m/s}^2$
$t = ?$

$y-y_0 = v_0 t + \frac{1}{2}at^2$
$0 = v_0 t + \frac{1}{2}at^2$
$t = -\frac{2v_0}{a} = -\frac{2(3.19 \text{ m/s})}{-9.80 \text{ m/s}^2} = \underline{0.651 \text{ s}}$

2-39
a) $v_{av} = \frac{\Delta y}{\Delta t} = \frac{y-y_0}{t}$

We need to use the constant acceleration formulas to find t. Take the origin of coordinates at the roof and take the positive y-direction to be upward.

$v_0 = +6.00 \text{ m/s}$
$y-y_0 = -12.0 \text{ m}$
$a = -9.80 \text{ m/s}^2$
$t = ?$

First find v:
$v^2 = v_0^2 + 2a(y-y_0)$
$v = -\sqrt{v_0^2 + 2a(y-y_0)}$
(We know that v is negative since the rock is traveling downward when it reaches the ground.)
$v = -\sqrt{(6.00 \text{ m/s})^2 + 2(-9.80 \text{ m/s}^2)(-12.0 \text{ m})} = -16.5 \text{ m/s}$

Now use $v = v_0 + at$ to solve for t:
$t = \frac{v-v_0}{a} = \frac{-16.5 \text{ m/s} - 6.00 \text{ m/s}}{-9.80 \text{ m/s}^2} = 2.296 \text{ s}$

Finally, $v_{av} = \frac{y-y_0}{t} = \frac{-12.0 \text{ m}}{2.296 \text{ s}} = -5.23 \text{ m/s}$.
The average velocity has magnitude 5.23 m/s and from the minus sign we see that it is directed downward.

b) $a_{av} = \frac{\Delta v}{\Delta t} = \frac{v-v_0}{t} = \frac{-16.5 \text{ m/s} - 6.00 \text{ m/s}}{2.296 \text{ s}} = -9.80 \text{ m/s}^2$

The average acceleration has magnitude 9.80 m/s² and is directed downward. (The acceleration is <u>constant</u> and equal to -9.80 m/s^2, so the average value must equal this constant value.)

2-41

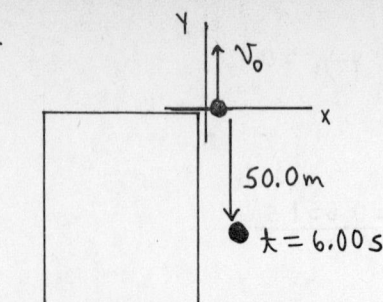

Take the positive y-direction to be upward, and the origin of coordinates at the cornice.

a) Consider the motion from the initial point to 6.00s later:
$v_0 = ?$
$t = 6.00s$
$y - y_0 = -50.0 m$
$a = -9.80 m/s^2$

$y - y_0 = v_0 t + \frac{1}{2} a t^2$

$v_0 = \frac{y-y_0}{t} - \frac{1}{2} a t = \frac{-50.0 m}{6.00s} - \frac{1}{2}(-9.80 m/s^2)(6.00s)$

$v_0 = -8.33 m/s + 29.4 m/s = \underline{21.1 m/s}$

b) Consider the motion from the initial point to the highest point. At the highest point $v = 0$.
$v = 0$
$a = -9.80 m/s^2$
$v_0 = +21.1 m/s$
$y - y_0 = ?$

$v^2 = v_0^2 + 2a(y-y_0)$

$y - y_0 = \frac{v^2 - v_0^2}{2a} = \frac{0^2 - (21.1 m/s)^2}{2(-9.80 m/s^2)} = \underline{+22.7 m}$

c) $v = 0$ at the maximum height.

d) In free-fall $a = 9.80 m/s^2$, downward at <u>all</u> points in the motion.

2-43

a) $v = 447 m/s$
$v_0 = 0$
$t = 1.80s$
$a = ?$

$v = v_0 + at$

$a = \frac{v - v_0}{t} = \frac{447 m/s - 0}{1.80s} = \underline{248 m/s^2}$

b) $\frac{a}{g} = \frac{248 m/s^2}{9.80 m/s^2} = \underline{25.3}$

c) $x - x_0 = v_0 t + \frac{1}{2} a t^2 = 0 + \frac{1}{2}(248 m/s^2)(1.80s)^2 = \underline{402 m}$

d) Calculate the acceleration, assuming that it is constant:
$t = 1.40s$
$v_0 = 283 m/s$
$v = 0$ (stops)
$a = ?$

$v = v_0 + at$

$a = \frac{v - v_0}{t} = \frac{0 - 283 m/s}{1.40s} = -202 m/s^2$

$\frac{a}{g} = \frac{-202 m/s^2}{9.80 m/s^2} = -20.6 \Rightarrow a = -20.6 g$

If the acceleration while the sled is stopping is constant then the magnitude of the acceleration is only 20.6 g. But if the acceleration is not constant, it is certainly possible that at some point the instantaneous acceleration could be as large as 40g.

2-47

$a = At - Bt^2$ with $A = 1.20 \text{ m/s}^3$ and $B = 0.120 \text{ m/s}^4$

a) $v = v_0 + \int_0^t a \, dt = v_0 + \int_0^t (At - Bt^2) \, dt = v_0 + \frac{1}{2}At^2 - \frac{1}{3}Bt^3$

At rest at $t=0 \Rightarrow v_0 = 0$, so
$v = \frac{1}{2}At^2 - \frac{1}{3}Bt^3 = \frac{1}{2}(1.20 \text{ m/s}^3)t^2 - \frac{1}{3}(0.120 \text{ m/s}^4)t^3$

$$\boxed{v = (0.60 \text{ m/s}^3)t^2 - (0.040 \text{ m/s}^4)t^3}$$

$x = x_0 + \int_0^t v \, dt = x_0 + \int_0^t (\frac{1}{2}At^2 - \frac{1}{3}Bt^3) \, dt = x_0 + \frac{1}{6}At^3 - \frac{1}{12}Bt^4$

At the origin at $t=0 \Rightarrow x_0 = 0$.
$x = \frac{1}{6}At^3 - \frac{1}{12}Bt^4 = \frac{1}{6}(1.20 \text{ m/s}^3)t^3 - \frac{1}{12}(0.120 \text{ m/s}^4)t^4$

$$\boxed{x = (0.20 \text{ m/s}^3)t^3 - (0.010 \text{ m/s}^4)t^4}$$

b) At the time t when v is maximum, $\frac{dv}{dt} = 0$. (Since $a = \frac{dv}{dt}$, the maximum velocity is when $a=0$. For earlier times a is positive so v is still increasing. For later times a is negative and v is decreasing.)

$a = \frac{dv}{dt} = 0 \Rightarrow At - Bt^2 = 0$

$t = \frac{A}{B} = \frac{1.20 \text{ m/s}^3}{0.120 \text{ m/s}^4} = 10.0 \text{ s}$

Then $v = (0.60 \text{ m/s}^3)t^2 - (0.040 \text{ m/s}^4)t^3$ gives
$v = (0.60 \text{ m/s}^3)(10.0 \text{ s})^2 - (0.040 \text{ m/s}^4)(10.0 \text{ s})^3 = 60 \text{ m/s} - 40 \text{ m/s} = \underline{20 \text{ m/s}}$

Problems

2-51

$a_{av} = \frac{\Delta v}{\Delta t} = \frac{v - v_0}{t}$

$v_0 = 0$ since the runner starts from rest.
$t = 4.0 \text{ s}$, but we need to calculate v, the speed of the runner at the end of the acceleration period.

For the last $9.1 \text{ s} - 4.0 \text{ s} = 5.1 \text{ s}$ the acceleration is zero and the runner travels a distance of $d_1 = (5.1 \text{ s}) v$ (using $x - x_0 = v_0 t + \frac{1}{2}at^2$).
During the acceleration phase of 4.0 s, where the velocity goes from 0 to v, the runner travels a distance
$d_2 = (\frac{v_0 + v}{2}) t = \frac{v}{2}(4.0 \text{ s}) = (2.0 \text{ s}) v$

The total distance traveled is 100 m, so $d_1 + d_2 = 100 \text{ m}$. This gives
$(5.1 \text{ s}) v + (2.0 \text{ s}) v = 100 \text{ m}$

2-51 (cont)

$$v = \frac{100 \text{ m}}{7.1 \text{ s}} = 14.08 \text{ m/s}$$

Now we can calculate a_{av}:

$$a_{av} = \frac{v - v_0}{t} = \frac{14.08 \text{ m/s} - 0}{4.0 \text{ s}} = \underline{3.5 \text{ m/s}^2}$$

2-53

Take the origin to be at Seward and the positive direction to be west.

a) average speed = $\frac{\text{distance traveled}}{\text{time}}$

The distance traveled (different from the net displacement $x - x_0$) is $76 \text{ km} + 34 \text{ km} = 110 \text{ km}$.

Find the total elapsed time by using $v_{av} = \frac{\Delta x}{\Delta t} = \frac{x - x_0}{t}$ to find t for each leg of the journey.

Seward to Auora: $t = \frac{x - x_0}{v_{av}} = \frac{76 \text{ km}}{88 \text{ km/h}} = 0.8636 \text{ h}$

Auora to York: $t = \frac{x - x_0}{v_{av}} = \frac{-34 \text{ km}}{-72 \text{ km/h}} = 0.4722 \text{ h}$

Total $t = 0.8636 \text{ h} + 0.4722 \text{ h} = 1.336 \text{ h}$.

$$\text{average speed} = \frac{110 \text{ km}}{1.336 \text{ h}} = \underline{82.3 \text{ km/h}}$$

b) $v_{av} = \frac{\Delta x}{\Delta t}$, where Δx is the displacement, not the total distance traveled.

For the whole trip he ends up $76 \text{ km} - 34 \text{ km} = 42 \text{ km}$ west of his starting point.

$$v_{av} = \frac{42 \text{ km}}{1.336 \text{ h}} = \underline{31.4 \text{ km/h}}$$

2-57

Let T be the time when you catch up with the cockroach. Take $x = 0$ to be at the $t = 0$ location of the roach and positive x to be in the direction of motion of the two objects.

roach	you
$v_0 = 1.50 \text{ m/s}$	$v_0 = 0.90 \text{ m/s}$
$a = 0$	$x_0 = -0.80 \text{ m}$
$x_0 = 0$	$x = 1.20 \text{ m}$
$x = 1.20 \text{ m}$	$t = T$
$t = T$	$a = ?$

Apply $x - x_0 = v_0 t + \frac{1}{2} a t^2$ to both objects:

2-57 (cont)

roach $1.20 \text{ m} = (1.50 \text{ m/s})T \Rightarrow T = 0.800 \text{ s}$

you $1.20 \text{ m} - (-0.80 \text{ m}) = (0.90 \text{ m/s})T + \frac{1}{2}aT^2$
$2.00 \text{ m} = (0.90 \text{ m/s})(0.800 \text{ s}) + \frac{1}{2}a(0.800 \text{ s})^2$
$2.00 \text{ m} = 0.72 \text{ m} + (0.320 \text{ s}^2)a$
$a = \underline{4.0 \text{ m/s}^2}$

2-61

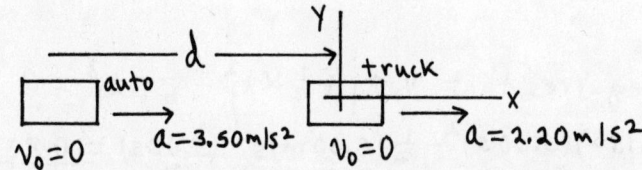

Take the origin of coordinates to be at the initial position of the truck. Let d be the distance that the auto initially is behind the truck, so $x_0(\text{auto}) = -d$ and $x_0(\text{truck}) = 0$. Let T be the time it takes the auto to catch the truck. Thus at time T the truck has undergone a displacement $x - x_0 = 60.0 \text{ m}$, so is at $x = x_0 + 60.0 \text{ m} = 60.0 \text{ m}$. The auto has caught the truck so at time T it is also at $x = 60.0 \text{ m}$.

a) Use the motion of the truck to calculate T:
$x - x_0 = 60.0 \text{ m}$ $x - x_0 = v_0 t + \frac{1}{2}at^2$
$v_0 = 0$ (starts from rest) $t = \sqrt{\frac{2(x-x_0)}{a}}$
$a = 2.20 \text{ m/s}^2$
$t = T$ $T = \sqrt{\frac{2(60.0 \text{ m})}{2.20 \text{ m/s}^2}} = \underline{7.39 \text{ s}}$

b) Use the motion of the auto to calculate d:
$x - x_0 = 60.0 \text{ m} + d$ $x - x_0 = v_0 t + \frac{1}{2}at^2$
$v_0 = 0$ $d + 60.0 \text{ m} = \frac{1}{2}(3.50 \text{ m/s}^2)(7.39 \text{ s})^2$
$a = 3.50 \text{ m/s}^2$
$t = 7.39 \text{ s}$ $d = 95.6 \text{ m} - 60.0 \text{ m} = \underline{35.6 \text{ m}}$

c) auto: $v = v_0 + at = 0 + (3.50 \text{ m/s}^2)(7.39 \text{ s}) = \underline{25.9 \text{ m/s}}$

   truck: $v = v_0 + at = 0 + (2.20 \text{ m/s}^2)(7.39 \text{ s}) = \underline{16.3 \text{ m/s}}$

d)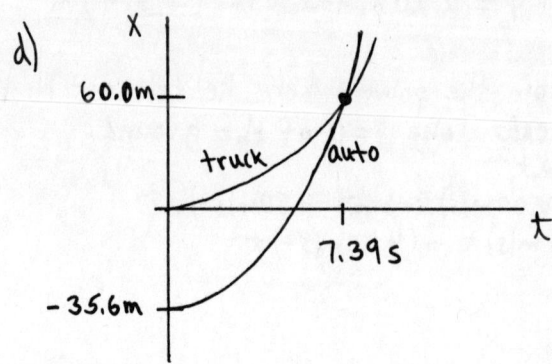

2-63

$a(t) = \alpha - \beta t$, with $\alpha = 3.00 \, m/s^2$ and $\beta = 2.00 \, m/s^3$

a) Find $v(t)$ and $x(t)$:

$v = v_0 + \int_0^t a \, dt = v_0 + \int_0^t (\alpha - \beta t) \, dt = v_0 + \alpha t - \frac{1}{2} \beta t^2$

$x = x_0 + \int_0^t v \, dt = x_0 + \int_0^t (v_0 + \alpha t - \frac{1}{2} \beta t^2) \, dt = x_0 + v_0 t + \frac{1}{2} \alpha t^2 - \frac{1}{6} \beta t^3$

At $t = 0$, $x = x_0$.

To have $x = x_0$ at $t_1 = 5.00 \, s$ requires that $v_0 t_1 + \frac{1}{2} \alpha t_1^2 - \frac{1}{6} \beta t_1^3 = 0$

$v_0 = \frac{1}{6} \beta t_1^2 - \frac{1}{2} \alpha t_1 = \frac{1}{6}(2.00 \, m/s^3)(5.00 \, s)^2 - \frac{1}{2}(3.00 \, m/s^2)(5.00 \, s) = \underline{0.833 \, m/s}$

b) With v_0 as calculated in part (a) and $t = 5.00 \, s$,

$v = v_0 + \alpha t - \frac{1}{2} \beta t^2 = 0.833 \, m/s + (3.00 \, m/s^2)(5.00 \, s) - \frac{1}{2}(2.00 \, m/s^3)(5.00 \, s)^2 = \underline{-9.17 \, m/s}$

2-65

Take positive y to be upward.

a) Consider the motion from when he applies the acceleration to when the shot leaves his hand.

$v_0 = 0$
$v = ?$
$a = 5.00 \, m/s^2$
$y - y_0 = 0.640 \, m$

$v^2 = v_0^2 + 2a(y - y_0)$

$v = +\sqrt{2a(y-y_0)} = \sqrt{2(5.0 \, m/s^2)(0.640 \, m)} = \underline{8.00 \, m/s}$

b) Consider the motion of the shot from the point where he releases it to its maximum height, where $v = 0$. Take $y = 0$ at the ground.

$y_0 = 2.20 \, m$
$y = ?$
$a = -9.80 \, m/s^2$ (free fall)
$v_0 = 8.00 \, m/s$ (from part (a))
$v = 0$ (at maximum height)

$v^2 = v_0^2 + 2a(y - y_0)$

$y - y_0 = \frac{v^2 - v_0^2}{2a} = \frac{0 - (8.00 \, m/s)^2}{2(-9.80 \, m/s^2)} = 3.27 \, m$

$y = 2.20 \, m + 3.27 \, m = \underline{5.47 \, m}$

c) Consider the motion of the shot from the point where he releases it to when it returns to the height of his head. Take $y = 0$ at the ground.

$y_0 = 2.20 \, m$
$y = 1.83 \, m$
$a = -9.80 \, m/s^2$
$v_0 = +8.00 \, m/s$
$t = ?$

$y - y_0 = v_0 t + \frac{1}{2} a t^2$

$1.83 \, m - 2.20 \, m = (8.00 \, m/s) t + \frac{1}{2}(-9.80 \, m/s^2) t^2$

$-0.37 \, m = (8.00 \, m/s) t - (4.90 \, m/s^2) t^2$

2-65 (cont)

Use the quadratic formula to solve for t:
$$4.90t^2 - 8.00t - 0.37 = 0$$
$$t = \frac{1}{9.80}\left(8.00 \pm \sqrt{(8.00)^2 - 4(4.90)(-0.37)}\right) = 0.816 \pm 0.861$$

t must be positive, so $t = 0.816s + 0.861s = \underline{1.68s}$

2-69

a) Calculate the speed of the diver when she reaches the water. Take the origin of coordinates to be at the platform, and take the $+y$-direction to be downward.

$y - y_0 = +24.4 m$
$a = +9.80 m/s^2$
$v_0 = 0$ (diver just steps off)
$v = ?$

$v^2 = v_0^2 + 2a(y-y_0)$
$v = +\sqrt{2a(y-y_0)}$
$v = +\sqrt{2(9.80 m/s^2)(24.4 m)}$
$v = +\underline{21.9 m/s}$

We know that v is positive because the diver is traveling downward when she reaches the water.

The announcer has exaggerated the final speed of the diver.

b) Use the same coordinates as in part (a). Calculate the initial upward velocity needed to give the diver a speed of 105 km/h when she reaches the water.

$v_0 = ?$
$v = 29.0 m/s$
$a = +9.80 m/s^2$
$y - y_0 = +24.4 m$

$v^2 = v_0^2 + 2a(y-y_0)$
$v_0 = -\sqrt{v^2 - 2a(y-y_0)} = -\sqrt{(29.0 m/s)^2 - 2(9.80 m/s^2)(24.4 m)}$
$v_0 = \underline{-19.0 m/s}$

(v_0 is negative since the direction of the initial velocity is upward.)

One way to decide if this speed is physically reasonable is to calculate the maximum height above the platform it would produce:

$v_0 = -19.0 m/s$
$v = 0$ (at maximum height)
$a = +9.80 m/s^2$
$y - y_0 = ?$

$v^2 = v_0^2 + 2a(y-y_0)$
$y - y_0 = \frac{v^2 - v_0^2}{2a} = \frac{0 - (-19.0 m/s)^2}{2(+9.80 m/s^2)} = -18.4 m$

This is not physically attainable.

2-73

Let $y = 0$ at the ground and let positive y be upward.

a) Let $t = 0$ when the football is at the window.

$v_0 = 5.00 m/s$
$v = 0$ (at maximum height)
$a = -9.80 m/s^2$
$y_0 = +15.0 m,\ y = ?$

$v^2 = v_0^2 + 2a(y-y_0)$
$y - y_0 = \frac{v^2 - v_0^2}{2a} = \frac{0 - (5.00 m/s)^2}{2(-9.80 m/s^2)} = 1.28 m$

2-73 (cont)

$$y = y_0 + 1.28\,m = 15.0\,m + 1.28\,m = \underline{16.3\,m}$$
(maximum height above ground)

b) Now let $t=0$ be when the football is at the ground. Use the motion from the ground to the window to find v_0, the speed of the football as it leaves the ground.

$v_0 = ?$
$a = -9.80\,m/s^2$
$v = +5.00\,m/s$
$y - y_0 = +15.0\,m$

$$v^2 = v_0^2 + 2a(y-y_0)$$
$$v_0 = +\sqrt{v^2 - 2a(y-y_0)} = \sqrt{(5.00\,m/s)^2 - 2(-9.80\,m/s^2)(+15.0\,m)}$$
$$v_0 = +17.9\,m/s$$

Now consider the motion from the ground to the maximum height.

$v_0 = +17.9\,m/s$
$t = ?$
$v = 0$ (at maximum height)
$a = -9.80\,m/s^2$

$$v = v_0 + at$$
$$t = \frac{v - v_0}{a} = \frac{0 - 17.9\,m/s}{-9.80\,m/s^2} = \underline{+1.83\,s}$$

2-75

a) It is very convenient to work in coordinates attached to the truck. Note that these coordinates move at constant velocity, relative to the earth. In these coordinates the truck is at rest, and the initial velocity of the car is $v_0 = 0$. Also, the car's acceleration in these coordinates is the same as in coordinates fixed to the earth.

First, let's calculate how far the car must travel relative to the truck:

[Diagram: initial position of car, relative to truck — 5m box, 25m gap, truck (20m, with 25m marks), 25m gap, 5m box. Total: 25m + 20m + 25m + 5m = 75m]

The car goes from $x_0 = -25.0\,m$ to $x = 50.0\,m$, so $x - x_0 = 75.0\,m$. Calculate the time that it takes the car to travel this distance:

$a = 0.600\,m/s^2$
$v_0 = 0$
$x - x_0 = 75.0\,m$
$t = ?$

$$x - x_0 = v_0 t + \tfrac{1}{2}at^2$$
$$t = \sqrt{\frac{2(x-x_0)}{a}} = \sqrt{\frac{2(75.0\,m)}{0.600\,m/s^2}} = \underline{15.8\,s}$$

b) Need how far the car travels relative to the earth, so go now to coordinates fixed in the earth. In these coordinates $v_0 = 20.0\,m/s$ for the car. Take the origin to be at the initial position of the car.

2-75 (cont)

$v_0 = 20.0 \text{ m/s}$
$a = 0.600 \text{ m/s}^2$
$t = 15.8 \text{ s}$
$x - x_0 = ?$

$x - x_0 = v_0 t + \frac{1}{2} a t^2$
$x - x_0 = (20.0 \text{ m/s})(15.8 \text{ s}) + \frac{1}{2}(0.600 \text{ m/s}^2)(15.8 \text{ s})^2$
$x - x_0 = 316 \text{ m} + 74.9 \text{ m} = \underline{391 \text{ m}}$

c) In coordinates fixed to the earth:
$v = v_0 + at$
$v = 20.0 \text{ m/s} + (0.600 \text{ m/s}^2)(15.8 \text{ s}) = \underline{29.5 \text{ m/s}}$

2-77

$x_A = \alpha t + \beta t^2$
$v_A = \frac{dx_A}{dt} = \alpha + 2\beta t$
$a_A = \frac{dv_A}{dt} = 2\beta$

$x_B = \gamma t^2 + \delta t^3$
$v_B = \frac{dx_B}{dt} = 2\gamma t + 3\delta t^2$
$a_B = \frac{dv_B}{dt} = 2\gamma + 6\delta t$

a) At $t = 0$, $v_A = \alpha$ and $v_B = 0$. So initially, car A moves ahead.

b) At the same point $\Rightarrow x_A = x_B$
$\alpha t + \beta t^2 = \gamma t^2 + \delta t^3$

One solution is $t = 0$, which says that they start from the same point.
Divide by $t \Rightarrow \alpha + \beta t = \gamma t + \delta t^2$
$\delta t^2 - (\gamma - \beta) t - \alpha = 0$

$t = \frac{1}{2\delta} \left((\beta - \gamma) + \sqrt{(\beta - \gamma)^2 + 4 \delta \alpha} \right)$ (t must be positive)

$t = \frac{1}{2(0.80)} \left(-0.60 + \sqrt{(0.60)^2 + 4(0.80)(4.00)} \right) = 1.89 \text{ s}$

So $x_A = x_B$ for $\underline{t = 0}$ and $\underline{t = 1.89 \text{ s}}$

c) $v_A = v_B \Rightarrow \alpha + 2\beta t = 2\gamma t + 3\delta t^2$
$3\delta t^2 + 2(\gamma - \beta) t - \alpha = 0$

$t = \frac{1}{6\delta} \left(2(\beta - \gamma) + \sqrt{4(\beta - \gamma)^2 + 12 \delta \alpha} \right) = \frac{1}{6(0.80)} \left(2(-0.60) + \sqrt{4(0.60)^2 + 12(0.80)(4.00)} \right)$
$t = \underline{1.06 \text{ s}}$

d) Distance from A to B is $x_B - x_A$. The rate of change of this distance is $\frac{d}{dt}(x_B - x_A)$. If this distance is not changing, $\frac{d}{dt}(x_B - x_A) = 0$. But this says $v_B - v_A = 0$. (The distance between A and B is neither decreasing nor decreasing at the instant when they have the same velocity.)
From part (c), this happens at $\underline{t = 1.06 \text{ s}}$.

CHAPTER 3

Exercises 1, 3, 5, 7, 11, 15, 19, 21, 25, 29, 33, 35

Problems 37, 43, 45, 49, 53, 55, 57, 59, 61, 63, 67

Exercises

3-1

a) $(V_{av})_x = \frac{\Delta x}{\Delta t} = \frac{x_2 - x_1}{t_2 - t_1} = \frac{-4.1m - 2.7m}{4.0s - 0} = \underline{-1.70 \, m/s}$

$(V_{av})_y = \frac{\Delta y}{\Delta t} = \frac{y_2 - y_1}{t_2 - t_1} = \frac{6.8m - 3.8m}{4.0s - 0} = \underline{+0.75 \, m/s}$

b) $\tan \alpha = \frac{(V_{av})_y}{(V_{av})_x} = \frac{0.75 \, m/s}{-1.70 \, m/s} = -0.441$

$\alpha = \underline{156°}$

$V_{av} = \sqrt{(V_{av})_x^2 + (V_{av})_y^2} = \sqrt{(-1.70 \, m/s)^2 + (0.75 \, m/s)^2} = \underline{1.86 \, m/s}$

3-3

a) First calculate the time t_{10}.

$\vec{r} = [1.5 \, cm + (2.0 \, cm/s^2)t^2]\hat{\imath} + (3.0 \, cm/s)t\hat{\jmath}$

At $t=0$, $\vec{r} = \vec{r}_0 = (1.5 \, cm)\hat{\imath}$

$\Delta \vec{r} = \vec{r} - \vec{r}_0 = (2.0 \, cm/s^2)t^2 \hat{\imath} + (3.0 \, cm/s)t \hat{\jmath}$

$|\Delta \vec{r}|^2 = [(2.0 \, cm/s^2)t^2]^2 + [(3.0 \, cm/s)t]^2$

At $t = t_{10}$, $|\Delta \vec{r}| = 10.0 \, cm$, so t_{10} is the solution of $100 = 4.0 t^4 + 9.0 t^2$

Let $\gamma = t^2$, then $4\gamma^2 + 9\gamma - 100 = 0$

quadratic formula gives $\gamma = \frac{1}{8}\left(-9 \pm \sqrt{(9)^2 + 4(4)(100)}\right) = \frac{1}{8}(-9 \pm 41)$

γ must be positive, so $\gamma = \frac{1}{8}(-9 + 41) = 4$

$t_{10} = \sqrt{\gamma} = \underline{2.0 \, s}$

Now can calculate the components of $\vec{V}_{av}$:

At $t=0$, $x_0 = 1.5 \, cm$, $y_0 = 0$.

At $t = t_{10} = 2.0 \, s$, $x_{10} = 9.5 \, cm$, $y_{10} = 6.0 \, cm$

$(V_{av})_x = \frac{\Delta x}{\Delta t} = \frac{x_{10} - x_0}{t_{10} - 0} = \frac{9.5 \, cm - 1.5 \, cm}{2.0 \, s} = \underline{4.0 \, cm/s}$

$(V_{av})_y = \frac{\Delta y}{\Delta t} = \frac{y_{10} - y_0}{t_{10} - 0} = \frac{6.0 \, cm - 0}{2.0 \, s} = \underline{3.0 \, cm/s}$

$V_{av} = \sqrt{(V_{av})_x^2 + (V_{av})_y^2} = \sqrt{(4.0 \, cm/s)^2 + (3.0 \, cm/s)^2} = \underline{5.0 \, cm/s}$

$\tan \alpha = \frac{(V_{av})_y}{(V_{av})_x} = \frac{3.0 \, cm/s}{4.0 \, cm/s} = 0.75 \Rightarrow \alpha = \underline{37°}$

28

3-3 (cont)

b) $\vec{v} = \frac{d\vec{r}}{dt} = (4.0\,cm/s^2)\,t\,\hat{\imath} + (3.0\,cm/s)\,\hat{\jmath}$

At $t = 0$, $v_x = 0$, $v_y = 3.0\,cm/s$ ⇒ $v = \underline{3.0\,cm/s}$, $\alpha = \underline{90°}$

At $t = t_{10} = 2.0\,s$, $v_x = 8.0\,cm/s$, $v_y = 3.0\,cm/s$
$v = \sqrt{v_x^2 + v_y^2} = \underline{8.5\,cm/s}$, $\tan\alpha = \frac{v_y}{v_x} = \frac{3.0\,cm/s}{8.0\,cm/s}$ ⇒ $\alpha = \underline{21°}$

c) The trajectory is a graph of y versus x.
$x = 1.5\,cm + (2.0\,cm/s^2)\,t^2$, $y = (3.0\,cm/s)\,t$
For values of t between 0 and 2.5 s, calculate x and y and plot y versus x.

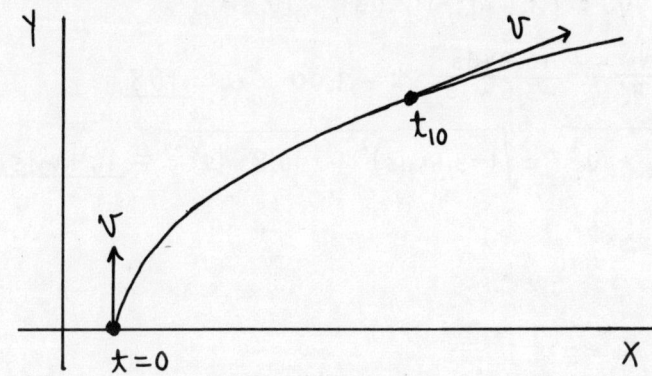

The sketch shows that the instantaneous velocity at any t is tangent to the trajectory.

3-5

a) $(a_{av})_x = \frac{\Delta v_x}{\Delta t} = \frac{v_{2x} - v_{1x}}{t_2 - t_1} = \frac{110\,m/s - 160\,m/s}{20.0\,s - 0} = \underline{-2.5\,m/s^2}$

$(a_{av})_y = \frac{\Delta v_y}{\Delta t} = \frac{v_{2y} - v_{1y}}{t_2 - t_1} = \frac{60\,m/s - (-80\,m/s)}{20.0\,s - 0} = \underline{+7.0\,m/s^2}$

b)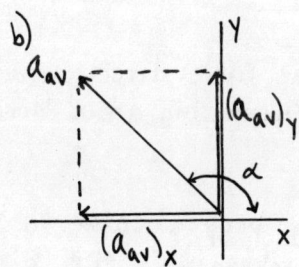

$\tan\alpha = \frac{(a_{av})_y}{(a_{av})_x} = \frac{7.0\,m/s^2}{-2.5\,m/s^2} = -2.8$

$\alpha = \underline{110°}$

$a_{av} = \sqrt{(a_{av})_x^2 + (a_{av})_y^2} = \sqrt{(-2.5\,m/s^2)^2 + (7.0\,m/s^2)^2}$

$a_{av} = \underline{7.4\,m/s^2}$

3-7

a)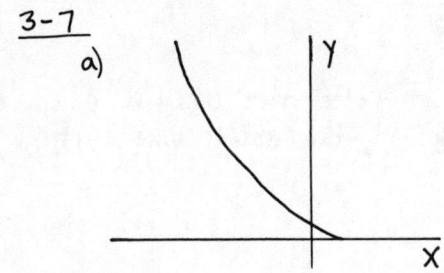

3-7 (cont)

b) $x = 2.0\text{ m} - \alpha t$, $\alpha = 3.6\text{ m/s}$
$y = \beta t^2$, $\beta = 1.8\text{ m/s}^2$

$v_x = \dfrac{dx}{dt} = -\alpha = -3.6\text{ m/s}$

$v_y = \dfrac{dy}{dt} = 2\beta t = (3.6\text{ m/s}^2)t$

$\Rightarrow \vec{v} = (-3.6\text{ m/s})\hat{i} + (3.6\text{ m/s}^2)t\,\hat{j}$

$a_x = \dfrac{dv_x}{dt} = 0$

$a_y = \dfrac{dv_y}{dt} = 2\beta = 3.6\text{ m/s}^2$

$\Rightarrow \vec{a} = (3.6\text{ m/s}^2)\hat{j}$

c) velocity

$t = 3.0\text{ s} \Rightarrow v_x = -3.6\text{ m/s}$, $v_y = (3.6\text{ m/s}^2)(3.0\text{ s}) = 10.8\text{ m/s}$

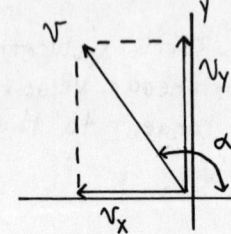

$\tan \alpha = \dfrac{v_y}{v_x} = \dfrac{10.8\text{ m/s}}{-3.6\text{ m/s}} = -3.00$, $\alpha = \underline{108°}$

$v = \sqrt{v_x^2 + v_y^2} = \sqrt{(-3.6\text{ m/s})^2 + (10.8\text{ m/s})^2} = \underline{11.4\text{ m/s}}$

acceleration

$t = 3.0\text{ s} \Rightarrow a_x = 0$, $a_y = 3.6\text{ m/s}^2$ (both acceleration components are independent of time)

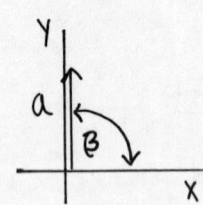

$a = 3.6\text{ m/s}^2$ and is in the $+y$-direction ($\beta = 90°$)

d)

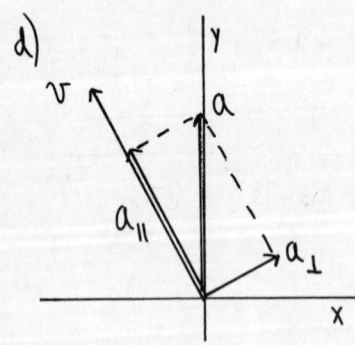

$\vec{a}$ has a component $a_{\parallel}$ in the same direction as $\vec{v}$, so we know that v is increasing (the bird is speeding up).

$\vec{a}$ also has a component $a_{\perp}$ perpendicular to $\vec{v}$ so the direction of $\vec{v}$ is changing; the bird is turning toward the $+y$-direction.

3-11

Take the positive y-direction to be upward. Take the origin of coordinates at the initial position of the book, at the point where it leaves the table top.

3-11 (cont)

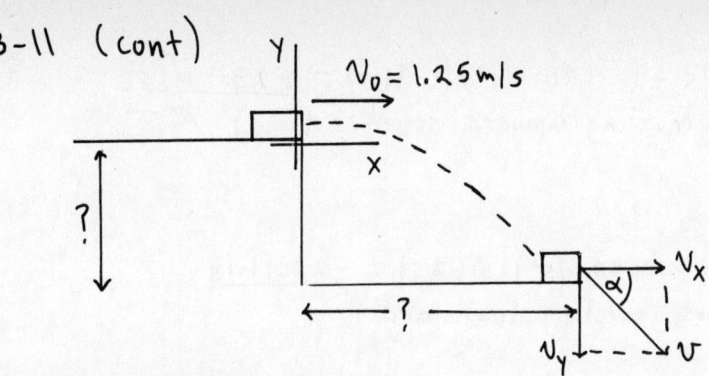

X-component
$a_x = 0$
$v_{ox} = 1.25 \text{ m/s}$
$t = 0.400 \text{ s}$

Y-component
$a_y = -9.80 \text{ m/s}^2$
$v_{oy} = 0$
$t = 0.400 \text{ s}$

a) $y - y_0 = ?$
$y - y_0 = v_{oy} t + \tfrac{1}{2} a_y t^2 = 0 + \tfrac{1}{2}(-9.80 \text{ m/s}^2)(0.400 \text{ s})^2 = -0.784 \text{ m}$
The table top is $\underline{0.784 \text{ m}}$ above the floor.

b) $x - x_0 = ?$
$x - x_0 = v_{ox} t + \tfrac{1}{2} a_x t^2 = (1.25 \text{ m/s})(0.400 \text{ s}) + 0 = \underline{0.500 \text{ m}}$

c) $v_x = v_{ox} + a_x t = \underline{1.25 \text{ m/s}}$ (The x-component of the velocity is constant, since $a_x = 0$.)
$v_y = v_{oy} + a_y t = 0 + (-9.80 \text{ m/s}^2)(0.400 \text{ s}) = \underline{-3.92 \text{ m/s}}$

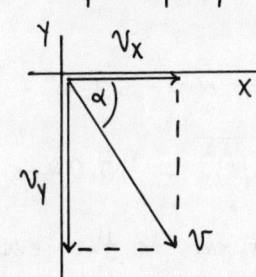

$\tan \alpha = \dfrac{v_y}{v_x} = \dfrac{-3.92 \text{ m/s}}{1.25 \text{ m/s}} = -3.14 \Rightarrow \alpha = -72.3°$

Direction of $\vec{v}$ is 72.3° below the horizontal.

$v = \sqrt{v_x^2 + v_y^2} = \sqrt{(1.25 \text{ m/s})^2 + (-3.92 \text{ m/s})^2} = \underline{4.11 \text{ m/s}}$

3-15

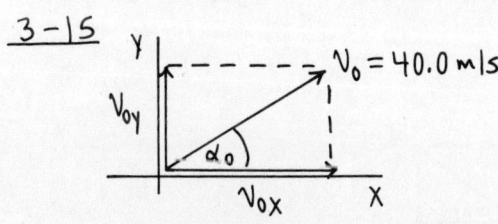

$v_{ox} = v_0 \cos \alpha_0 = (40.0 \text{ m/s}) \cos 53.1° = 24.0 \text{ m/s}$
$v_{oy} = v_0 \sin \alpha_0 = (40.0 \text{ m/s}) \sin 53.1° = 32.0 \text{ m/s}$

a) y-component (vertical motion)

$y - y_0 = +25.0 \text{ m}$
$v_{oy} = +32.0 \text{ m/s}$
$a_y = -9.80 \text{ m/s}^2$
$t = ?$

$y - y_0 = v_{oy} t + \tfrac{1}{2} a_y t^2$
$25.0 \text{ m} = (32.0 \text{ m/s}) t - (4.9 \text{ m/s}^2) t^2$
$(4.9 \text{ m/s}^2) t^2 - (32.0 \text{ m/s}) t + 25.0 \text{ m} = 0$
Apply the quadratic formula:
$t = \tfrac{1}{9.8} \left[32.0 \pm \sqrt{(32.0)^2 - 4(4.9)(25)} \right] \text{s} = (3.265 \pm 2.358) \text{ s}$

The ball is at a height of 25.0 m 0.91 s and 5.62 s after being thrown.

b) $\underline{t = 0.91 \text{ s}}$
$v_x = v_{ox} = \underline{+24.0 \text{ m/s}}$ (same at any t, since $a_x = 0$)

3-15 (cont)

$v_y = v_{oy} + a_y t = 32.0 \text{ m/s} + (-9.80 \text{ m/s}^2)(0.91 \text{ s}) = \underline{+23.1 \text{ m/s}}$

($v_y > 0$ means the baseball is traveling upward at this time)

$\underline{t = 5.62 \text{ s}}$
$v_x = v_{ox} = \underline{+24.0 \text{ m/s}}$
$v_y = v_{oy} + a_y t = 32.0 \text{ m/s} + (-9.80 \text{ m/s}^2)(5.62 \text{ s}) = \underline{-23.1 \text{ m/s}}$

($v_y < 0$ means the baseball is traveling downward)

c) $v_x = v_{ox} = 24.0 \text{ m/s}$

solve for v_y:
$v_y = ?$
$y - y_0 = 0$ (when ball returns to level from which thrown)
$a_y = -9.80 \text{ m/s}^2$
$v_{oy} = +32.0 \text{ m/s}$

$v_y^2 = v_{oy}^2 + 2a_y(y-y_0)$

$v_y = -v_{oy} = -32.0 \text{ m/s}$
(negative since must be traveling downward at this point)

Now that have the components can solve for the magnitude and direction of $\vec{v}$.

$\tan \alpha = \frac{v_y}{v_x} = \frac{-32.0 \text{ m/s}}{+24.0 \text{ m/s}} = -1.33 \Rightarrow \alpha = -53.1°$

$v = \sqrt{v_x^2 + v_y^2} = \sqrt{(24.0 \text{ m/s})^2 + (-32.0 \text{ m/s})^2} = 40.0 \text{ m/s}$

The velocity of the ball when it returns to the level from which it was thrown has magnitude $\underline{40.0 \text{ m/s}}$ (the same as the initial speed) and is directed at an angle of $53.1°$ below the horizontal.

3-19

Take the origin of coordinates at the roof and let the +y-direction be upward.

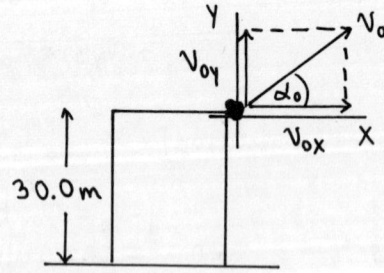

$v_{ox} = v_0 \cos \alpha_0 = 33.5 \text{ m/s}$
$v_{oy} = v_0 \sin \alpha_0 = 21.8 \text{ m/s}$

a) At the maximum height $v_y = 0$.
y-component
$a_y = -9.80 \text{ m/s}^2$
$v_y = 0$
$v_{oy} = +33.5 \text{ m/s}$
$y - y_0 = ?$

$v^2 = v_0^2 + 2a_y(y-y_0)$

$y - y_0 = \frac{v^2 - v_{oy}^2}{2a_y} = \frac{0^2 - (21.8 \text{ m/s})^2}{2(-9.80 \text{ m/s}^2)} = \underline{+24.2 \text{ m}}$

(measured from the roof)

3-19 (cont)

b) $v_x = v_{0x} = 33.5$ m/s (since $a_x = 0$)

y-component
$v_y = ?$
$a_y = -9.80$ m/s²
$y - y_0 = -30.0$ m (negative because at ground rock is below its initial position)
$v_{0y} = +21.8$ m/s

$v_y^2 = v_{0y}^2 + 2a_y(y-y_0)$
$v_y = -\sqrt{v_{0y}^2 + 2a_y(y-y_0)}$

(v_y is negative because at the ground the rock is traveling downward)

$v_y = -\sqrt{(21.8 \text{ m/s})^2 + 2(-9.80 \text{ m/s}^2)(-30.0 \text{ m})}$

$v_y = -32.6$ m/s

Then $v = \sqrt{v_x^2 + v_y^2} = \sqrt{(33.5 \text{ m/s})^2 + (-32.6 \text{ m/s})^2} = \underline{46.7 \text{ m/s}}$.

c) Use the vertical motion (y-component) to find the time the rock is in the air:

$t = ?$
$v_y = -32.6$ m/s (from part (b))
$a_y = -9.80$ m/s²
$v_{0y} = +21.8$ m/s

$v_y = v_{0y} + a_y t$
$t = \dfrac{v_y - v_{0y}}{a_y} = \dfrac{-32.6 \text{ m/s} - 21.8 \text{ m/s}}{-9.80 \text{ m/s}^2} = +5.55$ s

Can use this t to calculate the horizontal range:
$t = 5.55$ s
$x - x_0 = ?$
$a_x = 0$
$v_{0x} = 33.5$ m/s

$x - x_0 = v_{0x} t + \tfrac{1}{2} a_x t^2$
$x - x_0 = (33.5 \text{ m/s})(5.55 \text{ s}) = \underline{186 \text{ m}}$

3-21

Take the origin of coordinates at the point where the quarter leaves your hand, and take positive y to be upward.

$v_{0x} = v_0 \cos\alpha_0 = (6.4 \text{ m/s}) \cos 60° = 3.2$ m/s

$v_{0y} = v_0 \sin\alpha_0 = (6.4 \text{ m/s}) \sin 60° = 5.54$ m/s

a) Use the horizontal (x-component) of motion to solve for t, the time the quarter travels through the air:

x-component
$t = ?$
$x - x_0 = 2.1$ m
$v_{0x} = 3.2$ m/s
$a_x = 0$

$x - x_0 = v_{0x} t + \tfrac{1}{2} a_x t^2$
$t = \dfrac{x - x_0}{v_{0x}} = \dfrac{2.1 \text{ m}}{3.2 \text{ m/s}} = 0.656$ s

3-21 (cont)

Now find the vertical displacement of the quarter after this time:

y-component
$y - y_0 = ?$
$a_y = -9.80 \text{ m/s}^2$
$v_{0y} = +5.54 \text{ m/s}$
$t = 0.656 \text{ s}$

$y - y_0 = v_{0y} t + \frac{1}{2} a_y t^2$
$y - y_0 = (5.54 \text{ m/s})(0.656 \text{ s}) + \frac{1}{2}(-9.80 \text{ m/s}^2)(0.656 \text{ s})^2$
$y - y_0 = 3.64 \text{ m} - 2.11 \text{ m} = \underline{1.53 \text{ m}}$

b) $v_y = ?$
$t = 0.656 \text{ s}$
$a_y = -9.80 \text{ m/s}^2$
$v_{0y} = +5.54 \text{ m/s}$

$v_y = v_{0y} + a_y t$
$v_y = 5.54 \text{ m/s} + (-9.80 \text{ m/s}^2)(0.656 \text{ s}) = \underline{-0.89 \text{ m/s}}$

The minus sign indicates that the y-component of $\vec{v}$ is downward; at this point the quarter has passed through the highest point in its path and is on its way down.)

3-25

a)

$|\vec{v}|$ constant $\Rightarrow a_{tan} = \frac{d|\vec{v}|}{dt} = 0$

$a_{rad} = \frac{v^2}{R} = \frac{(8.00 \text{ m/s})^2}{14.0 \text{ m}} = 4.57 \text{ m/s}^2$

The resultant acceleration is $\underline{4.57 \text{ m/s}^2}$, upward.

b) $v = \frac{2\pi R}{T}$ (Eq. 3-29) $\Rightarrow T = \frac{2\pi R}{v} = \frac{2\pi (14.0 \text{ m})}{8.00 \text{ m/s}} = \underline{11.0 \text{ s}}$

3-29

Let W stand for the woman, G for the ground, and S for the sidewalk. Take the positive direction to be the direction in which the sidewalk is moving.

The velocities are $v_{W/G}$ (woman relative to the ground), $v_{W/S}$ (woman relative to the sidewalk), and $v_{S/G}$ (sidewalk relative to the ground).

Eq. (2-24) becomes $v_{W/G} = v_{W/S} + v_{S/G}$.

The time to reach the other end is given by $t = \frac{\text{distance traveled relative to ground}}{v_{W/G}}$.

a) $v_{S/G} = 1.0 \text{ m/s}$
$v_{W/S} = 2.0 \text{ m/s}$
$v_{W/G} = v_{W/S} + v_{S/G} = 2.0 \text{ m/s} + 1.0 \text{ m/s} = 3.0 \text{ m/s}$

$t = \frac{40.0 \text{ m}}{v_{W/G}} = \frac{40.0 \text{ m}}{3.0 \text{ m/s}} = \underline{13.3 \text{ s}}$

b) $v_{S/G} = 1.0 \text{ m/s}$
$v_{W/S} = -2.0 \text{ m/s}$
$v_{W/G} = v_{W/S} + v_{S/G} = -2.0 \text{ m/s} + 1.0 \text{ m/s} = -1.0 \text{ m/s}$

3-29 (cont)
$$t = \frac{-40.0 \text{ m}}{v_{W/G}} = \frac{-40.0 \text{ m}}{-1.0 \text{ m/s}} = \underline{40.0 \text{ s}}$$

(Now the woman travels opposite to the direction the sidewalk is moving; she gets on at the opposite end than she did in part (a).)

3-33
a) View the motion from above:

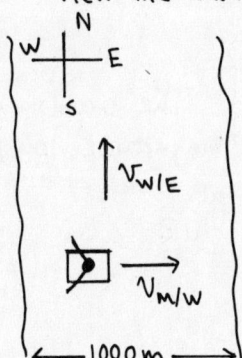

The velocity vectors in the problem are:
$\vec{v}_{M/E}$, the velocity of the man relative to the earth
$\vec{v}_{W/E}$, the velocity of the water relative to the earth
$\vec{v}_{M/W}$, the velocity of the man relative to the water

The rule for adding these velocities is
$$\vec{v}_{M/E} = \vec{v}_{M/W} + \vec{v}_{W/E}$$

The problem tells us that $\vec{v}_{W/E}$ has magnitude 2.4 m/s and direction due north. It also tells us that $\vec{v}_{M/W}$ has magnitude 3.5 m/s and direction due east.

The vector addition diagram is then

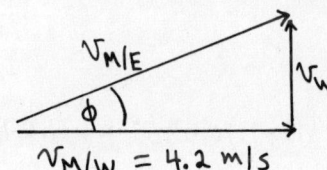

This diagram shows the vector addition $\vec{v}_{M/E} = \vec{v}_{M/W} + \vec{v}_{W/E}$ and also has $\vec{v}_{M/W}$ and $\vec{v}_{W/E}$ in their specified directions. Note that the vector addition diagram forms a right triangle.

The Pythagorean theorem applied to the vector addition diagram gives
$$v_{M/E}^2 = v_{M/W}^2 + v_{W/E}^2$$
$$v_{M/E} = \sqrt{v_{M/W}^2 + v_{W/E}^2} = \sqrt{(4.2 \text{ m/s})^2 + (2.4 \text{ m/s})^2} = 4.8 \text{ m/s}$$
$$\tan \phi = \frac{v_{W/E}}{v_{M/W}} = \frac{2.4 \text{ m/s}}{4.2 \text{ m/s}} = 0.571 \Rightarrow \phi = 30°$$

The velocity of the man relative to the earth has magnitude $\underline{4.8 \text{ m/s}}$ and direction $\underline{30° \text{ N of E}}$.

b) This is a tricky question. To cross the river the man must travel 1000 m due east relative to the earth. The man's velocity relative to the earth is $\vec{v}_{M/E}$. But, from the vector addition diagram the eastward component of $\vec{v}_{M/E}$ equals $v_{M/W} = 4.2 \text{ m/s}$.

Thus $t = \frac{x - x_0}{v_x} = \frac{1000 \text{ m}}{4.2 \text{ m/s}} = \underline{238 \text{ s}}$

3-33 (cont)

c) The northward component of $\vec{v}_{M/E}$ equals $v_{W/E} = 2.4 \text{ m/s}$. Therefore, in the 238 s it takes him to cross the river the distance north the man travels relative to the earth is

$$y - y_0 = v_y t = (2.4 \text{ m/s})(238 \text{ s}) = \underline{571 \text{ m}}$$

3-35

a) $\vec{v}_{P/A}$ is the velocity of the plane relative to the air. The problem states that $\vec{v}_{P/A}$ has magnitude 25 m/s and direction north.

$\vec{v}_{A/E}$ is the velocity of the air relative to the earth. The problem states that $\vec{v}_{A/E}$ is to the southwest (45° S of W) and has magnitude 10 m/s.

The relative velocity equation is $\vec{v}_{P/E} = \vec{v}_{P/A} + \vec{v}_{A/E}$

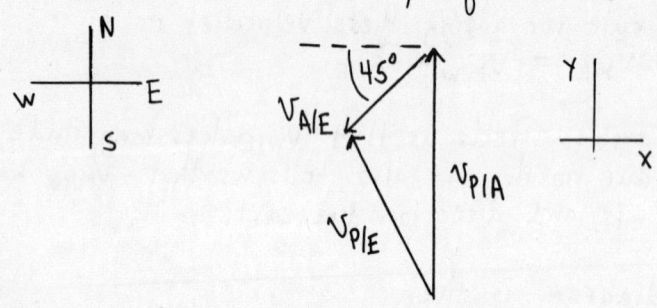

b) $(v_{P/A})_x = 0$, $(v_{P/A})_y = 25 \text{ m/s}$

$(v_{A/E})_x = -(10 \text{ m/s})\cos 45° = -7.07 \text{ m/s}$, $(v_{A/E})_y = -(10 \text{ m/s})\sin 45° = -7.07 \text{ m/s}$

$(v_{P/E})_x = (v_{P/A})_x + (v_{A/E})_x = 0 - 7.07 \text{ m/s} = \underline{-7.1 \text{ m/s}}$

$(v_{P/E})_y = (v_{P/A})_y + (v_{A/E})_y = 25 \text{ m/s} - 7.07 \text{ m/s} = \underline{17.9 \text{ m/s}}$

c)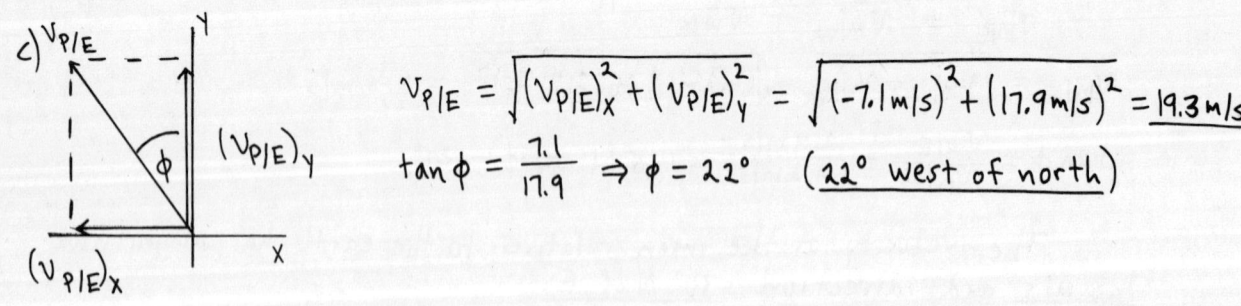

$v_{P/E} = \sqrt{(v_{P/E})_x^2 + (v_{P/E})_y^2} = \sqrt{(-7.1 \text{ m/s})^2 + (17.9 \text{ m/s})^2} = \underline{19.3 \text{ m/s}}$

$\tan \phi = \dfrac{7.1}{17.9} \Rightarrow \phi = 22°$ (22° west of north)

Problems

3-37

$$x = \alpha t, \quad y = 19.0 \text{ m} - \beta t^2$$

3-37 (cont)

a) At $t = 2.00\,s$, $x = (1.40\,m/s)(2.00\,s) = 2.80\,m$
$y = 19.0\,m - (0.600\,m/s^2)(2.00\,s)^2 = 16.6\,m$
$r = \sqrt{x^2 + y^2} = \sqrt{(2.80\,m)^2 + (16.6\,m)^2} = \underline{16.8\,m}$

b) $v_x = \frac{dx}{dt} = \alpha$, $v_y = \frac{dy}{dt} = -2\beta t$

At $t = 2.00\,s$, $v_x = 1.40\,m/s$, $v_y = -2(0.600\,m/s^2)(2.00\,s) = -2.40\,m/s$.

$v = \sqrt{v_x^2 + v_y^2} = \sqrt{(1.40\,m/s)^2 + (-2.40\,m/s)^2} = \underline{2.78\,m/s}$

$\tan\alpha = \frac{v_y}{v_x} = \frac{-2.40\,m/s}{1.40\,m/s} \Rightarrow \alpha = \underline{-59.7°}$

c) $a_x = \frac{dv_x}{dt} = 0$, $a_y = \frac{dv_y}{dt} = -2\beta = -1.20\,m/s^2$

$a = 1.20\,m/s^2$, at an angle of $\underline{-90.0°}$

d) $\vec{v} = \alpha\hat{i} - 2\beta t\hat{j}$, $\vec{a} = -2\beta\hat{j}$

$\vec{v}$ perpendicular to $\vec{a} \Rightarrow \vec{v} \cdot \vec{a} = 0$
$\vec{v} \cdot \vec{a} = +4\beta^2 t = 0 \Rightarrow \underline{t = 0}$

e) $\vec{r} = \alpha t\hat{i} + (19.0\,m - \beta t^2)\hat{j}$

$\vec{v}$ perpendicular to $\vec{r} \Rightarrow \vec{v} \cdot \vec{r} = 0$
$\vec{v} \cdot \vec{r} = \alpha^2 t - 2\beta t(19.0\,m - \beta t^2) = 0$
One solution is $t = 0$.
The other solution is t that satisfies $\alpha^2 = 2\beta(19.0\,m - \beta t^2)$
$19.0\,m - \beta t^2 = \frac{\alpha^2}{2\beta}$

$t = \sqrt{\frac{19.0\,m - \frac{\alpha^2}{2\beta}}{\beta}} = \sqrt{\frac{19.0\,m - \frac{(1.40\,m/s)^2}{2(0.600\,m/s^2)}}{0.600\,m/s^2}} = \underline{5.38\,s}$

At $t = 0$, $\vec{r} = (19.0\,m)\hat{j}$ ($r = 19.0\,m$, $\theta = 90.0°$)

At $t = 5.38\,s$, $\vec{r} = (1.40\,m/s)(5.38\,s)\hat{i} + (19.0\,m - (0.600\,m/s^2)(5.38\,s)^2)\hat{j}$
$\vec{r} = (7.53\,m)\hat{i} + (1.63\,m)\hat{j}$ ($r = 7.70\,m$, $\theta = 12.2°$)

f) The distance from the origin is given by $r = \sqrt{x^2 + y^2} = \sqrt{\alpha^2 t^2 + (19.0\,m - \beta t^2)^2}$

The minimum r occurs when $\frac{dr}{dt} = 0$, which gives $2\alpha^2 t - 4\beta t(19.0\,m - \beta t^2) = 0$

One solution is $t = 0$; $r = 19.0\,m$.

3-37 (cont)

The other solution comes from $\alpha^2 = 2\beta(19.0\text{m} - \beta t^2)$.
This is the same equation as in part (e), so $t = 5.38\text{s}$; $r = 7.70\text{m}$.

The minimum distance from the origin is 7.70m and occurs at $t = 5.38\text{s}$.

Note: $\frac{dr}{dt} = \frac{x}{r}\frac{dx}{dt} + \frac{y}{r}\frac{dy}{dt} = \frac{x}{r}v_x + \frac{y}{r}v_y = \frac{1}{r}(xv_x + yv_y) = \frac{1}{r}(\vec{r}\cdot\vec{v})$.
Thus $\frac{dr}{dt} = 0$ means $\vec{r}\cdot\vec{v} = 0$; this is why the same times are obtained in parts (f) and (e).

g)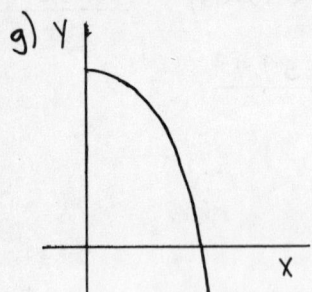

3-43

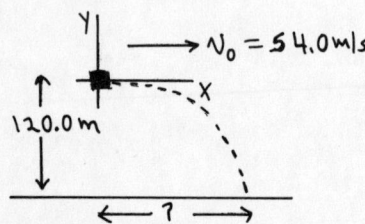

$v_0 = 54.0\text{m/s}$
120.0 m

Take the origin of coordinates at the point where the cannister is released. Take +y to be upward. The initial velocity of the cannister is the velocity of the plane, 54.0 m/s horizontally.

Use the vertical motion to find the time of fall:

y-component
$t = ?$
$v_{oy} = 0$
$a_y = -9.80 \text{m/s}^2$
$y - y_0 = -120.0\text{m}$
(When the cannister reaches the ground it is 120.0m below the origin.)

$y - y_0 = v_{oy}t + \frac{1}{2}a_y t^2$

$t = \sqrt{\frac{2(y-y_0)}{a_y}} = \sqrt{\frac{2(-120.0\text{m})}{-9.80\text{m/s}^2}} = 4.95 \text{ s}$

Find out how far the cannister travels horizontally in this time:

X-component
$x - x_0 = ?$
$a_x = 0$
$v_{ox} = 54.0 \text{m/s}$
$t = 4.95 \text{s}$

$x - x_0 = v_{ox}t + \frac{1}{2}a_x t^2$

$x - x_0 = (54.0\text{m/s})(4.95\text{s}) = \underline{267\text{m}}$

3-45

a)

Take the origin of coordinates at the point where the ball leaves the bat, and take +y to be upward.

$$v_{ox} = v_0 \cos\alpha_0$$
$$v_{oy} = v_0 \sin\alpha_0, \text{ but we don't know } v_0$$

Write down the equation for the horizontal displacement when the ball hits the ground and the corresponding equation for the vertical displacement. The time t is the same for both components, so this will give us two equations in two unknowns (v_0 and t).

y-component
$a_y = -9.80 \text{ m/s}^2$
$y - y_0 = -0.9 \text{ m}$
$v_{oy} = v_0 \sin 45°$

$$y - y_0 = v_{oy} t + \tfrac{1}{2} a_y t^2$$
$$-0.9 \text{ m} = (v_0 \sin 45°) t + \tfrac{1}{2}(-9.80 \text{ m/s}^2) t^2$$

X-component
$a_x = 0$
$x - x_0 = 188 \text{ m}$
$v_{ox} = v_0 \cos 45°$

$$x - x_0 = v_{ox} t + \tfrac{1}{2} a_x t^2$$
$$t = \frac{x - x_0}{v_{ox}} = \frac{188 \text{ m}}{v_0 \cos 45°}$$

Put this result into the y-component equation to eliminate t, and solve for v_0. (Note that $\sin 45° = \cos 45°$.)

$$-0.9 \text{ m} = (v_0 \sin 45°)\left(\frac{188 \text{ m}}{v_0 \cos 45°}\right) - (4.90 \text{ m/s}^2)\left(\frac{188 \text{ m}}{v_0 \cos 45°}\right)^2$$

$$4.90 \text{ m/s}^2 \left(\frac{188 \text{ m}}{v_0 \cos 45°}\right)^2 = 188 \text{ m} + 0.9 \text{ m} = 188.9 \text{ m}$$

$$\left(\frac{v_0 \cos 45°}{188 \text{ m}}\right)^2 = \frac{4.90 \text{ m/s}^2}{188.9 \text{ m}} \Rightarrow v_0 = \frac{188 \text{ m}}{\cos 45°}\sqrt{\frac{4.90 \text{ m/s}^2}{188.9 \text{ m}}} = \underline{42.8 \text{ m/s}}$$

b) Use the horizontal motion to find the time it takes the ball to reach the fence:

X-component
$x - x_0 = 116 \text{ m}$
$a_x = 0$
$v_{ox} = v_0 \cos 45° = (42.8 \text{ m/s}) \cos 45° = 30.3 \text{ m/s}$
$t = ?$

$$x - x_0 = v_{ox} t + \tfrac{1}{2} a_x t^2$$
$$t = \frac{x - x_0}{v_{ox}} = \frac{116 \text{ m}}{30.3 \text{ m/s}} = 3.83 \text{ s}$$

Find the vertical displacement of the ball at this t:

y-component
$y - y_0 = ?$
$a_y = -9.80 \text{ m/s}^2$
$v_{oy} = v_0 \sin 45° = 30.3 \text{ m/s}$
$t = 3.83 \text{ s}$

$$y - y_0 = v_{oy} t + \tfrac{1}{2} a_y t^2$$
$$y - y_0 = (30.3 \text{ m/s})(3.83 \text{ s}) + \tfrac{1}{2}(-9.80 \text{ m/s}^2)(3.83 \text{ s})^2$$
$$y - y_0 = 116.0 \text{ m} - 71.9 \text{ m} = +44.1 \text{ m, above the point where the ball was hit}$$

3-45 (cont)

The height of the ball above the ground is $44.1\,m + 0.90\,m = 45.0\,m$. It's height then above the top of the fence is $45.0\,m - 3.0\,m = \underline{42.0\,m}$.

3-49

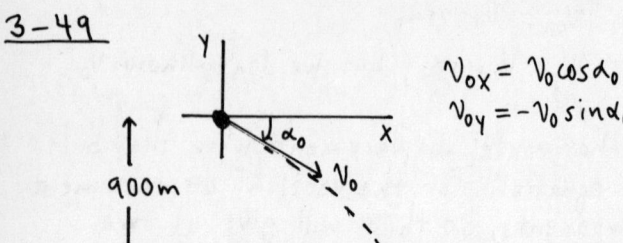

$v_{ox} = v_0 \cos\alpha_0$
$v_{oy} = -v_0 \sin\alpha_0$

Take the origin of coordinates to be at the point where the mailbag is released. Take $+y$ to be upward. The initial velocity of the mailbag equals the velocity of the plane.

a) Use the y-component. Consider the motion from the origin to where the bag strikes the ground.

$a_y = -9.80\,m/s^2$
$y - y_0 = -900\,m$
$t = 5.00\,s$
$v_{oy} = -v_0 \sin 40.9°$

$y - y_0 = v_{oy} t + \tfrac{1}{2} a_y t^2$
$v_{oy} = \dfrac{y - y_0}{t} - \tfrac{1}{2} a_y t$
$v_{oy} = \dfrac{-900\,m}{5.00\,s} - \tfrac{1}{2}(-9.80\,m/s^2)(5.00\,s) = -180\,m/s + 24.5\,m/s$
$v_{oy} = -155.5\,m/s$

Then $v_0 = -\dfrac{v_{oy}}{\sin 40.9°} = -\dfrac{-155.5\,m/s}{\sin 40.9°} = \underline{237\,m/s}$.

b) X-component

$a_x = 0$
$x - x_0 = ?$
$v_{ox} = v_0 \cos 40.9° = (237\,m/s) \cos 40.9°$
$v_{ox} = 179\,m/s$
$t = 5.00\,s$

$x - x_0 = v_{ox} t + \tfrac{1}{2} a_x t^2$
$x - x_0 = (179\,m/s)(5.00\,s) = \underline{896\,m}$

c) X-component
$a_x = 0 \Rightarrow v_x = v_{ox} = \underline{+179\,m/s}$

Y-component
$v_y = ?$
$v_{oy} = -155.5\,m/s$
$a_y = -9.80\,m/s^2$
$t = 5.00\,s$

$v_y = v_{oy} + a_y t$
$v_y = -155.5\,m/s + (-9.80\,m/s^2)(5.00\,s)$
$v_y = \underline{-204\,m/s}$

3-53

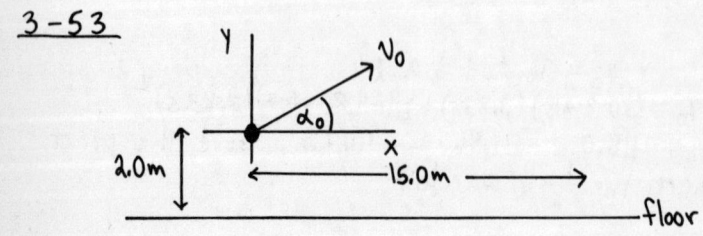

3-53 (cont)

a) First find the x- and y-components of the initial velocity.

X-component
$v_{0x} = ?$
$a_x = 0$
$t = 0.75 s$
$x - x_0 = 15.0 m$

$x - x_0 = v_{0x} t + \frac{1}{2} a_x t^2$

$v_{0x} = \frac{x-x_0}{t} = \frac{15.0 m}{0.75 s} = \underline{20.0 \, m/s}$

To find the y-component, use $\tan \alpha_0 = \frac{v_{0y}}{v_{0x}} \Rightarrow v_{0y} = v_{0x} \tan \alpha_0 = (20.0 \, m/s) \tan 19.0°$
$v_{0y} = \underline{6.9 \, m/s}$

Now can use the time in the air to find the components of the final velocity:

$v_x = v_{0x} + a_x t = v_{0x} = \underline{20.0 \, m/s}$, since $a_x = 0$

$v_y = v_{0y} + a_y t = +6.9 \, m/s + (-9.80 \, m/s^2)(0.75 s) = \underline{-0.4 \, m/s}$

b) $y - y_0 = ?$
$t = 0.75 s$
$a_y = -9.80 \, m/s^2$
$v_{0y} = +6.9 \, m/s$

$y - y_0 = v_{0y} t + \frac{1}{2} a_y t^2 = (6.9 \, m/s)(0.75 s) + \frac{1}{2}(-9.80 \, m/s^2)(0.75 s)^2$
$y - y_0 = 5.18 m - 2.76 m = 2.4 m$

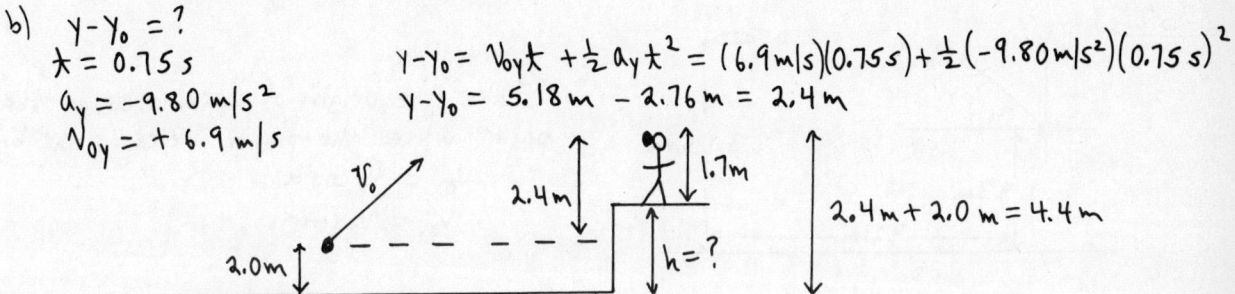

The impact is 2.4 m above the launch point and thus 4.4 m above the floor. The impact is 1.7 m above the stage, so the stage is a height $h = 4.4 m - 1.7 m = 2.7 m$ above the floor.

3-55

$v_{0x} = v_0 \cos \alpha_0$
$v_{0y} = v_0 \sin \alpha_0$

Take the origin of coordinates at the top of the ramp and take +y to be upward. The problem specifies that the object is displaced 40.0 m to the right when it is 15.0 m below the origin.

We don't know t, the time in the air, and we don't know v_0. Write down the equations for the horizontal and vertical displacements. Combine these two equations to eliminate one unknown.

Y-component
$y - y_0 = -15.0 \, m$
$a_y = -9.80 \, m/s^2$
$v_{0y} = v_0 \sin 53.0°$

$y - y_0 = v_{0y} t + \frac{1}{2} a_y t^2$
$\boxed{-15.0 \, m = (v_0 \sin 53°) t - (4.90 \, m/s^2) t^2}$

3-55 (cont)

X-component
$X - X_0 = 40.0 \text{ m}$
$a_x = 0$
$V_{ox} = V_0 \cos 53.0°$

$X - X_0 = V_{ox} t + \frac{1}{2} a_x t^2$

$\boxed{40.0 \text{ m} = (V_0 t) \cos 53.0°}$

The second equation says $V_0 t = \frac{40.0 \text{ m}}{\cos 53.0°} = 66.47 \text{ m}$.
Use this to replace $V_0 t$ in the first equation:
$-15.0 \text{ m} = (66.47 \text{ m}) \sin 53° - (4.90 \text{ m/s}^2) t^2$

$t = \sqrt{\frac{(66.47 \text{ m}) \sin 53° + 15.0 \text{ m}}{4.90 \text{ m/s}^2}} = \sqrt{\frac{68.09 \text{ m}}{4.90 \text{ m/s}^2}} = 3.728 \text{ s}$

Now that we have t we can use the x-component equation to solve for V_0:
$V_0 = \frac{40.0 \text{ m}}{t \cos 53.0°} = \frac{40.0 \text{ m}}{(3.728 \text{ s}) \cos 53.0°} = \underline{17.8 \text{ m/s}}$

3-57

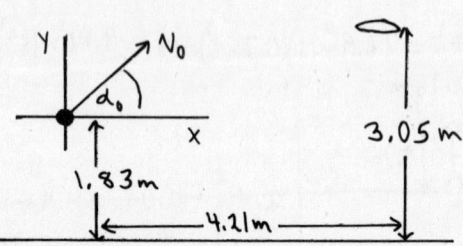

Take the origin of coordinates at the point where the player releases the ball.
$V_{ox} = V_0 \cos \alpha_0$
$V_{oy} = V_0 \sin \alpha_0$

Y-component
$V_y = 0$ (at the maximum height)
$V_{oy} = V_0 \sin \alpha_0 = (4.88 \text{ m/s}) \sin 35°$
$V_{oy} = 2.80 \text{ m/s}$
$a_y = -9.80 \text{ m/s}^2$
$Y - Y_0 = ?$

$V_y^2 = V_{oy}^2 + 2 a_y (y - y_0)$
$Y - Y_0 = \frac{V_y^2 - V_{oy}^2}{2 a_y} = \frac{0^2 - (2.80 \text{ m/s})^2}{2(-9.80 \text{ m/s}^2)}$
$Y - Y_0 = +0.400 \text{ m}$

The height above the floor then is $0.400 \text{ m} + 1.83 \text{ m}$
$= \underline{2.23 \text{ m}}$.

b) Use the vertical motion to find the time the ball is in the air:
Y-component
$Y - Y_0 = -1.83 \text{ m}$ (vertical displacement when the ball reaches the floor)
$V_{oy} = 2.80 \text{ m/s}$
$a_y = -9.80 \text{ m/s}^2$
$t = ?$

$Y - Y_0 = V_{oy} t + \frac{1}{2} a_y t^2$
$-1.83 \text{ m} = (2.80 \text{ m/s}) t - (4.9 \text{ m/s}^2) t^2$
$4.9 t^2 - 2.80 t - 1.83 = 0$

Quadratic formula gives
$t = \frac{1}{9.8} \left(2.80 \pm \sqrt{(2.80)^2 + 4(4.9)(1.83)} \right) \text{ s} = 0.286 \text{ s} \pm 0.675 \text{ s}$

t must be positive, so $t = 0.286 \text{ s} + 0.675 \text{ s} = \underline{0.961 \text{ s}}$

3-57 (cont)
Find the horizontal displacement for this t:

x-component
$x - x_0 = ?$
$v_{0x} = v_0 \cos \alpha_0 = (4.88 \text{ m/s}) \cos 35°$
$v_{0x} = 4.00 \text{ m/s}$
$a_x = 0$
$t = 0.961 \text{ s}$

$x - x_0 = v_{0x} t + \frac{1}{2} a_x t^2$
$x - x_0 = (4.00 \text{ m/s})(0.961 \text{ s}) = \underline{3.84 \text{ m}}$

c) We don't know either v_0 or the time t for the ball to reach the basket. Write down the equations for the horizontal and vertical displacements to get two equations for these two unknowns.

y-component
$a_y = -9.80 \text{ m/s}^2$
$v_{0y} = v_0 \sin 35°$
$y - y_0 = 3.05 \text{ m} - 1.83 \text{ m} = 1.22 \text{ m}$

$y - y_0 = v_{0y} t + \frac{1}{2} a_y t^2$
$\boxed{1.22 \text{ m} = (v_0 \sin 35°) t - (4.90 \text{ m/s}^2) t^2}$

x-component
$a_x = 0$
$v_{0x} = v_0 \cos 35°$
$x - x_0 = 4.21 \text{ m}$

$x - x_0 = v_{0x} t + \frac{1}{2} a_x t^2$
$\boxed{4.21 \text{ m} = (v_0 \cos 35°) t}$
$v_0 t = \frac{4.21 \text{ m}}{\cos 35°} = 5.139 \text{ m}$

Use this result in the y-component equation:
$1.22 \text{ m} = (5.139 \text{ m}) \sin 35° - (4.90 \text{ m/s}^2) t^2$
$(4.90 \text{ m/s}^2) t^2 = 1.728 \text{ m}$
$t = \sqrt{\frac{1.728 \text{ m}}{4.90 \text{ m/s}^2}} = 0.594 \text{ s}$

Then $v_0 = \frac{5.139 \text{ m}}{t} = \frac{5.139 \text{ m}}{0.594 \text{ s}} = \underline{8.65 \text{ m/s}}$.

d) We know that $v_0 = 8.65 \text{ m/s}$ from part (c).

y-component
$v_y = 0$ (at maximum height)
$a_y = -9.80 \text{ m/s}^2$
$v_{0y} = v_0 \sin 35° = (8.65 \text{ m/s}) \sin 35°$
$v_{0y} = 4.96 \text{ m/s}$

$v_y^2 = v_{0y}^2 + 2 a_y (y - y_0)$
$y - y_0 = \frac{v_y^2 - v_{0y}^2}{2 a_y} = \frac{0 - (4.96 \text{ m/s})^2}{2(-9.80 \text{ m/s}^2)} = +1.26 \text{ m}$

The maximum height above the floor is given by $1.83 \text{ m} + 1.26 \text{ m} = \underline{3.09 \text{ m}}$.

Also use the y-component to find the time t to reach the maximum height; can then use this t to find the horizontal displacement at this point.
$v_y = v_{0y} + a_y t$
$t = \frac{v_y - v_{0y}}{a_y} = \frac{0 - 4.96 \text{ m/s}}{-9.80 \text{ m/s}^2} = +0.506 \text{ s}$

3-57 (cont)

x-component

$x - x_0 = ?$
$a_x = 0$
$v_{0x} = v_0 \cos 35° = (8.65 \text{ m/s}) \cos 35°$
$v_{0x} = 7.09 \text{ m/s}$
$t = 0.506 \text{ s}$

$x - x_0 = v_{0x} t + \frac{1}{2} a_x t^2$
$x - x_0 = (7.09 \text{ m/s})(0.506 \text{ s}) = 3.59 \text{ m}$

This is the distance of the ball from the point where it was released. Its distance from the basket is then $4.21 \text{ m} - 3.59 \text{ m} = \underline{0.62 \text{ m}}$.

3-59

Use the equation derived in Example 3-9:
$$R = (v_0 \cos\alpha_0) \frac{2 v_0 \sin\alpha_0}{g}$$

Call the range R_1 when the angle is α_0 and R_2 when the angle is $90° - \alpha$.

$$R_1 = (v_0 \cos\alpha_0) \frac{2 v_0 \sin\alpha_0}{g}$$

$$R_2 = (v_0 \cos(90° - \alpha_0)) \frac{2 v_0 \sin(90° - \alpha_0)}{g}$$

The problem asks us to show that $R_1 = R_2$.

We can use the trig identities in Appendix B to show:
$\cos(90° - \alpha_0) = \cos(\alpha_0 - 90°) = \sin\alpha_0$
$\sin(90° - \alpha_0) = -\sin(\alpha_0 - 90°) = -(-\cos\alpha_0) = +\cos\alpha_0$

Thus $R_2 = (v_0 \sin\alpha_0) \frac{2 v_0 \cos\alpha_0}{g} = (v_0 \cos\alpha_0) \frac{2 v_0 \sin\alpha_0}{g} = R_1$.

3-61

$x = R \cos\omega t$, $y = R \sin\omega t$

a) $r = \sqrt{x^2 + y^2} = \sqrt{R^2 \cos^2\omega t + R^2 \sin^2\omega t} = \sqrt{R^2 (\sin^2\omega t + \cos^2\omega t)} = \sqrt{R^2} = R$,
since $\sin^2\omega t + \cos^2\omega t = 1$.

b) $v_x = \frac{dx}{dt} = -R\omega \sin\omega t$, $v_y = \frac{dy}{dt} = R\omega \cos\omega t$

$\vec{v} \cdot \vec{r} = v_x x + v_y y = (-R\omega \sin\omega t)(R \cos\omega t) + (R\omega \cos\omega t)(R \sin\omega t)$
$\vec{v} \cdot \vec{r} = R^2 \omega (-\sin\omega t \cos\omega t + \sin\omega t \cos\omega t) = 0$, so $\vec{v}$ is perpendicular to $\vec{r}$

c) $a_x = \frac{dv_x}{dt} = -R\omega^2 \cos\omega t = -\omega^2 x$
$a_y = \frac{dv_y}{dt} = -R\omega^2 \sin\omega t = -\omega^2 y$

$a = \sqrt{a_x^2 + a_y^2} = \sqrt{\omega^4 x^2 + \omega^4 y^2} = \omega^2 \sqrt{x^2 + y^2} = R\omega^2$

$\vec{a} = a_x \hat{i} + a_y \hat{j} = -\omega^2 (x\hat{i} + y\hat{j}) = -\omega^2 \vec{r}$; since ω^2 is positive this means the direction of $\vec{a}$ is opposite to the direction of $\vec{r}$.

3-61 (cont)

d) $v = \sqrt{v_x^2 + v_y^2} = \sqrt{R^2\omega^2 \sin^2\omega t + R^2\omega^2 \cos^2\omega t} = \sqrt{R^2\omega^2(\sin^2\omega t + \cos^2\omega t)}$

$v = \sqrt{R^2\omega^2} = R\omega$

e) $a = R\omega^2, \omega = \frac{v}{R} \Rightarrow a = R\frac{v^2}{R^2} = \frac{v^2}{R}$

3-63

The trick is to find the time it takes for the twins to meet. Since the pigeon flies at constant speed, the total distance the pigeon flies is just this speed times the time it flies.

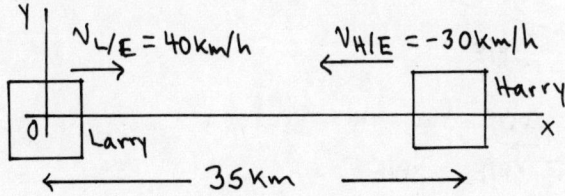

Work in a coordinate system attached to Larry's car. In these coordinates Harry undergoes a displacement of -35 km and the velocity of Harry is $v_{H/L}$, the velocity of Harry relative to Larry. Let $v_{L/E} = +40$ km/h be the velocity of Larry relative to the earth and let $v_{H/E} = -30$ km/h be the velocity of Harry relative to the earth (in the opposite direction). Then the relative velocity formula (Eq. 3-36) gives $v_{H/E} = v_{H/L} + v_{L/E}$, or
$v_{H/L} = v_{H/E} - v_{L/E} = -30$ km/h $- 40$ km/h $= -70$ km/h.

Then $x - x_0 = vt \Rightarrow t = \frac{x - x_0}{v} = \frac{-35 \text{km}}{-70 \text{km}} = +0.500 \text{h}.$

The distance the pigeon flies in this time is
distance = (speed)(time) = (60 km/h)(0.500 h) = __30 km__.

3-67

Select a coordinate system where $+y$ is north and $+x$ is east.

The velocity vectors in the problem are:
$\vec{v}_{P/E}$, the velocity of the plane relative to the earth
$\vec{v}_{P/A}$, the velocity of the plane relative to the air (the magnitude $v_{P/A}$ is the air speed of the plane and the direction of $\vec{v}_{P/A}$ is the compass course set by the pilot).
$\vec{v}_{A/E}$, the velocity of the air relative to the earth (the wind velocity).

The rule for combining relative velocities gives $\vec{v}_{P/E} = \vec{v}_{P/A} + \vec{v}_{A/E}$.

a) We are given the following information about the relative velocities:

3-67 (cont)

$\vec{v}_{P/A}$ has magnitude 220 km/h and its direction is west. In our coordinates it has components $(v_{P/A})_x = -220$ km/h and $(v_{P/A})_y = 0$.

From the displacement of the plane relative to the earth after 0.500 h, we find that $\vec{v}_{P/E}$ has components in our coordinate system of

$$(v_{P/E})_x = -\frac{180 \text{ km}}{0.500 \text{ h}} = -360 \text{ km/h (west)}$$

$$(v_{P/E})_y = -\frac{30 \text{ km}}{0.500 \text{ h}} = -60 \text{ km/h (south)}$$

With this information the diagram corresponding to the velocity addition equation is

We are asked to find $\vec{v}_{A/E}$, so solve for this vector:
$$\vec{v}_{P/E} = \vec{v}_{P/A} + \vec{v}_{A/E} \Rightarrow \vec{v}_{A/E} = \vec{v}_{P/E} - \vec{v}_{P/A}$$

The x-component of this equation gives
$$(v_{A/E})_x = (v_{P/E})_x - (v_{P/A})_x = -360 \text{ km/h} - (-220 \text{ km/h}) = -140 \text{ km/h}$$

The y-component of this equation gives
$$(v_{A/E})_y = (v_{P/E})_y - (v_{P/A})_y = -60 \text{ km/h}$$

Now that we have the components of $\vec{v}_{A/E}$ we can find its magnitude and direction.

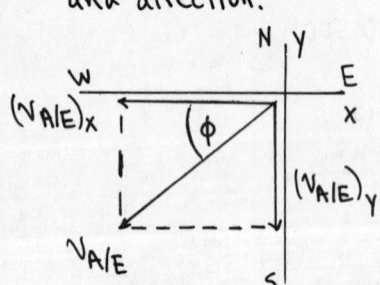

$$v_{A/E} = \sqrt{(v_{A/E})_x^2 + (v_{A/E})_y^2} = \sqrt{(-140 \text{ km/h})^2 + (-60 \text{ km/h})^2}$$

$$v_{A/E} = \underline{152 \text{ km/h}}$$

$$\tan\phi = \frac{60 \text{ km/h}}{140 \text{ km/h}} = 0.429 \Rightarrow \phi = 23.2°$$

The direction of the wind velocity is $\underline{23.2° \text{ S of W}}$.

b) The rule for combining the relative velocities is still $\vec{v}_{P/E} = \vec{v}_{P/A} + \vec{v}_{A/E}$, but some of these velocities have different values than in part (a).

$\vec{v}_{P/A}$ has magnitude 220 km/h but its direction is to be found.

$\vec{v}_{A/E}$ has magnitude 90 km/h and its direction is due south.

The direction of $\vec{v}_{P/E}$ is west; its magnitude is not given.

The vector diagram for $\vec{v}_{P/E} = \vec{v}_{P/A} + \vec{v}_{A/E}$ and the specified directions for the vectors is

3-67 (cont)

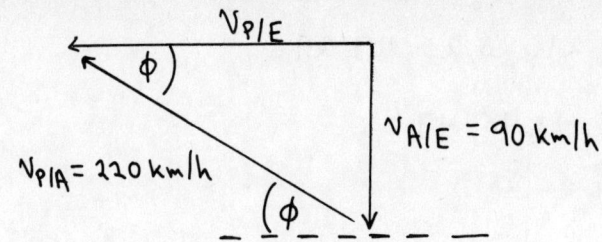

The vector addition diagram forms a right triangle.

$$\sin\phi = \frac{V_{A/E}}{V_{P/A}} = \frac{90 \text{ km/h}}{220 \text{ km/h}} = 0.409 \Rightarrow \phi = 24.1°$$

The pilot should set her course <u>24.1° N of W</u>.

CHAPTER 4

Exercises 5, 11, 13, 15, 17, 21, 23, 25, 27, 29

Problems 33, 35, 37, 41, 43, 45, 49

Exercises

4-5

Use a coordinate system where the +x-axis is in the direction of $\vec{F}_A$, the force applied by dog A.

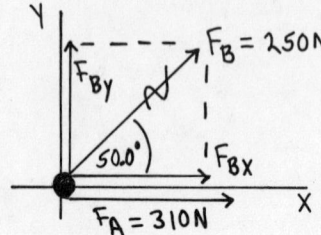

$F_{Ax} = +310 \text{ N}$
$F_{Ay} = 0$

$F_{Bx} = F_B \cos 50.0° = (250 \text{ N}) \cos 50.0° = +161 \text{ N}$
$F_{By} = F_B \sin 50.0° = (250 \text{ N}) \sin 50.0° = +192 \text{ N}$

$\vec{R} = \vec{F}_A + \vec{F}_B$
$R_x = F_{Ax} + F_{Bx} = +310 \text{ N} + 161 \text{ N} = +471 \text{ N}$
$R_y = F_{Ay} + F_{By} = 0 + 192 \text{ N} = +192 \text{ N}$

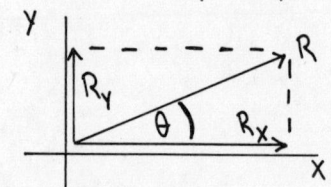

$R = \sqrt{R_x^2 + R_y^2} = \sqrt{(471 \text{ N})^2 + (192 \text{ N})^2} = \underline{509 \text{ N}}$

$\tan\theta = \dfrac{R_y}{R_x} = \dfrac{192 \text{ N}}{471 \text{ N}} = 0.408 \Rightarrow \theta = \underline{22.2°}$

4-11

a) During this time interval the acceleration is constant and equal to
$a_x = \dfrac{F_x}{m} = \dfrac{0.300 \text{ N}}{0.160 \text{ kg}} = 1.875 \text{ m/s}^2$

We can use the constant acceleration kinematic equations from Chapter 2.
$x - x_0 = v_{0x} t + \tfrac{1}{2} a_x t^2 = 0 + \tfrac{1}{2}(1.875 \text{ m/s}^2)(2.00 \text{ s})^2 = 3.75 \text{ m}$, so the puck is at $x = \underline{3.75 \text{ m}}$.
$v_x = v_{0x} + a_x t = 0 + (1.875 \text{ m/s}^2)(2.00 \text{ s}) = \underline{3.75 \text{ m/s}}$

b) In the time interval from $t = 2.00 \text{ s}$ to 5.00 s the force has been removed so the acceleration is zero. The speed stays constant at $v_x = 3.75 \text{ m/s}$. The distance the puck travels is $x - x_0 = v_{0x} t = (3.75 \text{ m/s})(5.00 \text{ s} - 3.00 \text{ s}) = 11.25 \text{ m}$. At the start of this interval the puck is at $x_0 = 3.75 \text{ m}$, so at the end of the interval it is at $x = x_0 + 11.25 \text{ m} = 15.0 \text{ m}$.

In the time interval from $t = 5.00 \text{ s}$ to 7.00 s the acceleration is again $a_x = 1.875 \text{ m/s}^2$. At the start of this interval $v_{0x} = 3.75 \text{ m/s}$ and $x_0 = 15.0 \text{ m}$.

4-11 (cont)

$$x - x_0 = v_{0x}t + \tfrac{1}{2}a_x t^2 = (3.75 \text{ m/s})(2.00 \text{ s}) + \tfrac{1}{2}(1.875 \text{ m/s}^2)(2.00 \text{ s})^2$$
$$x - x_0 = 7.50 \text{ m} + 3.75 \text{ m} = 11.25 \text{ m}.$$

Therefore, at $t = 7.00 \text{ s}$ the puck is at $x = x_0 + 11.25 \text{ m} = 15.0 \text{ m} + 11.25 \text{ m} = \underline{26.2 \text{ m}}$

$$v_x = v_{0x} + a_x t = 3.75 \text{ m/s} + (1.875 \text{ m/s}^2)(2.00 \text{ s}) = \underline{7.50 \text{ m/s}}$$

4-13

a) Constant velocity $\Rightarrow \vec{a} = 0 \Rightarrow \Sigma \vec{F} = 0$; the resultant force is zero; the vector sum of all the forces on the puck is zero.

b)

A •————————• B   Constant $\vec{v}$ says that the puck travels in a straight line.
  $\vec{v}$ $\vec{v}$

c) The new force produces a constant acceleration that is in the direction perpendicular to the line connecting points A and B. The path is a parabola, just as in projectile motion.

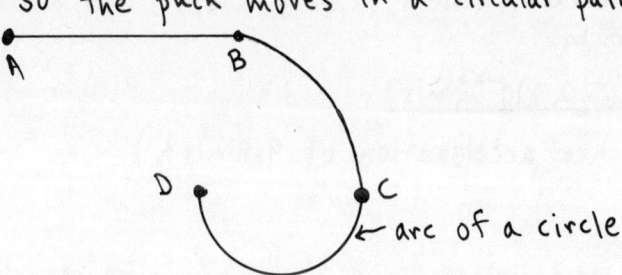

d) Now the acceleration is always perpendicular to the instantaneous velocity $\vec{v}$, so the puck moves in a circular path with constant speed.

(diagram showing points A, B, C, D with arc of a circle)

4-15

$w = mg$

The mass of the watermelon is constant, independent of its location. Its weight differs on earth and Mars. Use the information about the watermelon's weight on earth to calculate its mass:

$$w = mg \Rightarrow m = \frac{w}{g} = \frac{52.0 \text{ N}}{9.80 \text{ m/s}^2} = 5.31 \text{ kg}$$

On Mars, $m = \underline{5.31 \text{ kg}}$, the same as on earth. Thus the weight on Mars is $w = mg = (5.31 \text{ kg})(3.72 \text{ m/s}^2) = \underline{19.8 \text{ N}}$.

4-17

$F = ma$

We must use $w = mg$ to find the mass of the boulder:

$m = \dfrac{w}{g} = \dfrac{2800 \text{ N}}{9.80 \text{ m/s}^2} = 285.7 \text{ kg}$

Then $F = ma = (285.7 \text{ kg})(15.0 \text{ m/s}^2) = \underline{4290 \text{ N}}$.

4-21

a) The free-body diagram for the bottle is ↓ $w = mg$

The only force on the bottle is gravity.

b) w is the force of gravity that the earth exerts on the bottle. The reaction to this force is w', the gravity force that the bottle exerts on the earth. Note that these two equal and opposite forces produce very different accelerations because the bottle and the earth have very different masses.

4-23

The force of gravity that the earth exerts on her is her weight, $w = mg = (50 \text{ kg})(9.8 \text{ m/s}^2) = 490 \text{ N}$. By Newton's 3rd law, she exerts an equal and opposite force on the earth.

Apply $\Sigma \vec{F} = m\vec{a}$ to the earth, with $|\Sigma \vec{F}| = w = 490 \text{ N}$, but must use the mass of the earth for m.

$a = \dfrac{w}{m} = \dfrac{490 \text{ N}}{6.0 \times 10^{24} \text{ kg}} = \underline{8.2 \times 10^{-23} \text{ m/s}^2}$

(This is _much_ smaller than her acceleration of 9.8 m/s^2.)

4-25

The free-body diagram for the bucket is

T (the tension in the cord)

$w = mg$

$\Sigma F_y = ma_y$

$T - mg = ma$

$a = \dfrac{T - mg}{m} = \dfrac{60.0 \text{ N} - (4.80 \text{ kg})(9.80 \text{ m/s}^2)}{4.80 \text{ kg}} = \dfrac{60.0 \text{ N} - 47.04 \text{ N}}{4.80 \text{ kg}}$

$a = \underline{2.70 \text{ m/s}^2}$

4-27

Take the +y-direction to be downward since that is the direction of the person's acceleration.

The free-body diagram for the person is

4-27 (cont)

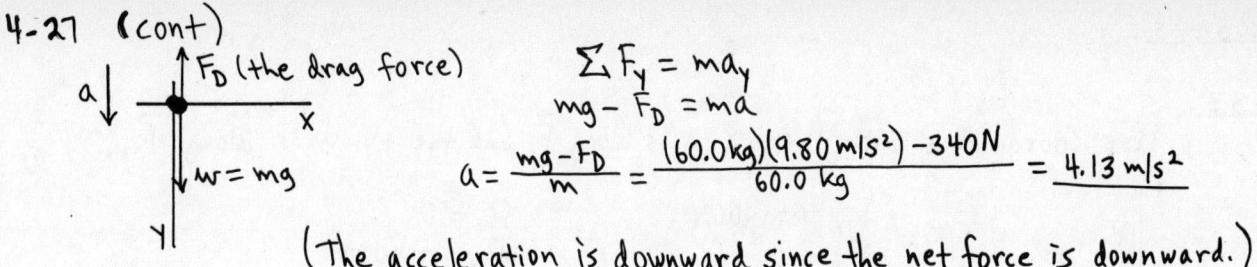

$$\Sigma F_y = ma_y$$
$$mg - F_D = ma$$

$$a = \frac{mg - F_D}{m} = \frac{(60.0\,\text{kg})(9.80\,\text{m/s}^2) - 340\,\text{N}}{60.0\,\text{kg}} = \underline{4.13\,\text{m/s}^2}$$

(The acceleration is downward since the net force is downward.)

4-29

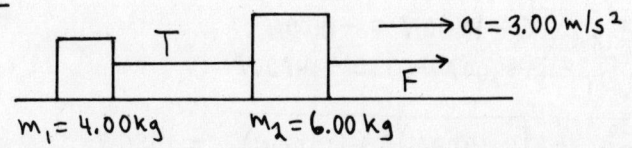

Since the crates are connected by a rope, they both have the same acceleration.

a) Consider the two crates and the rope connecting them as a single object of mass $m = m_1 + m_2 = 10.0\,\text{kg}$. The free-body diagram is

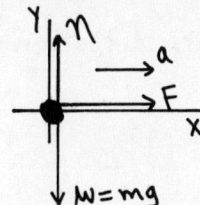

$$\Sigma F_x = ma_x$$
$$F = ma = (10.0\,\text{kg})(3.00\,\text{m/s}^2) = \underline{30.0\,\text{N}}$$

b) Consider the forces on the 4.00 kg crate:

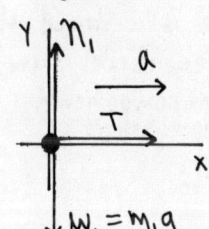

$$\Sigma F_x = ma_x$$
$$T = m_1 a = (4.00\,\text{kg})(3.00\,\text{m/s}^2) = \underline{12.0\,\text{N}}$$

As a check, can also consider the forces on the 6.00 kg crate:

$$\Sigma F_x = ma_x$$
$$F - T = m_2 a$$
$$T = F - m_2 a = 30.0\,\text{N} - (6.00\,\text{kg})(3.00\,\text{m/s}^2)$$
$$T = 30.0\,\text{N} - 18.0\,\text{N} = 12.0\,\text{N} \checkmark$$

Problems

4-33

Use coordinates with the +x-axis along $\vec{F_1}$ and the +y-axis along $\vec{R}$.

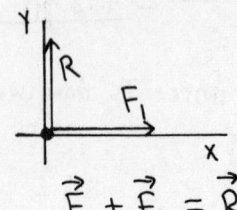

$F_{1x} = +1400 N$ $R_x = 0$
$F_{1y} = 0$ $R_y = +1400 N$

$\vec{F_1} + \vec{F_2} = \vec{R} \Rightarrow \vec{F_2} = \vec{R} - \vec{F_1}$
$F_{2x} = R_x - F_{1x} = 0 - 1400 N = -1400 N$
$F_{2y} = R_y - F_{1y} = +1400 N - 0 = +1400 N$

$F_2 = \sqrt{F_{2x}^2 + F_{2y}^2} = \sqrt{(-1400 N)^2 + (1400 N)^2} = 1980 N$

$\tan\theta = \dfrac{F_{2y}}{F_{2x}} = \dfrac{+1400 N}{-1400 N} = -1.00 \Rightarrow \theta = 135°$

The magnitude of $\vec{F_2}$ is __1980 N__ and its direction is __135°__ counterclockwise from the direction of $\vec{F_1}$.

4-35

$F_1 = 100 N$
$F_2 = 140 N$

If the box moves in the +x-direction it must have $a_y = 0$, so $\Sigma F_y = 0$. The smallest force the child can exert and still produce such motion is a force that makes the y-components of all three forces sum to zero, but that doesn't have any x-component.

Let $\vec{F_3}$ be the force exerted by the child.
$\Sigma F_y = 0 \Rightarrow F_{1y} + F_{2y} + F_{3y} = 0 \Rightarrow F_{3y} = -(F_{1y} + F_{2y})$

$F_{1y} = +F_1 \sin 60° = (100 N) \sin 60° = 86.6 N$
$F_{2y} = F_2 \sin(-30°) = -F_2 \sin 30° = -(140 N) \sin 30° = -70.0 N$
Then $F_{3y} = -(F_{1y} + F_{2y}) = -(86.8 N - 70.0 N) = -16.6 N$; $F_{3x} = 0$.

The smallest force the child can exert has magnitude __16.6 N__ and is directed at __90° clockwise__ from the +x-axis shown in the figure.

4-37

First use the information given about the height of the jump to calculate the speed he has at the instant his feet leave the ground. Use a coordinate system with the +y-axis upward and the origin at the position when his feet leave the ground.

4-37 (cont)

$v_y = 0$ (at the maximum height)
$v_{oy} = ?$
$a_y = -9.80 \text{ m/s}^2$
$y - y_0 = +1.2 \text{ m}$

$v_y^2 = v_{oy}^2 + 2a_y(y - y_0)$
$v_{oy} = \sqrt{-2a_y(y - y_0)}$
$v_{oy} = \sqrt{-2(-9.80 \text{ m/s}^2)(1.2 \text{ m})} = 4.85 \text{ m/s}$

Now consider the acceleration phase, from when he starts to jump until when his feet leave the ground. Use a coordinate system where the +y-axis is upward and the origin is at his position when he starts his jump. Calculate the average acceleration:

$(a_{av})_y = \dfrac{v_y - v_{oy}}{t} = \dfrac{4.85 \text{ m/s} - 0}{0.300 \text{ s}} = 16.17 \text{ m/s}^2$

Finally, find the average upward force that the ground must exert on him to produce this average upward acceleration. (Don't forget about the downward force of gravity.)

$(a_{av})_y \uparrow$ F_{av} (the average force the ground exerts on him)

$\downarrow mg = 890 \text{ N}$

$m = \dfrac{w}{g} = \dfrac{890 \text{ N}}{9.80 \text{ m/s}^2} = 90.8 \text{ kg}$

$\sum F_y = ma_y$
$F_{av} - mg = m(a_{av})_y$
$F_{av} = m(g + (a_{av})_y) = 90.8 \text{ kg}(9.80 \text{ m/s}^2 + 16.17 \text{ m/s}^2)$
$F_{av} = 2360 \text{ N}$

This is the average force exerted on him by the ground. But by Newton's 3rd law, the average force he exerts on the ground is equal and opposite, so is __2360 N__, __downward__.

4-41

a) $x = (9.0 \times 10^3 \text{ m/s}^2)t^2 - (8.0 \times 10^4 \text{ m/s}^3)t^3$
$x = 0$ at $t = 0$. When $t = 0.030 \text{ s}$, $x = (9.0 \times 10^3 \text{ m/s}^2)(0.030 \text{ s})^2 - (8.0 \times 10^4 \text{ m/s}^3)(0.030 \text{ s})^3$
$x = 5.9 \text{ m}$

The length of the barrel must be 5.9 m.

b) $v = \dfrac{dx}{dt} = (18.0 \times 10^3 \text{ m/s}^2)t - (24.0 \times 10^4 \text{ m/s}^3)t^2$
At $t = 0$, $v = 0$ (object starts from rest).
At $t = 0.030 \text{ s}$, when the object reaches the end of the barrel,
$v = (18.0 \times 10^3 \text{ m/s}^2)(0.030 \text{ s}) - (24.0 \times 10^4 \text{ m/s}^3)(0.030 \text{ s})^2 = \underline{320 \text{ m/s}}$

c) $\sum F_x = ma_x$, so find a.
$a = \dfrac{dv}{dt} = 18.0 \times 10^3 \text{ m/s}^2 - (48.0 \times 10^4 \text{ m/s}^3)t$

(i) at $t = 0$, $a = 18.0 \times 10^3 \text{ m/s}^2$ and $\sum F_x = (1.50 \text{ kg})(18.0 \times 10^3 \text{ m/s}^2) = \underline{2.7 \times 10^4 \text{ N}}$

(ii) at $t = 0.030 \text{ s}$, $a = 18.0 \times 10^3 \text{ m/s}^2 - (48.0 \times 10^4 \text{ m/s}^3)(0.030 \text{ s}) = 3.6 \times 10^3 \text{ m/s}^2$
and $\sum F_x = (1.50 \text{ kg})(3.6 \times 10^3 \text{ m/s}^2) = \underline{5.4 \times 10^3 \text{ N}}$

4-43

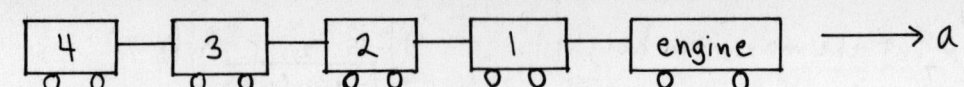

a) Consider all four cars together as one object. The horizontal force on this combined object is the force of the engine on the first car. The mass of all four cars together is 4m.

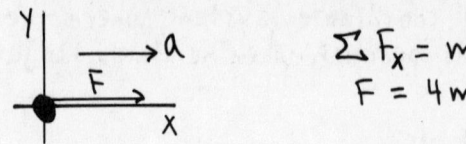

$\Sigma F_x = ma_x$
$F = 4ma$

b) Treat the last three cars together as one object. The horizontal force on this combined object is the force of the first car on the second car. The mass of these three cars together is 3m.

$\Sigma F_x = ma_x$
$F = 3ma$

c) Treat the last two cars together. Their total mass is 2m.

$\Sigma F_x = ma_x$
$F = 2ma$

d) By Newton's third law the force of the fourth car on the third car is equal and opposite to the force of the third car on the fourth. Free-body diagram for the fourth car:

$\Sigma F_x = ma_x$
$F = ma$

e) The forces would all be the same magnitude but would be in the opposite direction.

4-45

a) Take the +y-direction to be upward since that is the direction of the acceleration. The maximum upward acceleration is obtained from the maximum possible tension in the cables.

$\Sigma F_y = ma_y$
$T - mg = ma$
$a = \dfrac{T - mg}{m} = \dfrac{28{,}000\,N - (1800\,kg)(9.80\,m/s^2)}{1800\,kg} = \underline{5.76\,m/s^2}$

4-45 (cont)
b) What changes is the weight mg of the elevator.

$$a = \frac{T-mg}{m} = \frac{28{,}000\,N - (1800\,kg)(1.62\,m/s^2)}{1800\,kg} = \underline{13.9\,m/s^2}$$

4-49

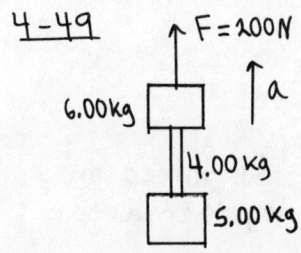

Note that in this problem the mass of the rope is given, and that it is not negligible compared to the other masses.

a) Treat the rope and the two blocks together as a single object, with mass $m = 6.00\,kg + 4.00\,kg + 5.00\,kg = 15.0\,kg$. Take $+y$ to be upward, since the acceleration is upward.

$$\Sigma F_y = ma_y$$
$$F - mg = ma$$
$$a = \frac{F - mg}{m} = \frac{200\,N - (15.0\,kg)(9.80\,m/s^2)}{15.0\,kg} = \frac{200\,N - 147\,N}{15.0\,kg}$$
$$a = \underline{3.53\,m/s^2}$$

b) Consider the forces on the top block (m = 6.00 kg), since the tension at the top of the rope (T_t) will be one of these forces.

$$\Sigma F_y = ma_y$$
$$F - mg - T_t = ma$$
$$T_t = F - m(g + a)$$
$$T_t = 200\,N - (6.00\,kg)(9.80\,m/s^2 + 3.53\,m/s^2) = \underline{120\,N}$$

__Alternatively__, can consider the forces on the combined object rope plus bottom block (m = 9.00 kg):

$$\Sigma F_y = ma_y$$
$$T_t - mg = ma$$
$$T_t = m(g+a) = 9.00\,kg\,(9.80\,m/s^2 + 3.53\,m/s^2) = 120\,N \checkmark$$

c) One way to do this is to consider the forces on the top half of the rope (m = 2.00 kg). Let T_m be the tension at the midpoint of the rope.

4-49 (cont)

$$\sum F_y = ma_y$$
$$T_t - T_m - mg = ma$$
$$T_m = T_t - m(a+g) = 120\,N - 2.00\,kg\,(3.53\,m/s^2 + 9.80\,m/s^2)$$
$$T_m = \underline{93.3\,N}$$

To check this answer we can alternatively consider the forces on the bottom half of the rope plus the lower block taken together as a combined object ($m = 2.00\,kg + 5.00\,kg = 7.00\,kg$):

$$\sum F_y = ma_y$$
$$T_m - mg = ma$$
$$T_m = m(g+a) = 7.00\,kg\,(9.80\,m/s^2 + 3.53\,m/s^2) = 93.3\,N,$$
which checks

CHAPTER 5

Exercises 3, 5, 7, 11, 13, 15, 17, 19, 21, 25, 27, 31, 35, 37, 39, 45, 47, 53, 55

Problems 63, 65, 67, 69, 73, 75, 77, 79, 83, 87, 93, 95, 97, 99, 101

Exercises

5-3

a) Force diagram for the person (T_1 and T_2 are the tension in each half of the rope):

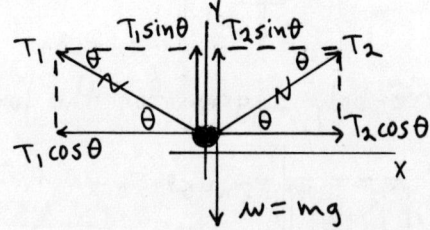

$\Sigma F_x = 0$
$T_2 \cos\theta - T_1 \cos\theta = 0$
$T_1 = T_2 = T$ (The tension is the same on both sides of the person.)

$\Sigma F_y = 0$
$T_1 \sin\theta + T_2 \sin\theta - mg = 0$
But $T_1 = T_2 = T$, so $2T\sin\theta = mg$

$T = \dfrac{mg}{2\sin\theta} = \dfrac{(81.6 \text{ kg})(9.80 \text{ m/s}^2)}{2 \sin 15.0°}$

$T = \underline{1540 \text{ N}}$

b) The relation $2T\sin\theta = mg$ still applies but now we are given that $T = 3.00 \times 10^4$ N (the breaking strength) and are asked to find θ.

$\sin\theta = \dfrac{mg}{2T} = \dfrac{(81.6 \text{ kg})(9.80 \text{ m/s}^2)}{2(3.00 \times 10^4 \text{ N})} = 0.0133 \Rightarrow \theta = \underline{0.764°}$

5-5

Force diagram for the wrecking ball:

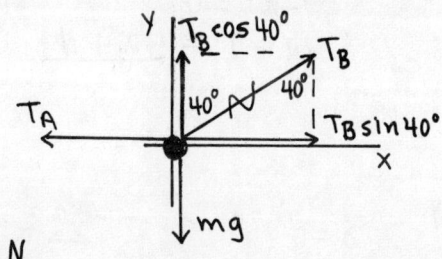

a) $\Sigma F_x = 0$
$T_B \sin 40° - T_A = 0$
$T_B = \dfrac{T_A}{\sin 40°} = \dfrac{580 \text{ N}}{\sin 40°} = \underline{902 \text{ N}}$

b) $\Sigma F_y = 0$
$T_B \cos 40° - mg = 0$
$m = \dfrac{T_B \cos 40°}{g} = \dfrac{(902 \text{ N})\cos 40°}{9.80 \text{ m/s}^2} = \underline{70.5 \text{ kg}}$

5-7

a)
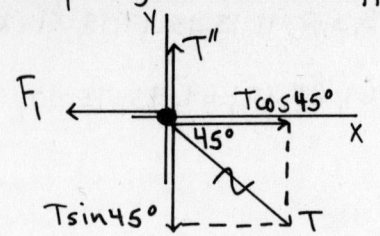

$T = 60.0$ N

Free-body diagram for the upper knot:

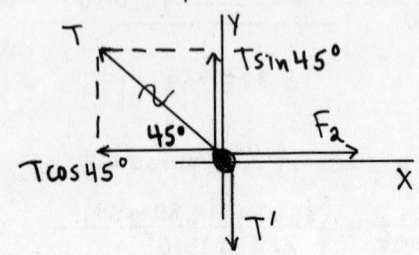

$\Sigma F_x = 0$
$T\cos 45° - F_1 = 0$
$F_1 = T\cos 45°$
$F_1 = (60.0\text{N})\cos 45°$
$F_1 = \underline{42.4 \text{ N}}$

Free-body diagram for the lower knot:

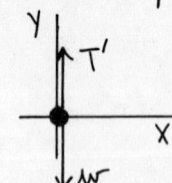

$\Sigma F_x = 0$
$F_2 - T\cos 45° = 0$
$F_2 = T\cos 45° = (60.0\text{N})\cos 45° = \underline{42.4 \text{ N}}$

b) Apply $\Sigma F_y = 0$ to the force diagram for the lower knot:
$\Sigma F_y = 0$
$T\sin 45° - T' = 0$
$T' = T\sin 45° = (60.0\text{N})\cos 45° = 42.4 \text{ N}$

Free-body diagram for the block:

$\Sigma F_y = 0$
$T' - w = 0$
$w = T' = \underline{42.4 \text{ N}}$

5-11

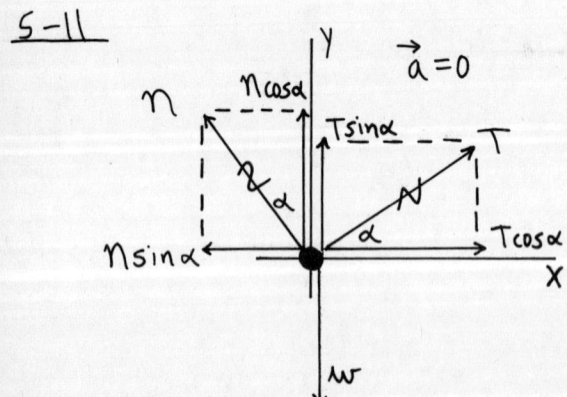

$\vec{a} = 0$

$\Sigma F_x = ma_x$
$T\cos\alpha - n\sin\alpha = 0$
$\boxed{T\cos\alpha = n\sin\alpha}$

$\Sigma F_y = ma_y$
$n\cos\alpha + T\sin\alpha - w = 0$
$\boxed{n\cos\alpha + T\sin\alpha = w}$

The first equation gives $n = T\dfrac{\cos\alpha}{\sin\alpha}$.

Use this in the second equation to eliminate n:

5-11 (cont)
$$\left(T\,\frac{\cos\alpha}{\sin\alpha}\right)\cos\alpha + T\sin\alpha = w$$
Multiply this equation by $\sin\alpha$:
$$T(\cos^2\alpha + \sin^2\alpha) = w\sin\alpha$$
$$\underline{T = w\sin\alpha} \quad (\text{since } \cos^2\alpha + \sin^2\alpha = 1)$$
Then $n = T\,\dfrac{\cos\alpha}{\sin\alpha} = w\sin\alpha\left(\dfrac{\cos\alpha}{\sin\alpha}\right) = \underline{w\cos\alpha}$

These results are the same as obtained in Example 5-3.

5-13

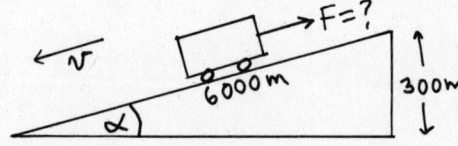

Let $\vec{F}$ be the resistive force.
$$\sin\alpha = \frac{300\,m}{6000\,m} = 0.0500$$

Free-body diagram for the car:

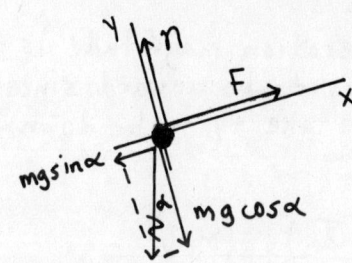

$\sum F_x = 0$ (constant speed $\Rightarrow \vec{a} = 0$)
$F - mg\sin\alpha = 0$
$F = mg\sin\alpha = (1500\,kg)(9.80\,m/s^2)(0.0500)$
$\underline{F = 735\,N}$

5-15

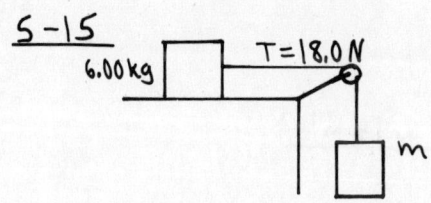

a) Free-body diagram for the 6.00 kg block:

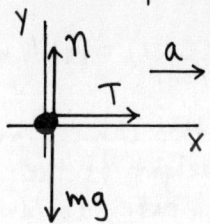

$\sum F_x = ma_x$
$T = ma$
$a = \dfrac{T}{m} = \dfrac{18.0\,N}{6.00\,kg} = \underline{3.00\,m/s^2}$

b) Free-body diagram for the hanging block: (Take $+y$ to be in the direction of the acceleration (downward). Since the blocks are connected by the rope, they have the same magnitude of acceleration.)

5-15 (cont)

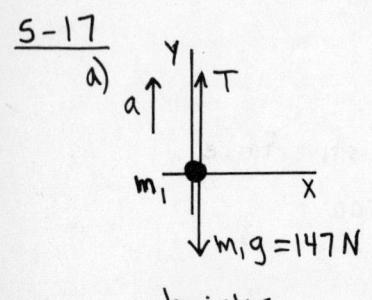

$\Sigma F_y = ma_y$
$mg - T = ma$
$m(g-a) = T$

$m = \dfrac{T}{g-a} = \dfrac{18.0\text{ N}}{9.80\text{ m/s}^2 - 3.00\text{ m/s}^2} = \underline{2.65\text{ kg}}$

5-17

a)

bricks: $a\uparrow$, $T\uparrow$, m_1, $m_1 g = 147\text{ N}\downarrow$

counterweight: $T\uparrow$, $a\downarrow$, m_2, $m_2 g = 274.4\text{ N}\downarrow$

b) Apply $\Sigma F_y = ma_y$ to each object. The acceleration magnitude is the same for the two objects. For the bricks take +y to be upward since $\vec{a}$ for the bricks is upward. For the counterweight take +y to be downward since $\vec{a}$ is downward.

bricks $\Sigma F_y = ma_y$
$\boxed{T - m_1 g = m_1 a}$

counterweight $\Sigma F_y = ma_y$
$\boxed{m_2 g - T = m_2 a}$

Add these two equations, to eliminate T:
$(m_2 - m_1) g = (m_1 + m_2) a$

$a = \left(\dfrac{m_2 - m_1}{m_1 + m_2}\right) g = \left(\dfrac{28.0\text{ kg} - 15.0\text{ kg}}{15.0\text{ kg} + 28.0\text{ kg}}\right)(9.80\text{ m/s}^2) = \underline{2.96\text{ m/s}^2}$

c) $T - m_1 g = m_1 a$ gives $T = m_1(a+g) = (15.0\text{ kg})(2.96\text{ m/s}^2 + 9.80\text{ m/s}^2) = \underline{191\text{ N}}$

As a check, calculate T using the other equation:
$m_2 g - T = m_2 a$ gives $T = m_2(g-a) = 28.0\text{ kg}(9.80\text{ m/s}^2 - 2.96\text{ m/s}^2) = 191\text{ N}$, which checks.

The tension is larger than the weight of the bricks; this causes the bricks to accelerate upward. The tension is less than the weight of the counterweight; this causes the counterweight to accelerate downward.

5-19

a) Consider the forces on the student. The reading on the scale equals the upward normal force exerted on the student. Also, the student is at rest relative to the elevator, so the acceleration of the student equals that of the elevator.

5-19 (cont)

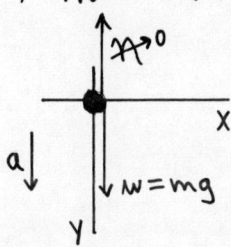

We know that $\vec{a}$ is upward since $n > w$.
$m = \frac{w}{g} = \frac{640N}{9.80 m/s^2} = 65.3 \text{ kg}$

$\Sigma F_y = ma_y$
$n - w = ma$
$a = \frac{n-w}{m} = \frac{800N - 640N}{65.3 \text{ kg}} = \underline{2.45 \text{ m/s}^2 \text{ (upward)}}$

b) Now $n < w$ so the student (and elevator) is accelerating downward. Take the $+y$-direction to be downward, in the direction of $\vec{a}$.

$\Sigma F_y = ma_y$
$w - n = ma$
$a = \frac{w-n}{m} = \frac{640N - 450N}{65.3 \text{ kg}} = \underline{2.91 \text{ m/s}^2 \text{ (downward)}}$

c) Now $n = 0$. $n < w$ so the acceleration is downward.

$\Sigma F_y = ma_y$
$mg = ma$
$a = g \text{ (downward)}$

The elevator and student are in free-fall; the elevator cable may have snapped.

5-21
a)

At rest $\Rightarrow a = 0$, so the forces on the lead sinker must sum to zero. The gravity force on the sinker is vertically downward, toward the center of the earth. The only other force on the sinker is the tension in the string. This force must be vertically upward in order for the two forces to sum to zero.

The sinker hangs such that the string is vertical.

b) <u>car moving up</u>

First consider the forces on the car, to calculate its acceleration. Take the x-axis parallel to the incline and the y-axis perpendicular to the incline.

5-21 (cont)

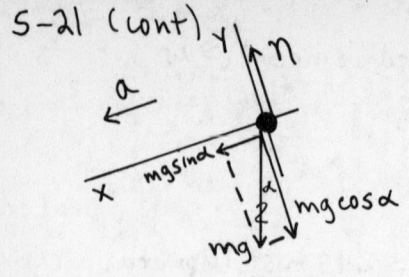

$$\Sigma F_x = ma_x$$
$$mg\sin\alpha = ma$$
$$\boxed{a = g\sin\alpha}$$

If the hanging sinker is at rest relative to the car then it also has acceleration $a = g\sin\alpha$ directed down the incline. Consider the forces on the sinker. Take the same x and y axes as for the car, so that $a_x = g\sin\alpha$ and $a_y = 0$. Assume that the string makes an angle β with the direction perpendicular to the ceiling of the car (the y-axis).

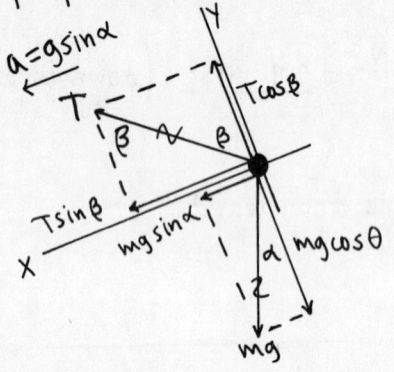

$$\Sigma F_x = ma_x$$
$$mg\sin\alpha + T\sin\beta = mg\sin\alpha$$

But this says $T\sin\beta = 0$, so $\beta = 0$ and the string is perpendicular to the ceiling of the car.

car moving down incline

The force diagrams for the car and for the sinker are the same as before, so again the sinker hangs such that the string is perpendicular to the ceiling.

There is <u>no deflection</u> during each stage of the motion.

5-25

a) constant speed $\Rightarrow a = 0$

Consider the free-body diagram for the box:

$$\Sigma F_y = ma_y$$
$$n - mg = 0$$
$$n = mg$$
$$f_k = \mu_k n = \mu_k mg$$

$$\Sigma F_x = ma_x$$
$$F - f_k = 0$$
$$F = f_k = \mu_k mg$$
$$F = (0.14)(8.00\,\text{kg})(9.80\,\text{m/s}^2)$$
$$F = \underline{11.0\,\text{N}}$$

5-25 (cont)

b)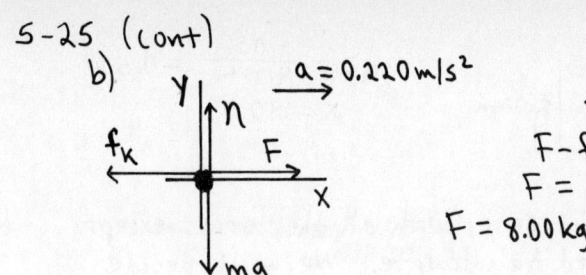

As in part (a), $f_k = \mu_k mg$
$\Sigma F_x = ma_x$
$F - f_k = ma$
$F = f_k + ma = \mu_k mg + ma = m(\mu_k g + a)$
$F = 8.00\text{kg}((0.14)(9.80\text{m/s}^2) + 0.220\text{m/s}^2) = \underline{12.7\text{ N}}$

c) The normal force $n = mg$ is reduced. This in turn reduces the friction force, so the magnitude of the force $\vec{F}$ required to move the box is less.

part (a): $F = \mu_k mg = (0.14)(8.00\text{kg})(1.62\text{m/s}^2) = \underline{1.8\text{ N}}$

part (b): $F = m(\mu_k g + a) = 8.00\text{kg}((0.14)(1.62\text{m/s}^2) + 0.220\text{m/s}^2) = \underline{3.6\text{ N}}$

5-27

a) constant speed $\Rightarrow a = 0$

Consider the free-body diagram for the crate. Let $\vec{F}$ be the horizontal force applied by the worker. The friction is kinetic friction since the crate is sliding along the surface.

$\Sigma F_y = ma_y$
$n - mg = 0$
$n = mg$
So $f_k = \mu_k n = \mu_k mg$

$\Sigma F_x = ma_x$
$F - f_k = 0$
$F = f_k = \mu_k mg = (0.20)(9.40\text{kg})(9.80\text{m/s}^2)$
$F = \underline{18.4\text{ N}}$

b) Now the only horizontal force on the crate is the kinetic friction force. Calculate the acceleration it produces. The friction force is $f_k = \mu_k mg$, just as in part (a).

$\Sigma F_x = ma_x$
$-f_k = ma$
$-\mu_k mg = ma$
$a = -\mu_k g = -(0.20)(9.80\text{m/s}^2) = -1.96\text{ m/s}^2$

Use the constant acceleration equations to find the time it takes this acceleration to reduce the speed from 4.50 m/s to zero:
$v = 0$
$v_0 = 4.50 \text{ m/s}$
$a = -1.96 \text{ m/s}^2$
$t = ?$

$v = v_0 + at$
$t = \dfrac{v - v_0}{a} = \dfrac{0 - 4.50\text{m/s}}{-1.96\text{m/s}^2} = \underline{2.3\text{ s}}$

63

5-31

constant speed
$\Rightarrow a = 0$

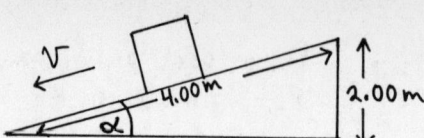

$\sin\alpha = \dfrac{2.00\,m}{4.00\,m} = 0.500$

$\alpha = 30.0°$

a) Consider the free-body diagram for the safe, with all the forces except for the applied force we are being asked to calculate. We must decide whether this force must be down the incline or up the incline, to make the total resultant force zero. Note that we know that the kinetic friction force f_k is directed up the incline, since the friction force opposes the motion. Use coordinates parallel and perpendicular to the incline.

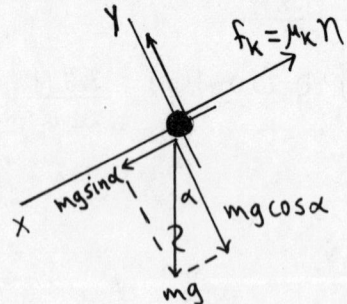

$mg\sin\alpha = (260\,kg)(9.80\,m/s^2)\sin 30.0° = 1274\,N$

$\sum F_y = ma_y$
$n - mg\cos\alpha = 0$
$n = mg\cos\alpha = (260\,kg)(9.80\,m/s^2)\cos 30.0° = 2207\,N$
$f_k = \mu_k n = (0.30)(2207\,N) = 662\,N$

$mg\sin\alpha > f_k$, so for the forces to balance ($a=0$) more force directed up the incline is needed; the safe must be __held back__ if it is to travel at constant speed.

b) Now we can add the applied force $\vec{F}$ to the free-body diagram, since we have determined its direction.

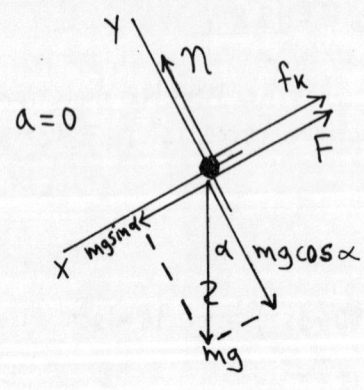

$\sum F_x = ma_x$
$mg\sin\alpha - F - f_k = 0$
$F = mg\sin\alpha - f_k$
$F = 1274\,N - 662\,N = \underline{612\,N}$

5-35

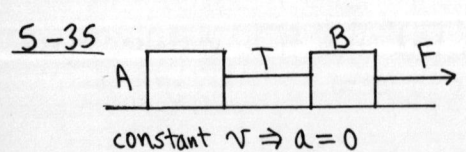

constant $v \Rightarrow a = 0$

free-body diagram for A:

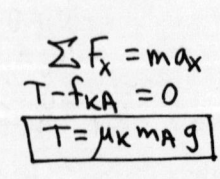

$\sum F_y = ma_y$
$n_A - m_A g = 0$
$n_A = m_A g$
$f_{kA} = \mu_k n_A = \mu_k m_A g$

$\sum F_x = ma_x$
$T - f_{kA} = 0$
$\boxed{T = \mu_k m_A g}$

64

5-35 (cont)
free-body diagram for B:

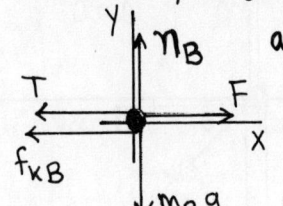

$\Sigma F_y = ma_y$
$n_B - m_B g = 0$
$n_B = m_B g$
$f_{kB} = \mu_k n_B = \mu_k m_B g$

$\Sigma F_x = ma_x$
$F - T - f_{kB} = 0$
$\boxed{F = T + \mu_k m_B g}$

Use the first equation to replace T in the second $\Rightarrow$ $F = \mu_k m_A g + \mu_k m_B g = \underline{\mu_k (m_A + m_B) g}$

5-37

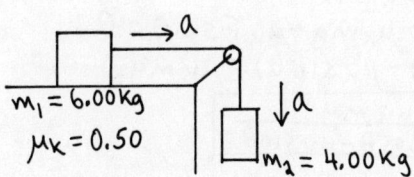

The magnitude of the acceleration is the same for both blocks. For each block take a positive coordinate direction to be the direction of the block's acceleration.

block on the table:

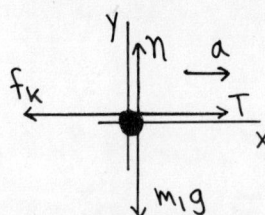

$\Sigma F_y = ma_y$
$n - m_1 g = 0$
$n = m_1 g$
$f_k = \mu_k n = \mu_k m_1 g$

$\Sigma F_x = ma_x$
$T - f_k = m_1 a$
$\boxed{T - \mu_k m_1 g = m_1 a}$

hanging block:

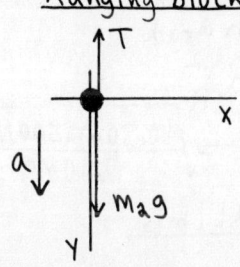

$\Sigma F_y = ma_y$
$m_2 g - T = m_2 a$
$\boxed{T = m_2 g - m_2 a}$

a) Use the second equation in the first
$\Rightarrow m_2 g - m_2 a - \mu_k m_1 g = m_1 a$
$(m_1 + m_2) a = (m_2 - \mu_k m_1) g$

$a = \dfrac{(m_2 - \mu_k m_1) g}{m_1 + m_2} = \dfrac{(4.00\,kg - (0.50)(6.00\,kg))(9.80\,m/s^2)}{6.00\,kg + 4.00\,kg} = \underline{0.98\,m/s^2}$

b) $T = m_2 g - m_2 a = m_2 (g - a) = 4.00\,kg\,(9.80\,m/s^2 - 0.98\,m/s^2) = \underline{35.3\,N}$

or, to check
$T - \mu_k m_1 g = m_1 a \Rightarrow T = m_1 (a + \mu_k g) = 6.00\,kg\,(0.98\,m/s^2 + (0.50)(9.80\,m/s^2))$
$T = 35.3\,N \checkmark$

5-39

a)

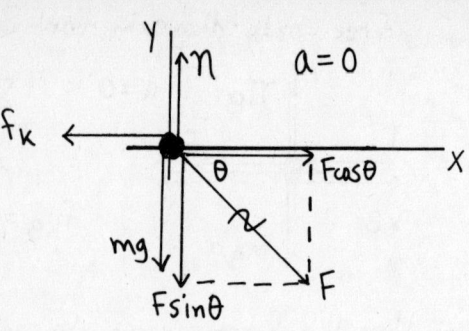

Free-body diagram for the crate:

constant $v \Rightarrow a = 0$
and friction is f_k

$\Sigma F_y = ma_y$
$n - mg - F\sin\theta = 0$
$n = mg + F\sin\theta$
$f_k = \mu_k n = \mu_k mg + \mu_k F\sin\theta$

$\Sigma F_x = ma_x$
$F\cos\theta - f_k = 0$
$F\cos\theta - \mu_k mg - \mu_k F\sin\theta = 0$
$F(\cos\theta - \mu_k \sin\theta) = \mu_k mg$
$$\boxed{F = \frac{\mu_k mg}{\cos\theta - \mu_k \sin\theta}}$$

b) "start the crate moving" means the same force diagram as in part (a), except that μ_k is replaced by μ_s.

Thus $F = \dfrac{\mu_s mg}{\cos\theta - \mu_s \sin\theta}$.

$F \to \infty$ if $\cos\theta - \mu_s \sin\theta = 0 \Rightarrow \mu_s = \dfrac{\cos\theta}{\sin\theta} = \dfrac{1}{\tan\theta}$

5-45

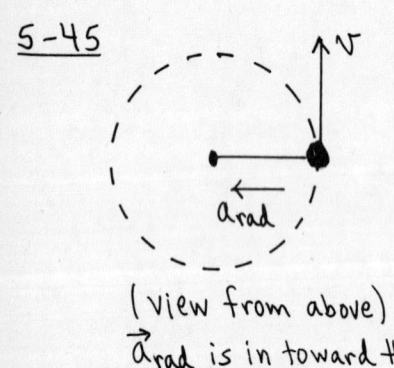

(view from above)
$\vec{a}_{rad}$ is in toward the center of the circular path

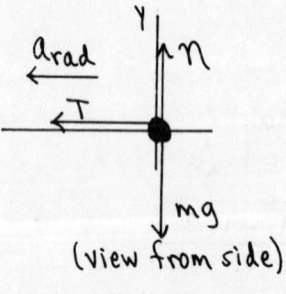

(view from side)
The maximum v produces $T = 500$ N.

$\Sigma F_x = ma_x$
$T = m\, a_{rad}$
$T = m\dfrac{v^2}{R}$
$v = \sqrt{\dfrac{RT}{m}} = \sqrt{\dfrac{(0.80\text{ m})(500\text{ N})}{0.90\text{ kg}}}$
$\underline{v = 21.1 \text{ m/s}}$

5-47

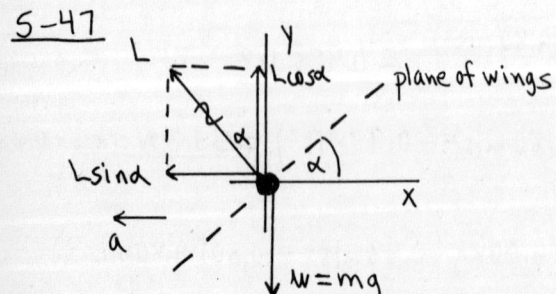

plane of wings

The lift force is perpendicular to the wings. The bank angle α is the angle between the plane of the wings and the horizontal.

For a constant-altitude turn the acceleration is horizontal, in toward the center of the circular path.

5-47 (cont)

Use coordinates where x is horizontal and y is vertical. Then $a_y = 0$.

$\Sigma F_y = ma_y$
$L\cos\alpha - w = 0$
$L = 3.8w \Rightarrow 3.8\cancel{w}\cos\alpha = \cancel{w}$
$\cos\alpha = \frac{1}{3.8} \Rightarrow \underline{\alpha = 75°}$

5-53

At each point $\vec{a}_{rad}$ is directed toward the center of the circular path.

a) Force diagram for the pilot at the top of the path:

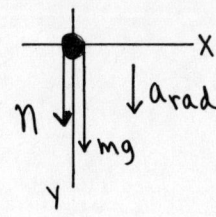

"the pilot feels weightless" means that the vertical normal force n exerted on the pilot by the chair on which she sits is zero.

$\Sigma F_y = ma_y$
$\cancel{m}g = \cancel{m}a_{rad}$
$g = \frac{v^2}{R} \Rightarrow v = \sqrt{gR} = \sqrt{(9.80\text{m/s}^2)(250\text{m})} = 49.5\text{m/s}$

$v = (49.5\text{m/s})\left(\frac{1\text{km}}{10^3\text{m}}\right)\left(\frac{3600\text{s}}{1\text{h}}\right) = \underline{178\text{ km/h}}$

b) Force diagram for the pilot at the bottom of the path. Note that the vertical normal force exerted on her by the chair on which she sits is now upward.

$\Sigma F_y = ma_y$
$n - mg = m\frac{v^2}{R}$
$n = mg + m\frac{v^2}{R}$; this normal force is her apparent weight

$w = 500\text{N} \Rightarrow m = \frac{w}{g} = 51.0\text{kg}$
$v = (250\text{km/h})\left(\frac{1\text{h}}{3600\text{s}}\right)\left(\frac{10^3\text{m}}{1\text{km}}\right) = 69.4\text{m/s}$

Thus $n = 500\text{N} + 51.0\text{kg}\frac{(69.4\text{m/s})^2}{250\text{m}} = \underline{1480\text{N}}$

5-55

Consider the free-body diagram for the water when the pail is at the top of its circular path. The radial acceleration is in toward the center of the circle so at this point it is downward. n is the downward normal force exerted on the water by the bottom of the pail.

5-55 (cont)

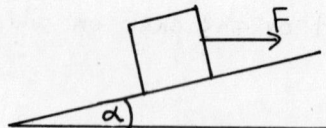

$$\Sigma F_y = ma_y$$
$$n + mg = m\frac{v^2}{R}$$

At the minimum speed the water is just ready to lose contact with the bottom of the pail, so at this speed $n \to 0$. (Note that the force n cannot be upward.)

$$n \to 0 \Rightarrow mg = m\frac{v^2}{R}$$
$$v = \sqrt{gR} = \sqrt{(9.80 \text{ m/s}^2)(0.700 \text{ m})} = \underline{2.62 \text{ m/s}}$$

Problems

5-63

No friction.
Constant speed $\Rightarrow a = 0$.

Since $a = 0$ it is equally convenient to use
(1) coordinate axes parallel and perpendicular to the incline;
(2) coordinate axes that are horizontal and vertical.
We will work the problem both ways.

coordinate axes parallel and perpendicular to the incline

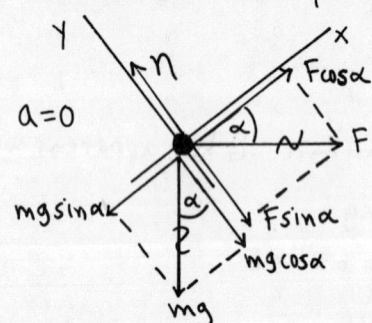

$$\Sigma F_x = ma_x$$
$$F\cos\alpha - mg\sin\alpha = 0$$
$$F = mg\frac{\sin\alpha}{\cos\alpha}$$
$$\boxed{F = mg\tan\alpha}$$

coordinates that are horizontal and vertical

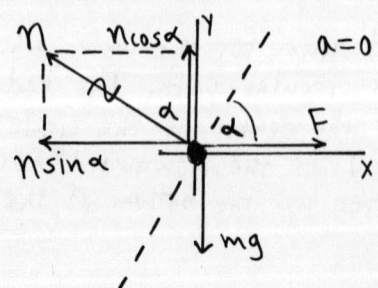

$$a = 0$$

$$\Sigma F_x = ma_x$$
$$F - n\sin\alpha = 0$$

$$\Sigma F_y = ma_y$$
$$n\cos\alpha - mg = 0$$
$$n = \frac{mg}{\cos\alpha}$$

Combine these two equations to eliminate n:
$$F = n\sin\alpha = \left(\frac{mg}{\cos\alpha}\right)\sin\alpha$$

$F = mg\tan\alpha$, the same as before

5-65

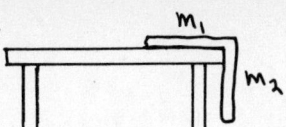

Let m_1 be the mass of that part of the rope that is on the table, and let m_2 be the mass of that part of the rope that is hanging over the edge.
($m_1 + m_2 = m$, the total mass of the rope)

Since the mass of the rope is not being neglected, the tension in the rope varies along the length of the rope. Let T be the tension in the rope at that point that is at the edge of the table.

Free-body diagram for the hanging section of the rope:

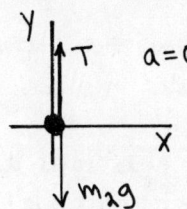

$\sum F_y = ma_y$
$T - m_2 g = 0$
$\boxed{T = m_2 g}$

Free-body diagram for the part of the rope that is on the table:

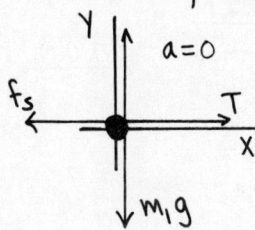

$\sum F_y = ma_y$
$n - m_1 g = 0$
$n = m_1 g$

When the maximum amount of rope hangs over the edge the static friction has its maximum value:
$f_s = \mu_s n = \mu_s m_1 g$

$\sum F_x = ma_x$
$T - f_s = 0$
$\boxed{T = \mu_s m_1 g}$

Use the first equation to replace T:
$m_2 g = \mu_s m_1 g$
$m_2 = \mu_s m_1$

The fraction that hangs over is $\dfrac{m_2}{m} = \dfrac{m_2}{m_1 + m_2} = \dfrac{\mu_s m_1}{m_1 + \mu_s m_1} = \dfrac{\mu_s}{1 + \mu_s}$

5-67
a)

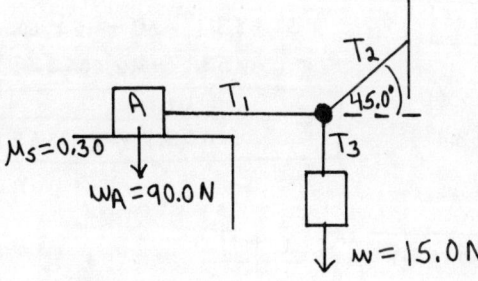

Free-body diagram for the hanging block:

$T_3 \quad a=0 \qquad \sum F_y = ma_y$
$\qquad\qquad\qquad T_3 - w = 0$
$\qquad\qquad\qquad \underline{T_3 = 15.0 \text{ N}}$

Free-body diagram for the knot:

$\sum F_y = ma_y$
$T_2 \sin 45.0° - T_3 = 0$
$T_2 = \dfrac{T_3}{\sin 45.0°} = \dfrac{15.0 \text{ N}}{\sin 45.0°}$
$\underline{T_2 = 21.2 \text{ N}}$

$\sum F_x = ma_x$
$T_2 \cos 45.0° - T_1 = 0$
$T_1 = T_2 \cos 45.0°$
$\underline{T_1 = 15.0 \text{ N}}$

5-67 (cont)
Free-body diagram for block A:

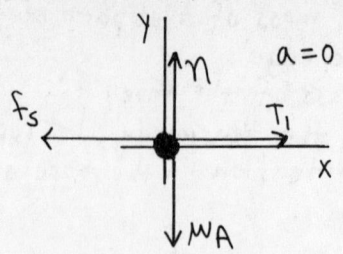

$\Sigma F_x = ma_x$
$T_1 - f_s = 0$
$f_s = T_1 = \underline{15.0 \text{ N}}$

Note: $\Sigma F_y = ma_y$
$n - w_A = 0$
$n = w_A = 90.0 \text{ N}$

$\mu_s n = (0.30)(90.0 \text{ N}) = 27.0 \text{ N}$
$f_s < \mu_s n$; for this value of w the static friction force can hold the blocks in place.

b) We have all the same free-body diagrams and force equations as in part (a), but now the static friction force has its largest possible value, $f_s = \mu_s n = 27.0 \text{ N}$. Then $T_1 = f_s = \underline{27.0 \text{ N}}$.

$T_2 \cos 45.0° - T_1 = 0 \Rightarrow T_2 = \dfrac{T_1}{\cos 45.0°} = \dfrac{27.0 \text{ N}}{\cos 45.0°} = \underline{38.2 \text{ N}}$
$T_2 \sin 45.0° - T_3 = 0 \Rightarrow T_3 = T_2 \sin 45.0° = (38.2 \text{ N}) \sin 45.0° = \underline{27.0 \text{ N}}$
$T_3 - w = 0 \Rightarrow w = T_3 = \underline{27.0 \text{ N}}$

5-69

Consider the forces on the scrub brush. Note that the normal force exerted by the wall is horizontal, since it is perpendicular to the wall. The kinetic friction force exerted by the wall is parallel to the wall and opposes the motion, so it is vertically downward.

$\Sigma F_x = ma_x$
$n - F \cos 53.1° = 0$
$\boxed{n = F \cos 53.1°}$

Then $f_k = \mu_k n$
$f_k = \mu_k F \cos 53.1°$

$\Sigma F_y = ma_y$
$F \sin 53.1° - w - f_k = 0$
$F \sin 53.1° - w - \mu_k F \cos 53.1° = 0$
$F(\sin 53.1° - \mu_k \cos 53.1°) = w$
$\boxed{F = \dfrac{w}{\sin 53.1° - \mu_k \cos 53.1°}}$

a) $F = \dfrac{w}{\sin 53.1° - \mu_k \cos 53.1°} = \dfrac{8.00 \text{ N}}{\sin 53.1° - (0.30) \cos 53.1°} = \underline{12.9 \text{ N}}$

b) $n = F \cos 53.1° = (12.9 \text{ N}) \cos 53.1° = \underline{7.75 \text{ N}}$

5-73

Block B is pulled to the left at constant speed $\Rightarrow$ block A moves to the right at constant speed and $a = 0$ for each block.

5-73 (cont)

Free-body diagram for block A:

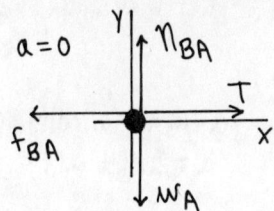

n_{BA} is the normal force that B exerts on A.
$f_{BA} = \mu_k n_{BA}$ is the kinetic friction force that B exerts on A. Block A moves to the right relative to B, and f_{BA} opposes this motion $\Rightarrow f_{BA}$ is to the left.
Note also that F acts just on B, not on A.

$\Sigma F_y = ma_y$
$n_{BA} - w_A = 0$
$n_{BA} = 2.70 \text{ N}$
$f_{BA} = \mu_k n_{BA} = (0.25)(2.70 \text{ N}) = 0.675 \text{ N}$

$\Sigma F_x = ma_x$
$T - f_{BA} = 0$
$T = f_{BA} = 0.675 \text{ N}$

Free-body diagram for block B:

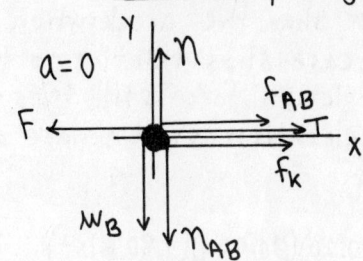

n_{AB} is the normal force that block A exerts on B. By Newton's third law n_{AB} and n_{BA} are equal and opposite $\Rightarrow n_{AB} = 2.70 \text{ N}$.

f_{AB} is the kinetic friction force that A exerts on B. Block B moves to the left relative to A and f_{AB} opposes this motion $\Rightarrow f_{AB}$ is to the right.

Also, $f_{AB} = \mu_k n_{AB} = (0.25)(2.70 \text{ N}) = 0.675 \text{ N}$. But also note that f_{AB} and f_{BA} are a third law action-reaction pair, so they must be equal and opposite and this is indeed what our calculation gives.

n and f_k are the normal and the friction force exerted by the floor on block B; $f_k = \mu_k n$. Note that block B moves to the left relative to the floor and f_k opposes this motion, so f_k is to the right.

$\Sigma F_y = ma_y$
$n - w_B - n_{AB} = 0$
$n = w_B + n_{AB} = 5.40 \text{ N} + 2.70 \text{ N} = 8.10 \text{ N}$
Then $f_k = \mu_k n = (0.25)(8.10 \text{ N}) = 2.025 \text{ N}$.

$\Sigma F_x = ma_x$
$f_{AB} + T + f_k - F = 0$
$F = T + f_{AB} + f_k$
$F = 0.675 \text{ N} + 0.675 \text{ N} + 2.025 \text{ N} = \underline{3.38 \text{ N}}$

5-75
a)

$\to v_0$

$\to a = 1.80 \text{ m/s}^2$

First calculate the <u>maximum</u> acceleration that the static friction force can give to the case:

5-75 (cont)

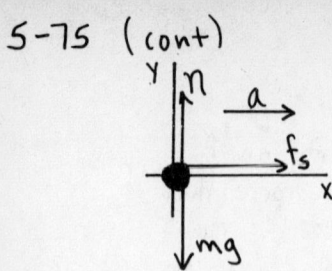

The static friction force is to the right since it tries to make the case move with the truck. The maximum value it can have is $f_s = \mu_s n$.

$\sum F_y = ma_y$ $\sum F_x = ma_x$ $a = \mu_s g = (0.30)(9.80 \text{ m/s}^2)$
$n - mg = 0$ $f_s = ma$ $a = 2.94 \text{ m/s}^2$
$n = mg$ $\mu_s \cancel{m} g = \cancel{m} a$
$f_s = \mu_s n = \mu_s mg$

The truck's acceleration is less than this so the case doesn't slip relative to the truck; the case's acceleration is $a = 1.80 \text{ m/s}^2$ (eastward).

Then $f_s = ma = (30.0 \text{ kg})(1.80 \text{ m/s}^2) = \underline{54.0 \text{ N}}$, eastward.

b) Now the acceleration of the truck is greater than the acceleration static friction can give the case. Therefore, the case slips relative to the truck and the friction is kinetic friction. The friction force still tries to keep the case moving with the truck, so the acceleration of the case and the friction force are both westward.

$\sum F_y = ma_y$ $f_k = \mu_k mg = (0.20)(30.0 \text{ kg})(9.80 \text{ m/s}^2)$
$n - mg = 0$ $f_k = \underline{58.8 \text{ N}}$, westward
$n = mg$

Note: $f_k = ma \Rightarrow a = \dfrac{f_k}{m} = \dfrac{58.8 \text{ N}}{30.0 \text{ kg}} = 1.96 \text{ m/s}^2$. The magnitude of the acceleration of the case is less than that of the truck and the case slides toward the front of the truck.

5-77

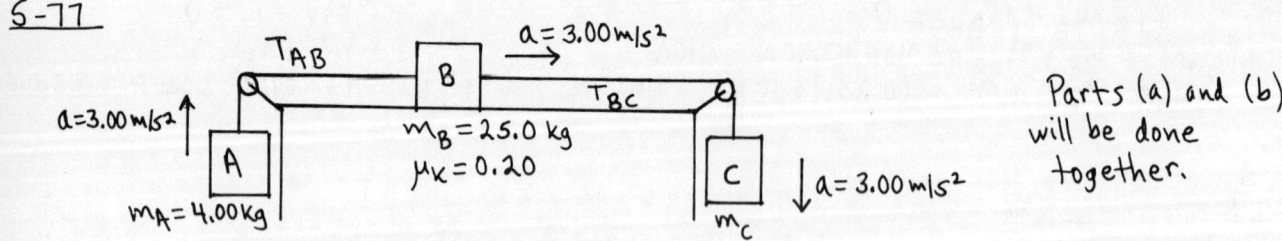

Parts (a) and (b) will be done together.

Consider the forces on each block. Note that each block has the same magnitude of acceleration, but in different directions. For each block let the direction of $\vec{a}$ be a positive coordinate direction.

5-77 (cont)

Block A:
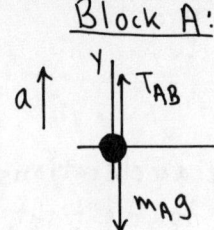

$\Sigma F_y = m a_y$
$T_{AB} - m_A g = m_A a$
$T_{AB} = m_A(a+g) = 4.00 \text{ kg}(3.00 \text{ m/s}^2 + 9.80 \text{ m/s}^2) = \underline{51.2 \text{ N}}$

Block B:
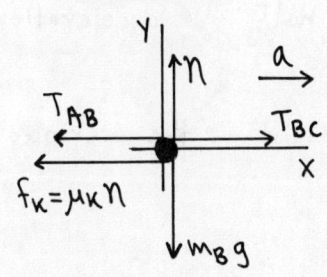

$\Sigma F_y = m a_y$
$n - m_B g = 0$
$n = m_B g$

$f_k = \mu_k n = \mu_k m_B g$
$f_k = (0.20)(25.0 \text{ kg})(9.80 \text{ m/s}^2)$
$f_k = 49.0 \text{ N}$

$\Sigma F_x = m a_x$
$T_{BC} - T_{AB} - f_k = m_B a$
$T_{BC} = T_{AB} + f_k + m_B a$
$T_{BC} = 51.2 \text{ N} + 49.0 \text{ N} + (25.0 \text{ kg})(3.00 \text{ m/s}^2)$
$T_{BC} = \underline{175 \text{ N}}$

Block C:
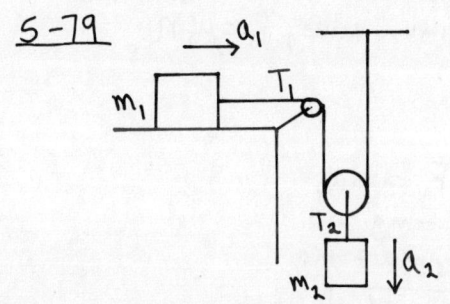

$\Sigma F_y = m a_y$
$m_C g - T_{BC} = m_C a$
$m_C(g-a) = T_{BC}$
$m_C = \dfrac{T_{BC}}{g-a} = \dfrac{175 \text{ N}}{9.80 \text{ m/s}^2 - 3.00 \text{ m/s}^2} = \underline{25.7 \text{ kg}}$

5-79

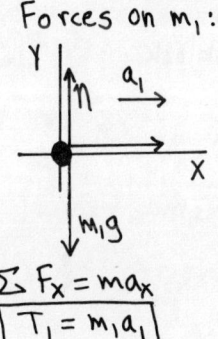

Forces on m_1:

$\Sigma F_x = m a_x$
$\boxed{T_1 = m_1 a_1}$

Forces on m_2:
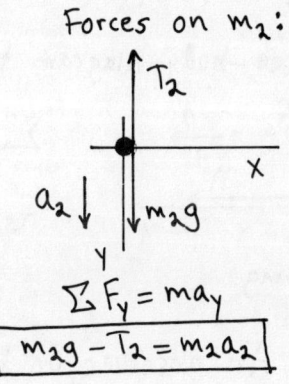

$\Sigma F_y = m a_y$
$\boxed{m_2 g - T_2 = m_2 a_2}$

This gives us two equations, but there are 4 unknowns ($T_1, T_2, a_1,$ and a_2) so two more equations are required.

Free-body diagram for the moveable pulley (mass m):

$\Sigma F_y = m a_y$
$mg + T_2 - 2T_1 = ma$

But our pulleys have negligible mass
$\Rightarrow mg = ma = 0$
$\Rightarrow \boxed{T_2 = 2T_1}$

5-79 (cont)

Combine these three equations to eliminate T_1 and T_2:
$$m_2 g - T_2 = m_2 a_2 \Rightarrow m_2 g - 2T_1 = m_2 a_2$$
Then $T_1 = m_1 a_1 \Rightarrow \boxed{m_2 g - 2 m_1 a_1 = m_2 a_2}$

There are still two unknowns, a_1 and a_2. But the accelerations a_1 and a_2 are related. In any time interval, if m_1 moves to the right a distance d, then in the same time m_2 moves downward a distance $d/2$. One of the constant acceleration kinematic equations says $x - x_0 = v_0 t + \frac{1}{2} a t^2$, so if m_2 moves half the distance it must have half the acceleration of m_1: $a_2 = \frac{1}{2} a_1$, or $\boxed{a_1 = 2 a_2}$.

This is the additional equation we need. Use it in the previous boxed equation $\Rightarrow m_2 g - 2 m_1 (2 a_2) = m_2 a_2$
$$a_2 (4 m_1 + m_2) = m_2 g$$

$$a_2 = \frac{m_2 g}{4 m_1 + m_2} \quad \text{and} \quad a_1 = 2 a_2 = \frac{2 m_2 g}{4 m_1 + m_2}.$$

5-83

The cart and the block have the same acceleration. The normal force exerted by the cart on the block is perpendicular to the front of the cart, so is horizontal and to the right.

The friction force on the block is directed so as to hold the block up against the downward pull of gravity. We want to calculate the minimum a required, so take static friction to have its maximum value, $f_s = \mu_s n$.

Free-body diagram for the block:

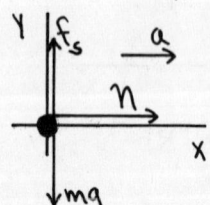

$\sum F_x = m a_x$
$n = m a$
$f_s = \mu_s n = \mu_s m a$

$\sum F_y = m a_y$
$f_s - m g = 0$
$\mu_s m a = m g$
$\boxed{a = g / \mu_s}$

An observer on the cart sees the block pinned there, with no reason for a horizontal force on it because the block is at rest relative to the cart. Therefore, such an observer concludes that $n = 0$ and thus $f_s = 0$, and he doesn't understand what holds the block up against the downward force of gravity. The reason for this difficulty is that $\sum \vec{F} = m \vec{a}$ does not apply in a coordinate frame attached to the cart. This reference frame is accelerated, and hence is not inertial.

5-87

a) Force diagram for the two blocks taken as a single combined object:

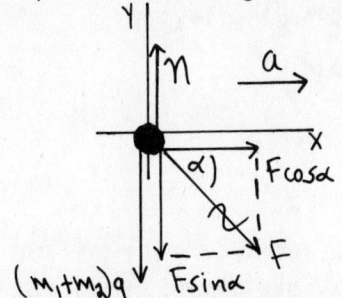

$\sum F_x = ma_x$
$F\cos\alpha = (m_1 + m_2)a$
$a = \dfrac{F\cos\alpha}{m_1 + m_2}$

$\sum F_y = ma_y$
$n - F\sin\alpha - (m_1 + m_2)g = 0$
$n = F\sin\alpha + (m_1 + m_2)g$

b) We now know the acceleration of each block if they move together. Now apply $\sum \vec{F} = m\vec{a}$ to each block.

bottom block:

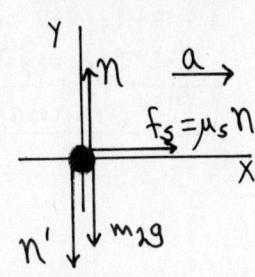

n is the normal force applied by the surface, and was calculated in part (a).
n' is the normal force applied by the top block.
Note that $\vec{F}$ acts only on the top block so doesn't appear in this force diagram.

$\sum F_x = ma_x$
$f_s = m_2 a$
$\mu_s n' = m_2 a$

$\sum F_y = ma_y$
$n - n' - m_2 g = 0$
$n' = n - m_2 g$

In $n' = n - m_2 g$ use the value of n calculated in part (a):
$n' = n - m_2 g = F\sin\alpha + m_1 g + m_2 g - m_2 g = F\sin\alpha + m_1 g$

Use this result, and also $a = F\cos\alpha/(m_1 + m_2)$ from part (a), in $\mu_s n' = m_2 a$:
$\mu_s F\sin\alpha + \mu_s m_1 g = m_2 F\cos\alpha/(m_1 + m_2)$

Solve for F:
$(\mu_s \sin\alpha (m_1 + m_2) - m_2 \cos\alpha) F = -\mu_s m_1 g (m_1 + m_2)$

$F = \dfrac{\mu_s m_1 (m_1 + m_2) g}{m_2 \cos\alpha - \mu_s \sin\alpha (m_1 + m_2)}$, as was to be shown.

An alternative approach is to apply $\sum \vec{F} = m\vec{a}$ to the top block:

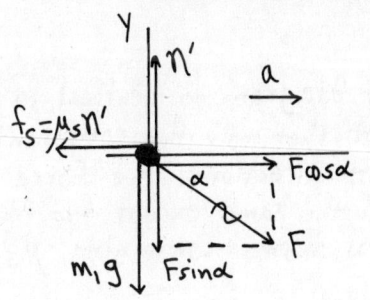

$\sum F_x = ma_x$
$F\cos\alpha - f_s = m_1 a$
$F\cos\alpha - \mu_s n' = m_1 a$

$\sum F_y = ma_y$
$n' - m_1 g - F\sin\alpha = 0$
$n' = m_1 g + F\sin\alpha$

Combine these two equations to eliminate n':
$F\cos\alpha - \mu_s m_1 g - \mu_s F\sin\alpha = m_1 a$

Use the value of a calculated in part (a):
$F\cos\alpha - \mu_s m_1 g - \mu_s F\sin\alpha = \dfrac{m_1 F\cos\alpha}{m_1 + m_2}$

5-87 (cont)

Solve for F:
$$F((m_1+m_2)\cos\alpha - \mu_s(m_1+m_2)\sin\alpha - m_1\cos\alpha) = \mu_s m_1(m_1+m_2)g$$

$$F = \frac{\mu_s m_1(m_1+m_2)g}{m_2\cos\alpha - \mu_s(m_1+m_2)\sin\alpha}, \text{ which is the same result.}$$

5-93

a) To keep the car from sliding up the banking the static friction force is directed down the incline. At the maximum speed the static friction force has its maximum value, $f_s = \mu_s n$.

Free-body diagram for the car:

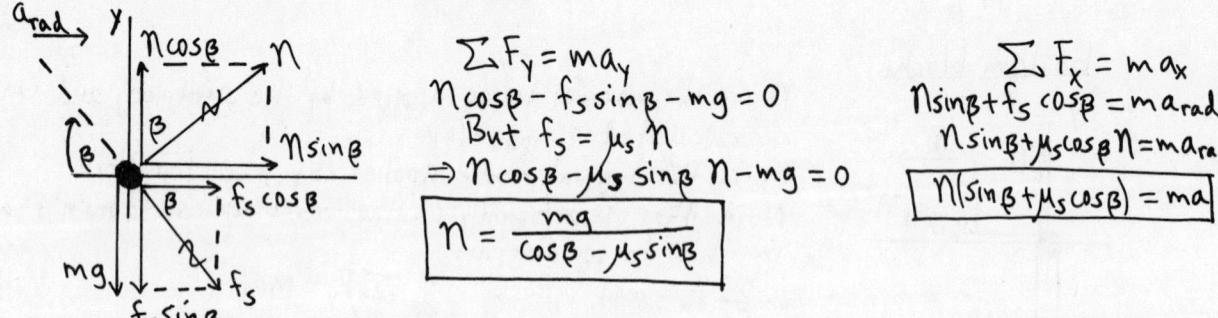

$\sum F_y = ma_y$
$n\cos\beta - f_s\sin\beta - mg = 0$
But $f_s = \mu_s n$
$\Rightarrow n\cos\beta - \mu_s \sin\beta \, n - mg = 0$

$$\boxed{n = \frac{mg}{\cos\beta - \mu_s\sin\beta}}$$

$\sum F_x = ma_x$
$n\sin\beta + f_s\cos\beta = ma_{rad}$
$n\sin\beta + \mu_s\cos\beta \, n = ma_{rad}$

$$\boxed{n(\sin\beta + \mu_s\cos\beta) = ma}$$

Use the first equation to replace n:

$$\left(\frac{mg}{\cos\beta - \mu_s\sin\beta}\right)(\sin\beta + \mu_s\cos\beta) = ma_{rad}$$

$$a_{rad} = \left(\frac{\sin\beta + \mu_s\cos\beta}{\cos\beta - \mu_s\sin\beta}\right)g = \left(\frac{\sin 25° + (0.35)\cos 25°}{\cos 25° - (0.35)\sin 25°}\right)(9.80 \text{ m/s}^2) = 9.56 \text{ m/s}^2$$

$$a_{rad} = \frac{v^2}{R} \Rightarrow v = \sqrt{a_{rad} R} = \sqrt{(9.56 \text{ m/s}^2)(50 \text{ m})} = \underline{21.9 \text{ m/s}}$$

b) To keep the car from sliding <u>down</u> the banking the static friction force is directed up the incline. At the minimum speed the static friction force has its maximum value $f_s = \mu_s n$.

Free-body diagram for the car:

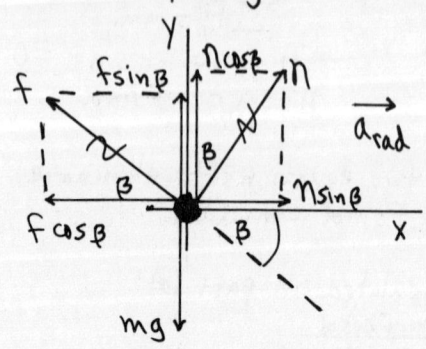

The free-body diagram is identical to that in part (a) except that now the components of f_s have opposite directions. The force equations are all the same except for the opposite sign for terms containing μ_s.

5-93 (cont)

Thus, $a_{rad} = \left(\dfrac{\sin\beta - \mu_s \cos\beta}{\cos\beta + \mu_s \sin\beta}\right)g = \left(\dfrac{\sin 25° - (0.35)\cos 25°}{\cos 25° + (0.35)\sin 25°}\right)(9.80\,m/s^2) = 0.980\,m/s^2$

$v = \sqrt{a_{rad} R} = \sqrt{(0.980\,m/s^2)(50\,m)} = \underline{7.0\,m/s}$

5-95

$v = (0.60\,rev/s)\left(\dfrac{2\pi R}{1\,rev}\right) = (0.60\,rev/s)\left(\dfrac{2\pi(2.5\,m)}{1\,rev}\right) = 9.425\,m/s$

a)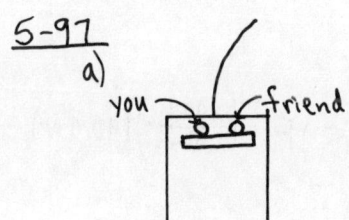

The person is held up against gravity by the static friction force exerted on him by the wall. The acceleration of the person is a_{rad}, directed in toward the axis of rotation.

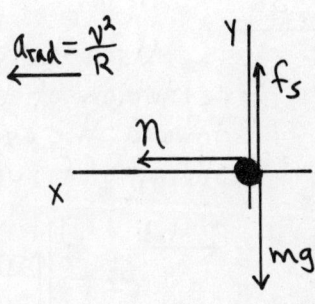

b) To calculate the minimum μ_s required, take f_s to have its maximum value, $f_s = \mu_s n$.

$\sum F_y = ma_y$
$f_s - mg = 0$
$\boxed{\mu_s n = mg}$

$\sum F_x = ma_x$
$\boxed{n = m\dfrac{v^2}{R}}$

Combine these two equations to eliminate n:

$\mu_s m \dfrac{v^2}{R} = mg$

$\mu_s = \dfrac{Rg}{v^2} = \dfrac{(2.5\,m)(9.80\,m/s^2)}{(9.425\,m/s)^2} = \underline{0.28}$

c) No, the mass of the person divided out of the equation for μ_s.

5-97

a) Turn to the right. In the absence of sufficient friction your friend doesn't make the turn completely and you move to the right toward your friend.

b) The maximum radius of the turn is the one that makes a_{rad} just equal to the maximum acceleration that static friction can give to your friend, and for this f_s has its maximum value $f_s = \mu_s n$.

Free-body diagram for your friend, as viewed by someone standing behind the car:

5-97 (cont)

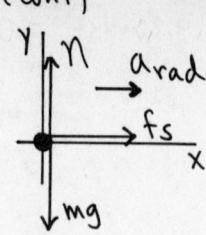

$\Sigma F_y = ma_y$
$n - mg = 0$
$n = mg$

$\Sigma F_x = ma_x$
$f_s = ma_{rad}$
$\mu_s n = m\frac{v^2}{R}$
$\mu_s \cancel{m}g = \cancel{m}\frac{v^2}{R}$
$R = \frac{v^2}{\mu_s g} = \frac{(25 m/s)^2}{(0.40)(9.80 m/s^2)} = \underline{159 \, m}$

5-99

Use the information given about Kathy to find the time t for one revolution of the merry-go-round. Her acceleration is a_{rad}, directed in toward the axle. Let F_1 be the horizontal force that keeps her from sliding off. Let her speed be v_1 and let R_1 be her distance from the axis.

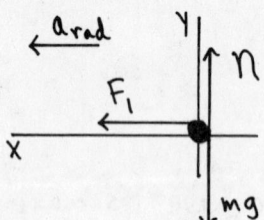

$\Sigma F_x = ma_x$
$F_1 = ma_{rad}$
$F_1 = m\frac{v_1^2}{R_1}$
$v_1 = \sqrt{\frac{R_1 F_1}{m}}$

The time for one revolution is $t = \frac{2\pi R_1}{v_1} = 2\pi R_1 \sqrt{\frac{m}{R_1 F_1}}$.

Karen goes around once in the same time but her speed (v_2) and the radius of her circular path (R_2) are different.

$v_2 = \frac{2\pi R_2}{t} = 2\pi R_2 \left(\frac{1}{2\pi R_1}\right)\sqrt{\frac{R_1 F_1}{m}} = \frac{R_2}{R_1}\sqrt{\frac{R_1 F_1}{m}}$

Free-body diagram for Karen:

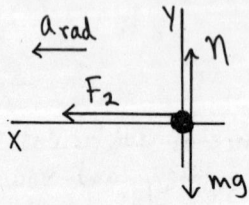

$\Sigma F_x = ma_x$
$F_2 = ma_{rad}$
$F_2 = m\frac{v_2^2}{R_2} = \frac{\cancel{m}}{R_2}\frac{R_2^2}{R_1^2}\frac{R_1 F_1}{\cancel{m}} = \left(\frac{R_2}{R_1}\right)F_1 = \left(\frac{4.00 m}{2.00 m}\right)(90.0 N)$
$F_2 = \underline{180.0 \, N}$

5-101

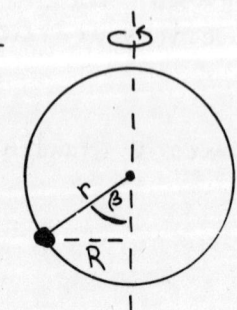

The bead moves in a circle of radius $R = r\sin\beta$.

The normal force exerted on the bead by the hoop is radially inward.

5-101 (cont)

Free-body diagram for the bead:

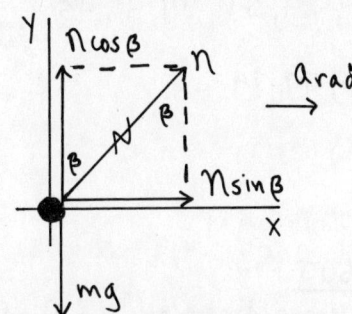

$\sum F_y = ma_y$
$n\cos\beta - mg = 0$
$n = \dfrac{mg}{\cos\beta}$

$\sum F_x = ma_x$
$n\sin\beta = ma_{rad}$

Combine these two equations to eliminate n

$\left(\dfrac{mg}{\cos\beta}\right)\sin\beta = m\,a_{rad}$

$$\boxed{\dfrac{\sin\beta}{\cos\beta} = \dfrac{a_{rad}}{g}}$$

$a_{rad} = \dfrac{v^2}{R}$ and $v = \dfrac{2\pi R}{T} \Rightarrow a_{rad} = \dfrac{4\pi^2 R}{T^2}$, where T is the time for one revolution

$R = r\sin\beta \Rightarrow a_{rad} = \dfrac{4\pi^2 r \sin\beta}{T^2}$

Use this in the above equation:

$\dfrac{\sin\beta}{\cos\beta} = \dfrac{4\pi^2 r \sin\beta}{T^2 g}$

This equation is satisfied by $\sin\beta = 0$ ($\beta = 0°$)

or $\dfrac{1}{\cos\beta} = \dfrac{4\pi^2 r}{T^2 g} \Rightarrow \boxed{\cos\beta = \dfrac{T^2 g}{4\pi^2 r}}$

a) 3.00 rev/s $\Rightarrow T = \dfrac{1}{3.00}\,s = 0.3333\,s$

$\cos\beta = \dfrac{(0.3333s)^2(9.80\,m/s^2)}{4\pi^2(0.100\,m)} = 0.2758 \Rightarrow \underline{\beta = 74.0°}$

b) This would mean $\beta = 90°$. But $\cos 90° = 0$ so this requires $T \to 0$. β approaches $90°$ as the hoop rotates very fast, but $\beta = 90°$ is not possible.

c) 1.00 rev/s $\Rightarrow T = 1.00\,s$

The $\cos\beta = \dfrac{T^2 g}{4\pi^2 r}$ equation then says $\cos\beta = \dfrac{(1.00s)^2(9.80\,m/s^2)}{4\pi^2(0.100\,m)} = 2.48$, which is not possible.

The only way to have the $\sum \vec{F} = m\vec{a}$ equations satisfied is for $\sin\beta = 0$. This means $\beta = 0°$; the bead sits at the bottom of the hoop.

CHAPTER 6

Exercises 1, 3, 7, 11, 15, 17, 21, 23, 25, 27, 29, 31, 37, 41, 45

Problems 47, 49, 55, 57, 59, 61, 65, 71, 73, 77, 79

Exercises

6-1

a) $W_F = (F\cos\phi)s = (3.00\text{ N}\cos 0°)(1.20\text{ m}) = \underline{+3.60\text{ J}}$
(The force and displacement vectors are in the same direction and the work done is positive.)

b) $W_f = (f\cos\phi)s = (0.600\text{ N}\cos 180°)(1.20\text{ m}) = \underline{-0.720\text{ J}}$
(The force and displacement vectors are in opposite directions and the work done is negative.)

6-3

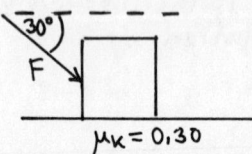

constant speed
$\Rightarrow a = 0$

$\mu_k = 0.30$

a) Free-body diagram for the crate:

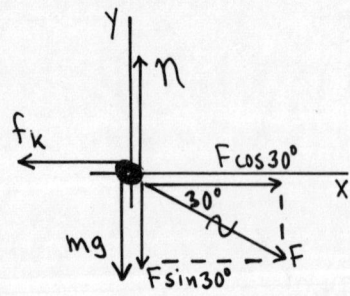

$\Sigma F_y = ma_y$
$n - mg - F\sin 30° = 0$
$n = mg + F\sin 30°$

$f_k = \mu_k n = \mu_k mg + F\mu_k \sin 30°$

$\Sigma F_x = ma_x$
$F\cos 30° - f_k = 0$
$F\cos 30° - \mu_k mg - \mu_k \sin 30° F = 0$

$F = \dfrac{\mu_k mg}{\cos 30° - \mu_k \sin 30°}$

$F = \dfrac{0.30(25.0\text{ kg})(9.80\text{ m/s}^2)}{\cos 30° - (0.30)\sin 30°} = \underline{103\text{ N}}$

b) $W_F = (F\cos\phi)s = (103\text{ N}\cos 30°)(6.0\text{ m}) = \underline{533\text{ J}}$

($F\cos 30°$ is the horizontal component of $\vec{F}$; the work done by $\vec{F}$ is the displacement times the component of $\vec{F}$ in the direction of the displacement.)

c) We have an expression for f_k from part (a):
$f_k = \mu_k(mg + F\sin 30°) = (0.30)[(25.0\text{ kg})(9.80\text{ m/s}^2) + (103\text{ N})(\sin 30°)] = 89.0\text{ N}$

$\phi = 180°$ since f_k is opposite to the displacement.

$\Rightarrow W_f = (f_k \cos\phi)s = (89.0\text{ N})(\cos 180°)(6.0\text{ m}) = \underline{-533\text{ J}}$

6-3 (cont)

d) The normal force is perpendicular to the displacement so $\phi = 90°$ and $W_n = 0$.

The gravity force (the weight) is perpendicular to the displacement so $\phi = 90°$ and $W_w = 0$.

e) $W_{tot} = W_F + W_f + W_n + W_w = +533 \text{ J} + (-533 \text{ J}) = 0$

6-7

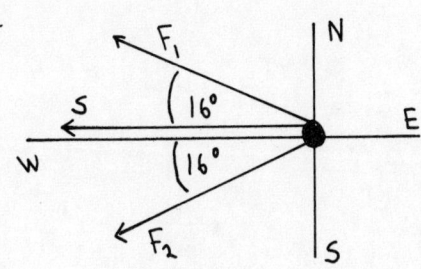

$W = Fs \cos\phi$

$W_1 = F_1 s \cos\phi_1 = (1.50 \times 10^6 \text{N})(0.65 \times 10^3 \text{m}) \cos 16°$
$W_1 = 9.37 \times 10^8 \text{ J}$

$W_2 = F_2 s \cos\phi_2 = W_1$

$W_{tot} = W_1 + W_2 = 2(9.37 \times 10^8 \text{ J}) = \underline{1.87 \times 10^9 \text{ J}}$

6-11

a) $W = K_2 - K_1$

$K_1 = \tfrac{1}{2} m v_1^2, \quad K_2 = \tfrac{1}{2} m v_2^2$

$v_2 = \tfrac{1}{3} v_1 \Rightarrow K_2 = \tfrac{1}{2} m (\tfrac{1}{3} v_1)^2 = \tfrac{1}{9}(\tfrac{1}{2} m v_1^2) = \tfrac{1}{9} K_1$

$W = K_2 - K_1 = \tfrac{1}{9} K_1 - K_1 = -\tfrac{8}{9} K_1$

b) K depends only on the magnitude of $\vec{v}$ not on its direction, so the answer for W in part (a) does **not** depend on the final direction of the electron's motion.

6-15

a) $W_{tot} = K_2 - K_1 \Rightarrow K_2 = W_{tot} + K_1$

$K_1 = \tfrac{1}{2} m v_1^2 = \tfrac{1}{2}(6.00 \text{ kg})(4.00 \text{ m/s})^2 = 48.0 \text{ J}$

The only force that does work on the wagon is the 10.0 N force. This force is in the direction of the displacement so $\phi = 0°$ and the force does positive work:

$W = (F \cos\phi) s = (10.0 \text{ N})(\cos 0°)(4.0 \text{ m}) = 40.0 \text{ J}$

Then $K_2 = W_{tot} + K_1 = 40.0 \text{ J} + 48.0 \text{ J} = 88.0 \text{ J}$.

$K_2 = \tfrac{1}{2} m v_2^2 \Rightarrow v_2 = \sqrt{\dfrac{2 K_2}{m}} = \sqrt{\dfrac{2(88.0 \text{ J})}{6.00 \text{ kg}}} = \underline{5.42 \text{ m/s}}$

6-15 (cont)

b) [free body diagram: n up, mg down, F right, a right]

$\sum F_x = ma_x$
$F = ma$
$a = \frac{F}{m} = \frac{10.0\,N}{6.00\,kg} = 1.67\,m/s^2$

$v_2^2 = v_1^2 + 2a(x-x_0)$
$v_2 = \sqrt{v_1^2 + 2a(x-x_0)} = \sqrt{(4.00\,m/s)^2 + 2(1.67\,m/s^2)(4.0\,m)} = \underline{5.42\,m/s}$

This agrees with the result calculated in part (a).

6-17

$W_{tot} = K_2 - K_1$

$K_1 = \tfrac{1}{2}mv_1^2 = \tfrac{1}{2}(0.420\,kg)(2.00\,m/s)^2 = 0.84\,J$
$K_2 = \tfrac{1}{2}mv_2^2 = \tfrac{1}{2}(0.420\,kg)(6.00\,m/s)^2 = 7.56\,J$
$W_{tot} = K_2 - K_1 = 7.56\,J - 0.84\,J = 6.72\,J$

The 40.0 N force is the only force doing work on the ball, so it must do 6.72 J of work.

$W = (F\cos\phi)s \Rightarrow s = \frac{W}{F\cos\phi} = \frac{6.72\,J}{40.0\,N \cos 0°} = \underline{0.17\,m}$

6-21

a) $W_{tot} = K_2 - K_1$

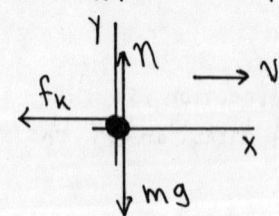

$f_k = \mu_k n = \mu_k mg$

f_k is the only force that does work, so $W_{tot} = W_f$
f_k is opposite to the displacement so $\phi = 180°$ and
$W_f = f(\cos\phi)s = \mu_k mg(\cos 180°)s = -\mu_k mgs$

$K_2 = 0$ (car stops)
$K_1 = \tfrac{1}{2}mv_0^2$

$W_{tot} = K_2 - K_1 \Rightarrow -\mu_k \cancel{m}gs = -\tfrac{1}{2}\cancel{m}v_0^2 \Rightarrow s = \frac{v_0^2}{2\mu_k g}$

b) $90\,km/h = (90\,km/h)\left(\frac{0.2778\,m/s}{1\,km/h}\right) = 25.0\,m/s$

$60\,km/h = (60\,km/h)\left(\frac{0.2778\,m/s}{1\,km/h}\right) = 16.67\,m/s$

6-21 (cont)

$$S = \frac{v_0^2}{2\mu_k g} \Rightarrow 2\mu_k g = \frac{v_0^2}{S} = \frac{(25.0 \text{ m/s})^2}{98.3 \text{ m}} = 6.358 \text{ m/s}^2$$

Then for $v_0 = 16.67 \text{ m/s}$, $S = \frac{v_0^2}{2\mu_k g} = \frac{(16.67 \text{ m/s})^2}{6.358 \text{ m/s}^2} = \underline{43.7 \text{ m}}$

6-23

Use the information given to calculate the force constant k of the spring:

$F = kx \Rightarrow k = \frac{F}{x} = \frac{120 \text{ N}}{0.040 \text{ m}} = 3000 \text{ N/m}$

a) $F = kx = (3000 \text{ N/m})(0.010 \text{ m}) = \underline{30 \text{ N}}$

$F = kx = (3000 \text{ N/m})(-0.080 \text{ m}) = -240 \text{ N}$ (magnitude $\underline{240 \text{ N}}$)

b) $W = \frac{1}{2}kx^2 = \frac{1}{2}(3000 \text{ N/m})(0.010 \text{ m})^2 = \underline{0.15 \text{ J}}$

$W = \frac{1}{2}kx^2 = \frac{1}{2}(3000 \text{ N/m})(-0.080 \text{ m})^2 = \underline{9.6 \text{ J}}$

6-25

a) [Free body diagram: n up, w=mg down, F_{spring} left, $f_s = \mu_s n$ right, $a=0$]

$\Sigma F_y = ma_y$
$n - mg = 0$
$n = mg$
$f_s = \mu_s mg$

$\Sigma F_x = ma_x$
$f_s - F_{spring} = 0$
$\mu_s mg - kd = 0$
$\mu_s = \frac{kd}{mg} = \frac{(20.0 \text{ N/m})(0.086 \text{ m})}{(0.100 \text{ kg})(9.80 \text{ m/s}^2)}$

$\underline{\mu_s = 1.76}$

b) Use the results of part (a) to calculate d, the amount the spring is stretched when the glider stops instantaneously:

$\mu_s mg = kd$

$d = \frac{\mu_s mg}{k} = \frac{(0.60)(0.100 \text{ kg})(9.80 \text{ m/s}^2)}{20.0 \text{ N/m}} = 0.0294 \text{ m}$

Now apply the work-energy theorem to the motion of the glider:
$W_{tot} = K_2 - K_1$

$K_1 = \frac{1}{2}mv_1^2$, $K_2 = 0$ (instantaneously stops)

$W_{tot} = W_{spring} + W_{fric} = -\frac{1}{2}kd^2 - \mu_k mg d$ (as in Example 6-8)
$W_{tot} = -\frac{1}{2}(20.0 \text{ N/m})(0.0294 \text{ m})^2 - 0.47(0.100 \text{ kg})(9.80 \text{ m/s}^2)(0.0294 \text{ m}) = -0.02218 \text{ J}$

Then $W_{tot} = K_2 - K_1$ gives $-0.02218 \text{ J} = -\frac{1}{2}mv_1^2$.

$v_1 = \sqrt{\frac{2(0.02218 \text{ J})}{0.100 \text{ kg}}} = \underline{0.67 \text{ m/s}}$

6-27

The magnitude of the work done by F_x equals the area under the F_x versus x curve. The work is positive when F_x and the displacement are in the same direction; it is negative when they are in opposite directions.

a) F_x is positive and the displacement Δx is positive, so $W > 0$.
$$W = \tfrac{1}{2}(2N)(2.0m) + (2N)(1.0m) = \underline{4.0\,J}$$

b) During this displacement $F_x = 0$, so $\underline{W = 0}$.

c) F_x is negative, Δx is positive, so $W < 0$.
$$W = -\tfrac{1}{2}(1N)(2.0m) = \underline{-1.0\,J}$$

d) The work is the sum of the answers to parts (a), (b), and (c), so
$$W = 4.0\,J + 0 - 1.0\,J = \underline{3.0\,J}.$$

6-29

a) $W_{tot} = K_2 - K_1 \Rightarrow K_2 = K_1 + W_{tot}$

$K_1 = 0$ (released with no initial velocity), $K_2 = \tfrac{1}{2}mv_2^2$

The only force doing work is the spring force. Eq. (6-10) gives the work done *on* the spring to move its end from x_1 to x_2. The force the spring exerts on an object attached to it is $F = -kx$, so the work the spring force does is $W_{spr} = -(\tfrac{1}{2}kx_2^2 - \tfrac{1}{2}kx_1^2) = \tfrac{1}{2}kx_1^2 - \tfrac{1}{2}kx_2^2$. Here $x_1 = -0.375\,m$ and $x_2 = 0$. Thus $W_{spr} = \tfrac{1}{2}(4000\,N/m)(-0.375\,m)^2 - 0 = 281\,J$.

$K_2 = K_1 + W_{tot} = 0 + 281\,J = 281\,J$

$K_2 = \tfrac{1}{2}mv_2^2 \Rightarrow v_2 = \sqrt{\dfrac{2K_2}{m}} = \sqrt{\dfrac{2(281\,J)}{80.0\,kg}} = \underline{2.65\,m/s}$

b) $K_2 = K_1 + W_{tot}$

$K_1 = 0$
$W_{tot} = W_{spr} = \tfrac{1}{2}kx_1^2 - \tfrac{1}{2}kx_2^2$
Now $x_2 = -0.200\,m$, so
$W_{spr} = \tfrac{1}{2}(4000\,N/m)(-0.375\,m)^2 - \tfrac{1}{2}(4000\,N/m)(-0.200\,m)^2 = 281\,J - 80\,J = 201\,J$

$K_2 = 0 + 201\,J = 201\,J$

$K_2 = \tfrac{1}{2}mv_2^2 \Rightarrow v_2 = \sqrt{\dfrac{2K_2}{m}} = \sqrt{\dfrac{2(201\,J)}{80.0\,kg}} = \underline{2.24\,m/s}$

6-31

a)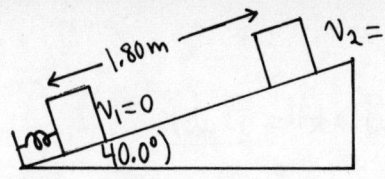

$W_{tot} = K_2 - K_1 = 0$

$W_{tot} = W_{spr} + W_w = 0 \Rightarrow W_{spr} = -W_w$

(The spring does positive work on the glider since the spring force is directed up the incline, the same as the direction of the displacement.)

$W_w = (w\cos\phi)s = (mg\cos 130.0°)s = (0.0750\,kg)(9.80\,m/s^2)$
$\cdot(\cos 130.0°)(1.80\,m)$

$W_w = -0.850\,J$

(The component of w parallel to the incline is directed down the incline, opposite to the displacement, so gravity does negative work.)

$W_{spr} = -W_w = +0.850\,J$

$W_{spr} = \tfrac{1}{2}kx^2$, so $x = \sqrt{\dfrac{2W_{spr}}{k}} = \sqrt{\dfrac{2(0.850\,J)}{640\,N/m}} = \underline{0.0515\,m}$

b)

$W_{tot} = K_2 - K_1$
$K_2 = K_1 + W_{tot}$
$K_1 = 0$

$W_{tot} = W_{spr} + W_w$

From part (a), $W_{spr} = 0.850\,J$ and

$W_w = (mg\cos 130.0°)s = (0.0750\,kg)(9.80\,m/s^2)(\cos 130.0°)(0.80\,m) = -0.378\,J$

Then $K_2 = W_{spr} + W_w = +0.850\,J - 0.378\,J = \underline{+0.472\,J}$.

(Kinetic energy must always be positive.)

6-37

Find the total mass that can be lifted:

$P_{av} = \dfrac{\Delta W}{\Delta t} = \dfrac{mgh}{t}$, so $m = \dfrac{P_{av}\,t}{gh}$

$P_{av} = (30\,hp)\left(\dfrac{746\,W}{1\,hp}\right) = 2.238 \times 10^4\,W$

$m = \dfrac{P_{av}\,t}{gh} = \dfrac{(2.238 \times 10^4\,W)(15.0\,s)}{(9.80\,m/s^2)(20.0\,m)} = 1.71 \times 10^3\,kg$

The mass of passengers is $1.71 \times 10^3\,kg - 600\,kg = 1.11 \times 10^3\,kg$.

The number of passengers is $\dfrac{1.11 \times 10^3\,kg}{65.0\,kg} = 17.1$;

17 passengers can ride.

6-41

a) $F = ma$
$v = v_0 + at$ and $v_0 = 0$, so $v = at$.
The instantaneous power is $P = Fv = (ma)(at) = ma^2 t$.

b) P is proportional to a^2.
Triple $a \Rightarrow$ increase P by a factor of $\underline{9}$.

c) $\frac{P}{t} = ma^2 =$ constant, so $\frac{P_1}{t_1} = \frac{P_2}{t_2}$.
$P_2 = P_1 \left(\frac{t_2}{t_1}\right) = 30\text{W}\left(\frac{10.0\text{s}}{5.0\text{s}}\right) = \underline{60\text{ W}}$

6-45

a) $P = \vec{F} \cdot \vec{v} = Fv \Rightarrow F = \frac{P}{v}$

$P = (6.00 \text{ hp})\left(\frac{746 \text{ W}}{1 \text{ hp}}\right) = 4476 \text{ W}$
$v = (60.0 \text{ km/h})\left(\frac{1 \times 10^3 \text{ m}}{1 \text{ km}}\right)\left(\frac{1 \text{ h}}{3600 \text{ s}}\right) = 16.67 \text{ m/s}$
$F = \frac{P}{v} = \frac{4476 \text{ W}}{16.67 \text{ m/s}} = \underline{269 \text{ N}}$

b) The power required is the 6.00 hp of part (a) plus the power P_g required to lift the car against gravity.

$\tan\alpha = \frac{10\text{m}}{100\text{m}} = 0.10 \Rightarrow \alpha = 5.71°$

The vertical component of the velocity of the car is $v\sin\alpha = (16.67 \text{ m/s}) \sin 5.71° = 1.658 \text{ m/s}$.

Then $P_g = F(v\sin\alpha) = mgv\sin\alpha = (1400 \text{ kg})(9.80 \text{ m/s}^2)(1.658 \text{ m/s}) = 2.27 \times 10^4 \text{ W}$
$P_g = 2.27 \times 10^4 \text{ W}\left(\frac{1 \text{ hp}}{746 \text{ W}}\right) = 30.5 \text{ hp}$

The total power required is $6.00 \text{ hp} + 30.5 \text{ hp} = \underline{36.5 \text{ hp}}$.

c) The power is <u>reduced</u> by the rate at which gravity does work:

$\tan\alpha = 0.0200 \Rightarrow \alpha = 1.146°$

$P_g = mg(v\sin\alpha) = (1400 \text{ kg})(9.80 \text{ m/s}^2)(16.67 \text{ m/s}) \sin 1.146° = 4.574 \times 10^3 \text{ W}\left(\frac{1 \text{ hp}}{746 \text{ W}}\right) = 6.132 \text{ hp}$

The power required from the engine is then $6.00 \text{ hp} - 6.132 \text{ hp} = \underline{-0.13 \text{ hp}}$.
(The engine must produce a force up the incline to prevent the car from speeding up.)

d) No power is needed from the engine if gravity does work at the rate of
$P_g = 6.00 \text{ hp} = 4476 \text{ W}$.
$P_g = mgv\sin\alpha \Rightarrow \sin\alpha = \frac{P_g}{mgv} = \frac{4476 \text{ W}}{(1400 \text{ kg})(9.80 \text{ m/s}^2)(16.67 \text{ m/s})} = 0.01957$
$\alpha = 1.121° \Rightarrow \tan\alpha = 0.0196$; a 1.96% grade.

Problems

6-47

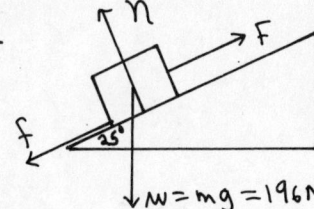

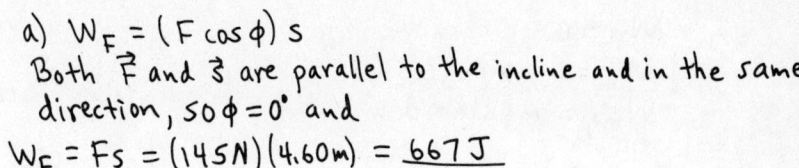

a) $W_F = (F \cos\phi) s$
Both $\vec{F}$ and $\vec{s}$ are parallel to the incline and in the same direction, so $\phi = 0°$ and
$W_F = Fs = (145N)(4.60m) = \underline{667 J}$

b) $W_w = (w \cos\phi) s$

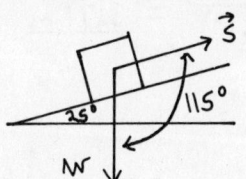

$\phi = 115°$, so $W_w = (196N)(\cos 115°)(4.60m) = -381 J$

Alternatively, the component of w parallel to the incline is $w \sin 25°$. This component is down the incline so its angle with $\vec{s}$ is $\phi = 180°$.
$W_{w\sin 25°} = (196N \sin 25°)(\cos 180°)(4.60m) = -381 J.$
The other component of w, $w \cos 25°$, is perpendicular to $\vec{s}$ and hence does no work. Thus $W_w = W_{w\sin 25°} = -381 J$, which agrees with the above.

c) The normal force is perpendicular to the displacement ($\phi = 90°$), so $W_n = 0$.

d) $n = w \cos 25°$ so $f_k = \mu_k n = \mu_k w \cos 25° = (0.30)(196N) \cos 25° = 53.3 N$
$W_f = (f_k \cos\phi) s = (53.3 N)(\cos 180°)(4.60m) = \underline{-245 J}$

e) $W_{tot} = W_F + W_w + W_n + W_f = 667 J - 381 J + 0 - 245 J = \underline{41 J}$

f) $W_{tot} = K_2 - K_1$, $K_1 = 0 \Rightarrow K_2 = W_{tot}$
$\frac{1}{2} m v_2^2 = W_{tot}$
$v_2 = \sqrt{\dfrac{2 W_{tot}}{m}} = \sqrt{\dfrac{2(41 J)}{20.0 kg}} = \underline{2.0 m/s}$

6-49

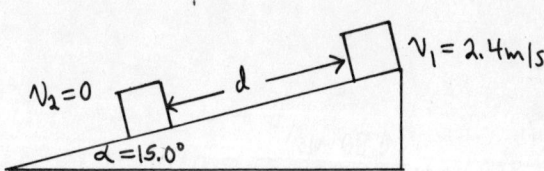

Apply $W_{tot} = K_2 - K_1$ to the motion of the package from point 1 to point 2.
$K_2 = 0$, $K_1 = \frac{1}{2} m v_1^2$

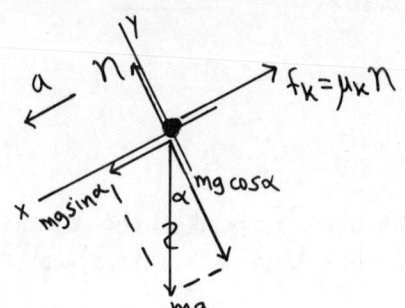

$\Sigma F_y = m a_y$
$n - mg \cos\alpha = 0$
$n = mg \cos\alpha$
$f_k = \mu_k n = \mu_k mg \cos\alpha$

87

6-49 (cont)

$$W_{tot} = W_n + W_f + W_{mg}$$

$W_n = 0$
$W_f = -\mu_k mg \cos\alpha \, d$
$W_{mg} = mg \sin\alpha \, d$

Then $W_{tot} = K_2 - K_1$ gives $mg(\sin\alpha - \mu_k \cos\alpha)d = -\frac{1}{2}m v_1^2$

$$d = \frac{-v_1^2}{2g(\sin\alpha - \mu_k \cos\alpha)} = \frac{-(2.4 \text{ m/s})^2}{2(9.80 \text{ m/s}^2)(\sin 15.0° - (0.35)\cos 15.0°)} = \underline{3.71 \text{ m}}$$

6-55

$F = \alpha x^3$, with $\alpha = 5.00 \text{ N/m}^3$.

a) $x = 1.00 \text{ m} \Rightarrow F = (5.00 \text{ N/m}^3)(1.00 \text{ m})^3 = 5.00 \text{ N}$
Force directed toward origin $\Rightarrow \vec{F}$ is in $-x$-direction; $\vec{F} = -(5.00 \text{ N})\hat{i}$.

b) $x = 2.00 \text{ m} \Rightarrow F = (5.00 \text{ N/m}^3)(2.00 \text{ m})^3 = 40.0 \text{ N}$; $\vec{F} = -(40.0 \text{ N})\hat{i}$.

c) $W = \int_{x_1}^{x_2} F_x \, dx$

$F_x = -\alpha x^3$, minus since $\vec{F}$ is toward the origin.

$W = \int_{x_1}^{x_2} (-\alpha x^3) \, dx = -\frac{\alpha}{4}\left(x^4 \Big|_{x_1}^{x_2}\right) = -\frac{\alpha}{4}(x_2^4 - x_1^4)$

$W = -\frac{1}{4}(5.00 \text{ N/m}^3)((2.00 \text{ m})^4 - (1.00 \text{ m})^4) = \underline{-18.8 \text{ J}}$

The work done is negative since the object moves away from the origin and the force is in the opposite direction, toward the origin.

6-57

a) Free-body diagram for the block:

$a_{rad} = \frac{v^2}{R}$

$\sum F_x = ma_x$

$T = m\frac{v^2}{R} = (0.0800 \text{ kg})\frac{(0.80 \text{ m/s})^2}{0.30 \text{ m}} = \underline{0.171 \text{ N}}$

b) $T = m\frac{v^2}{R} = (0.0800 \text{ kg})\frac{(2.40 \text{ m/s})^2}{0.10 \text{ m}} = \underline{4.61 \text{ N}}$

c) The tension changes as the distance of the block from the hole changes. We could use $W = \int_{x_1}^{x_2} F_x \, dx$ to calculate the work. But a much simpler approach is to use $W_{tot} = K_2 - K_1$.

6-57 (cont)

The only force doing work on the block is the tension in the cord, so $W_{tot} = W_T$.

$$K_1 = \tfrac{1}{2}mv_1^2 = \tfrac{1}{2}(0.0800 \text{ kg})(0.80 \text{ m/s})^2 = 0.0256 \text{ J}$$
$$K_2 = \tfrac{1}{2}mv_2^2 = \tfrac{1}{2}(0.0800 \text{ kg})(2.40 \text{ m/s})^2 = 0.2304 \text{ J}$$

$W_{tot} = K_2 - K_1 \Rightarrow W_{tot} = 0.2304 \text{ J} - 0.0256 \text{ J} = \underline{0.205 \text{ J}}$

This is the amount of work done by the person who pulled the cord.

6-59

a) $x(t) = \alpha t^2 + \beta t^3$
$v(t) = \frac{dx}{dt} = 2\alpha t + 3\beta t^2$

$t = 4.00 \text{ s} \Rightarrow v = 2(2.00 \text{ m/s}^2)(4.00 \text{ s}) + 3(0.200 \text{ m/s}^3)(4.00 \text{ s})^2 = \underline{25.6 \text{ m/s}}$

b) $a(t) = \frac{dv}{dt} = 2\alpha + 6\beta t$
$F = ma = m(2\alpha + 6\beta t)$

$t = 4.00 \text{ s} \Rightarrow F = 6.00 \text{ kg} \left(2(2.00 \text{ m/s}^2) + 6(0.200 \text{ m/s}^3)(4.00 \text{ s})\right) = \underline{52.8 \text{ N}}$

c) $W_{tot} = K_2 - K_1$

At $t_1 = 0$, $v_1 = 0$ so $K_1 = 0$.
$W_{tot} = W_F$
$K_2 = \tfrac{1}{2}mv_2^2 = \tfrac{1}{2}(6.00 \text{ kg})(25.6 \text{ m/s})^2 = 1970 \text{ J}$

$W_{tot} = K_2 - K_1 \Rightarrow W_F = \underline{1970 \text{ J}}$

6-61

a) $W_{tot} = K_2 - K_1$

$K_1 = \tfrac{1}{2}mv_1^2 = \tfrac{1}{2}(80.0 \text{ kg})(6.00 \text{ m/s})^2 = 1440 \text{ J}$
$K_2 = \tfrac{1}{2}mv_2^2 = \tfrac{1}{2}(80.0 \text{ kg})(2.50 \text{ m/s})^2 = 250 \text{ J}$

$W_{tot} = 250 \text{ J} - 1440 \text{ J} = \underline{-1190 \text{ J}}$

b) Neglecting friction, work is done by you (with the force you apply to the pedals) and by gravity: $W_{tot} = W_{you} + W_{gravity}$.

The gravity force is $w = mg = (80.0 \text{ kg})(9.80 \text{ m/s}^2) = 784 \text{ N}$, downward.
The displacement is 5.20 m, upward. Thus $\phi = 180°$ and
$W_{gravity} = (F \cos\phi)s = (784 \text{ N})(5.20 \text{ m}) \cos 180° = -4077 \text{ J}$

6-61 (cont)

$W_{tot} = W_{you} + W_{gravity}$

$\Rightarrow W_{you} = W_{tot} - W_{gravity} = -1190\,J - (-4077\,J) = \underline{+2890\,J}$

6-65

$W_{tot} = K_2 - K_1$

$K_1 = 0, \; K_2 = 0$

$W_{tot} = W_{fric} + W_{spr}$

$W_{spr} = \frac{1}{2} k X^2$, where $X = 0.200\,m$ (Spring force is in direction of motion of block, so it does positive work.)

$W_{fric} = -\mu_k mgd$

$W_{tot} = K_2 - K_1$ gives $\frac{1}{2}kX^2 - \mu_k mgd = 0$

$d = \dfrac{kX^2}{2\mu_k mg} = \dfrac{(250\,N/m)(0.200\,m)^2}{2(0.30)(1.50\,kg)(9.80\,m/s^2)} = \underline{1.13\,m}$, measured from the point where the block was released

(The block travels $1.13\,m - 0.20\,m = 0.93\,m$ from its initial position, where it was before it was forced against the spring.)

6-71

Apply $W_{tot} = K_2 - K_1$ to the system consisting of both blocks.

$K_1 = \frac{1}{2} m_A v_1^2 + \frac{1}{2} m_B v_1^2 = \frac{1}{2}(m_A + m_B) v_1^2$

$K_2 = 0$

The tension T in the rope does positive work on block B and the same magnitude of negative work on block A, so T does no net work on the system.

Gravity does work $W_{mg} = m_A g d$ on block A, where $d = 2.95\,m$. (Block B moves horizontally, so no work is done on it by gravity.)

Friction does work $W_{fric} = -\mu_k m_B g d$ on block B.

Thus $W_{tot} = W_{mg} + W_{fric} = m_A g d - \mu_k m_B g d$

$W_{tot} = K_2 - K_1$ gives $m_A g d - \mu_k m_B g d = -\frac{1}{2}(m_A + m_B) v_1^2$

$\mu_k = \dfrac{m_A}{m_B} + \dfrac{\frac{1}{2}(m_A + m_B) v_1^2}{m_B g d} = \dfrac{6.00\,kg}{8.00\,kg} + \dfrac{\frac{1}{2}(6.00\,kg + 8.00\,kg)(2.00\,m/s)^2}{(8.00\,kg)(9.80\,m/s^2)(2.95\,m)} = \underline{0.871}$

6-73

a) As in Example 6-11, $W = mgh$.

We need the mass of blood lifted; we are given the volume
$$V = (7500 \text{L}) \left(\frac{1 \times 10^{-3} \text{ m}^3}{1 \text{ L}} \right) = 7.50 \text{ m}^3.$$

$m = \text{density} \times \text{volume} = (1.05 \times 10^3 \text{ kg/m}^3)(7.50 \text{ m}^3) = 7.875 \times 10^3 \text{ kg}$

Then
$$W = mgh = (7.875 \times 10^3 \text{ kg})(9.80 \text{ m/s}^2)(1.63 \text{ m}) = \underline{1.26 \times 10^5 \text{ J}}$$

b) $P_{av} = \frac{\Delta W}{\Delta t} = \frac{1.26 \times 10^5 \text{ J}}{(24 \text{h})(3600 \text{s/h})} = \underline{1.46 \text{ W}}$

6-77

The power output is $P_{av} = 2000 \text{ MW} = 2.00 \times 10^9 \text{ W}$.

$P_{av} = \frac{\Delta W}{\Delta t}$ and 92% of the work done on the water by gravity is converted to electrical power output, so in 1.00s the amount of work done on the water by gravity is
$$W = \frac{P_{av} \Delta t}{0.92} = \frac{(2.00 \times 10^9 \text{ W})(1.00 \text{s})}{0.92} = 2.174 \times 10^9 \text{ J}$$

$W = mgh$, so the mass of water flowing over the dam in 1.00s must be
$$m = \frac{W}{gh} = \frac{2.174 \times 10^9 \text{ J}}{(9.80 \text{ m/s}^2)(170 \text{ m})} = 1.30 \times 10^6 \text{ kg}$$

$\text{density} = \frac{m}{V} \Rightarrow V = \frac{m}{\text{density}} = \frac{1.30 \times 10^6 \text{ kg}}{1.00 \times 10^3 \text{ kg/m}^3} = \underline{1.30 \times 10^3 \text{ m}^3}$

6-79

a) $P = F_{tot} \, v$, with $F_{tot} = F_{roll} + F_{air}$

$F_{air} = \frac{1}{2} C A \rho v^2 = \frac{1}{2}(1.0)(0.463 \text{ m}^2)(1.2 \text{ kg/m}^3)(12.0 \text{ m/s})^2 = 40.0 \text{ N}$
$F_{roll} = \mu_r n = \mu_r w = (0.0045)(490 \text{ N} + 118 \text{ N}) = 2.74 \text{ N}$

$P = (F_{roll} + F_{air}) v = (2.74 \text{ N} + 40.0 \text{ N})(12.0 \text{ m/s}) = \underline{513 \text{ W}}$

b) $F_{air} = \frac{1}{2} C A \rho v^2 = \frac{1}{2}(0.88)(0.366 \text{ m}^2)(1.2 \text{ kg/m}^3)(12.0 \text{ m/s})^2 = 27.8 \text{ N}$
$F_{roll} = \mu_r n = \mu_r w = (0.0030)(490 \text{ N} + 88 \text{ N}) = 1.73 \text{ N}$

$P = (F_{roll} + F_{air}) v = (1.73 \text{ N} + 27.8 \text{ N})(12.0 \text{ m/s}) = \underline{354 \text{ W}}$

c) $F_{air} = \frac{1}{2} C A \rho v^2 = \frac{1}{2}(0.88)(0.366 \text{ m}^2)(1.2 \text{ kg/m}^3)(6.0 \text{ m/s})^2 = 6.96 \text{ N}$
$F_{roll} = \mu_r w = 1.73 \text{ N}$ (unchanged)

$P = (F_{roll} + F_{air}) v = (1.73 \text{ N} + 6.96 \text{ N})(6.0 \text{ m/s}) = \underline{52.1 \text{ W}}$

CHAPTER 7

Exercises 3, 7, 11, 15, 17, 19, 21, 23, 27, 31, 33, 35, 37

Problems 41, 45, 47, 49, 51, 53, 55, 61, 63, 65, 69

Exercises

7-3

a)

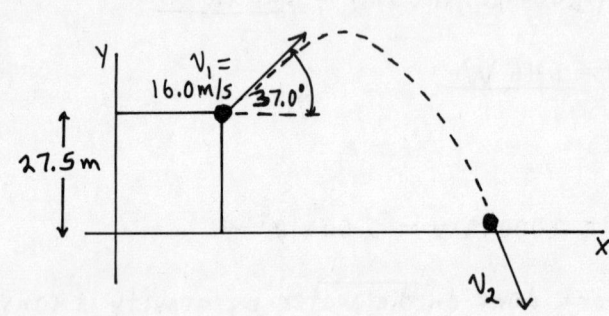

$K_1 + U_1 + W_{other} = K_2 + U_2$

$W_{other} = 0$ (The only force on the ball while it is in the air is gravity.)
$K_1 = \frac{1}{2}mv_1^2$
$K_2 = \frac{1}{2}mv_2^2$
$U_1 = mgy_1, \quad y_1 = 27.5m$
$U_2 = mgy_2 = 0$, since $y_2 = 0$ for our choice of coordinates

$\frac{1}{2}\cancel{m}v_1^2 + \cancel{m}gy_1 = \frac{1}{2}\cancel{m}v_2^2$
$v_2 = \sqrt{v_1^2 + 2gy_1} = \sqrt{(16.0 m/s)^2 + 2(9.80 m/s^2)(27.5 m)} = \underline{28.2 \, m/s}$

Note that the projection angle of 37.0° doesn't enter into the calculation. The kinetic energy depends only on the magnitude of the velocity; it is independent of the direction of the velocity.

b) Nothing changes in the calculation. The expression derived in part (a) for v_2 is independent of the angle, so $v_2 = 28.2 \, m/s$, the same as before.

c) The ball travels a shorter distance in part (b), so in that case air resistance will have less effect.

7-7

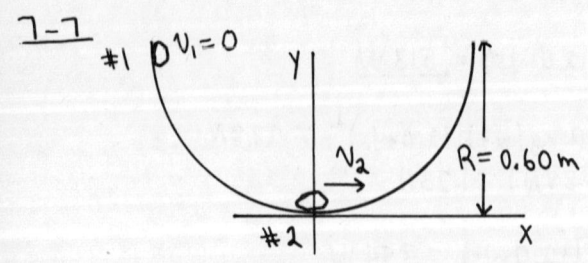

$y_1 = R, \quad y_2 = 0$

The forces on the object are gravity, the normal force, and friction. The normal force is at all points in the motion perpendicular to the displacement $\Rightarrow W_n = 0$.
Hence $W_{other} = W_f$, the work done by friction.

$K_1 + U_1 + W_{other} = K_2 + U_2$

$K_1 = \frac{1}{2}mv_1^2 = 0 \quad , \quad K_2 = \frac{1}{2}mv_2^2$
$U_1 = mgy_1 = (0.10 kg)(9.80 m/s^2)(0.60 m) = 0.588 J \quad , \quad U_2 = mgy_2 = 0$

7-7 (cont)

$W_{other} = W_f = -0.22 \text{ J}$

Thus $0 + U_1 + W_f = K_2 + 0$

$0.588 \text{ J} - 0.22 \text{ J} = \frac{1}{2}(0.10) v_2^2$

$v_2 = \sqrt{\frac{2(0.368 \text{ J})}{0.10 \text{ kg}}} = \underline{2.71 \text{ m/s}}$

7-11

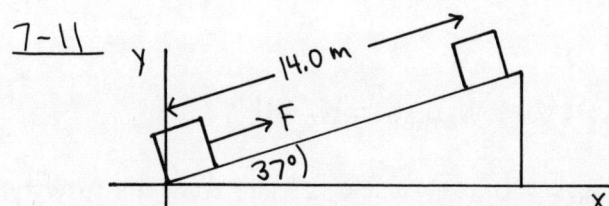

$y_1 = 0$

$y_2 = (14.0 \text{ m}) \sin 37° = 8.43 \text{ m}$

a) $W_F = (F \cos \phi) s = (120 \text{ N})(\cos 0°)(14.0 \text{ m}) = \underline{1680 \text{ J}}$

b) Use the free-body diagram for the oven to calculate the normal force n; then the friction force can be calculated from $f_k = \mu_k n$. For this calculation use coordinates parallel and perpendicular to the incline.

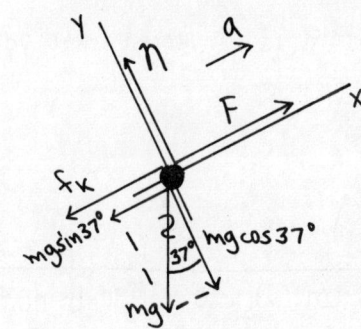

$\Sigma F_y = m a_y$
$n - mg \cos 37° = 0$
$n = mg \cos 37°$
$f_k = \mu_k n = \mu_k mg \cos 37° = (0.25)(12.0 \text{ kg})(9.80 \text{ m/s}^2) \cos 37°$
$f_k = 23.5 \text{ N}$

$W_f = (f_k \cos \phi) s = (23.5 \text{ N})(\cos 180°)(14.0 \text{ m}) = \underline{-329 \text{ J}}$

c) $\Delta U = U_2 - U_1 = mg(y_2 - y_1) = (12.0 \text{ kg})(9.80 \text{ m/s}^2)(8.43 \text{ m} - 0) = \underline{991 \text{ J}}$

d) $K_1 + U_1 + W_{other} = K_2 + U_2$
$\Delta K = K_2 - K_1 = U_1 - U_2 + W_{other}$
$\Delta K = W_{other} - \Delta U$

$W_{other} = W_F + W_f = 1680 \text{ J} - 329 \text{ J} = 1351 \text{ J}$
$\Delta U = 991 \text{ J}$
$\Rightarrow \Delta K = 1351 \text{ J} - 991 \text{ J} = \underline{360 \text{ J}}$

e) We can use the free-body diagram that is in part (b):
$\Sigma F_x = m a_x$
$F - f_k - mg \sin 37° = ma$
$a = \frac{F - f_k - mg \sin 37°}{m} = \frac{120 \text{ N} - 23.5 \text{ N} - (12.0 \text{ kg})(9.80 \text{ m/s}^2) \sin 37°}{12.0 \text{ kg}} = \underline{2.144 \text{ m/s}^2}$

7-11 (cont)

$v_1 = 0$
$a = 2.144 \text{ m/s}^2$
$x - x_0 = 14.0 \text{ m}$
$v_2 = ?$

$v_2^2 = v_1^{2\,0} + 2a(x-x_0)$
$v_2 = \sqrt{2a(x-x_0)} = \sqrt{2(2.144 \text{ m/s}^2)(14.0 \text{ m})} = 7.748 \text{ m/s}$

Then $\Delta K = K_2 - K_1^{\,0} = \frac{1}{2}mv_2^2 = \frac{1}{2}(12.0 \text{ kg})(7.748 \text{ m/s})^2 = \underline{360 \text{ J}}$; this agrees with the result calculated in part (d) using energy methods.

7-15

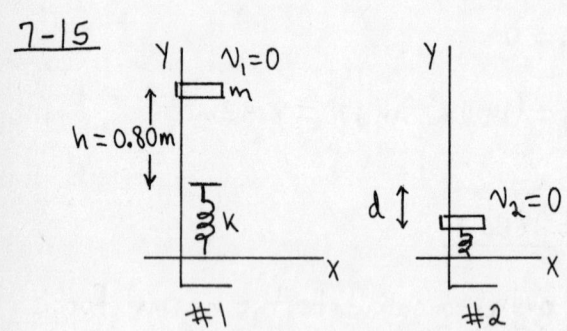

$K_1 + U_1 + W_{other} = K_2 + U_2$

$W_{other} = 0$ (only work is that done by gravity and spring force)
$K_1 = 0, \quad K_2 = 0$

$y = 0$ at final position of book
$U_1 = mg(h+d), \quad U_2 = \frac{1}{2}kd^2$

$0 + mg(h+d) + 0 = \frac{1}{2}kd^2$

(the original gravitational potential energy of the system is converted into potential energy of compressed spring)

$\frac{1}{2}kd^2 - mgd - mgh = 0$

$d = \frac{1}{k}\left(mg \pm \sqrt{(mg)^2 + 4(\frac{1}{2}k)(mgh)}\right)$

d must be positive, so $d = \frac{1}{k}\left(mg + \sqrt{(mg)^2 + 2kmgh}\right)$

$d = \frac{1}{1960 \text{ N/m}}\left((1.20 \text{ kg})(9.80 \text{ m/s}^2) + \sqrt{((1.20 \text{ kg})(9.80 \text{ m/s}^2))^2 + 2(1960 \text{ N/m})(1.20 \text{ kg})(9.80 \text{ m/s}^2)(0.80 \text{ m})}\right)$

$d = 0.0060 \text{ m} + 0.0982 \text{ m} = \underline{0.10 \text{ m}}$

7-17

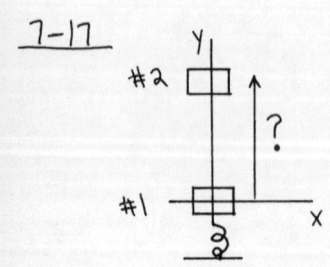

$K_1 + U_1 + W_{other} = K_2 + U_2$

The spring force and gravity are the only forces doing work on the brick so $W_{other} = 0$ and $U = U_{grav} + U_{el}$.
Brick released from rest $\Rightarrow K_1 = 0$.
At the maximum height $v_2 = 0$ so $K_2 = 0$.
$U_1 = U_{1,el} + U_{1,grav}$
$y_1 = 0 \Rightarrow U_{1,grav} = 0$
$U_{1,el} = \frac{1}{2}kx_1^2 = \frac{1}{2}(1500 \text{ N/m})(0.20 \text{ m})^2 = 30.0 \text{ J}$

(Here x_1 refers to the amount the spring is stretched or compressed when the brick is at position #1; it is <u>not</u> the x-coordinate of the brick in the coordinate system shown in the sketch.)

94

7-17 (cont)

$U_2 = U_{2,grav} + U_{2,el}$

$U_{2,grav} = mgy_2$, where y_2 is the height we are solving for
$U_{2,el} = 0$ since now the spring is no longer compressed

Putting all this into $K_1 + U_1 + W_{other} = K_2 + U_2$ gives
$U_{1,el} = U_{2,grav}$

The description in terms of energy is very simple: the elastic potential energy originally stored in the spring is converted to gravitational potential energy of the system.

$$30.0 J = mgy_2$$

$$y_2 = \frac{30.0 J}{mg} = \frac{30.0 J}{(1.60 kg)(9.80 m/s^2)} = \underline{1.91 m}$$

7-19

a) $K_1 + U_1 + W_{other} = K_2 + U_2$

$1 \Rightarrow$ the glider is at its initial position, where $x_1 = 0.100 m$ and $v_1 = 0$
$2 \Rightarrow$ the glider is at $x = 0$

$K_1 = 0$ (released from rest), $K_2 = \frac{1}{2} m v_2^2$
$U_1 = \frac{1}{2} k x_1^2$, $U_2 = 0$, $W_{other} = 0$ (only the spring force does work)

$$\frac{1}{2} k x_1^2 = \frac{1}{2} m v_2^2$$

(the initial potential energy of the stretched spring is converted entirely into kinetic energy of the glider)

$$v_2 = x_1 \sqrt{\frac{k}{m}} = (0.100 m)\sqrt{\frac{5.00 N/m}{0.200 kg}} = \underline{0.500 m/s}$$

b) The maximum speed occurs at $x = 0$, so the same equation applies.
$\frac{1}{2} k x_1^2 = \frac{1}{2} m v_2^2$

$$x_1 = v_2 \sqrt{\frac{m}{k}} = 2.00 m/s \sqrt{\frac{0.200 kg}{5.00 N/m}} = \underline{0.400 m}$$

7-21

a) $1 \Rightarrow$ the glider is at its initial position, where $x_1 = 0.100 m$ and $v_1 = 0$
$2 \Rightarrow$ the glider is at $x_2 = 0.080 m$

$K_1 + U_1 + W_{other} = K_2 + U_2$

$K_1 = 0$, $K_2 = \frac{1}{2} m v_2^2$
$U_1 = \frac{1}{2} k x_1^2 = \frac{1}{2}(5.00 N/m)(0.100 m)^2 = 0.025 J$
$U_2 = \frac{1}{2} k x_2^2 = \frac{1}{2}(5.00 N/m)(0.080 m)^2 = 0.016 J$
$W_{other} = W_{fric} = -\mu_k mg (x_1 - x_2) = -(0.10)(0.200 kg)(9.80 m/s^2)(0.020 m) = -0.00392 J$

7-21 (cont)

Thus $0.025\,J - 0.00392\,J = \tfrac{1}{2}mv_2^2 + 0.016\,J$

$\tfrac{1}{2}mv_2^2 = 0.00508\,J$

$v_2 = \sqrt{\dfrac{2(0.00508\,J)}{0.200\,kg}} = \underline{0.23\,m/s}$

b) Same analysis as in part (a), but now $x_2 = 0$.

$K_1 = 0,\ K_2 = \tfrac{1}{2}mv_2^2$
$U_1 = 0.025\,J,\ U_2 = 0$
$W_{other} = -\mu_k mg(x_1 - x_2) = -0.10\,(0.200\,kg)(9.80\,m/s^2)(0.100\,m) = -0.0196\,J$

Thus $0.025\,J - 0.0196\,J = \tfrac{1}{2}mv_2^2$

$\tfrac{1}{2}mv_2^2 = 0.0054\,J \Rightarrow v_2 = \sqrt{\dfrac{2(0.0054\,J)}{0.200\,kg}} = \underline{0.23\,m/s}$

7-23

a) Choose point 1 as in Example 7-11 and let that be the origin, so $y_1 = 0$. Let point 2 be 1.50 m below point 1, so $y_2 = -1.50\,m$.

$K_1 + U_1 + W_{other} = K_2 + U_2$

$K_1 = \tfrac{1}{2}mv_1^2 = \tfrac{1}{2}(2000\,kg)(25\,m/s)^2 = 625,000\,J,\quad U_1 = 0$
$W_{other} = -f|y_2| = -(17,000\,N)(1.50\,m) = -25,500\,J$
$K_2 = \tfrac{1}{2}mv_2^2$
$U_2 = U_{2,grav} + U_{2,el} = mgy_2 + \tfrac{1}{2}ky_2^2 = (2000\,kg)(9.80\,m/s^2)(-1.50\,m) + \tfrac{1}{2}(1.41\times 10^5\,N/m)(1.50\,m)^2$
$U_2 = -29,400\,J + 158,625\,J = +129,225\,J$

Thus $625,000\,J - 25,500\,J = \tfrac{1}{2}mv_2^2 + 129,225\,J$

$\tfrac{1}{2}mv_2^2 = 470,275\,J$

$v_2 = \sqrt{\dfrac{2(470,275\,J)}{2000\,kg}} = \underline{21.7\,m/s}$

b) Free-body diagram for the elevator:

$F_{spr} = kd$, where d is the distance the spring is compressed
$\Sigma F_y = ma_y$
$f_k + F_{spr} - mg = ma$
$f_k + kd - mg = ma$

$a = \dfrac{f_k + kd - mg}{m} = \dfrac{17,000\,N + (1.41\times 10^5\,N/m)(1.50\,m) - (2000\,kg)(9.80\,m/s^2)}{2000\,kg}$

$a = \underline{104\,m/s^2}$

(a is calculated to be positive, so the acceleration is upward)

7-27

$$W = \int_1^2 \vec{F} \cdot d\vec{l}, \quad \vec{F} = -\alpha x^2 \hat{i}$$

a) $d\vec{l} = dy\,\hat{j}$ (x is constant; the displacement is in the $+y$-direction)
$\vec{F} \cdot d\vec{l} = 0$ (since $\hat{i} \cdot \hat{j} = 0$) and $W = 0$

b) $d\vec{l} = dx\,\hat{i}$
$\vec{F} \cdot d\vec{l} = (-\alpha x^2 \hat{i}) \cdot (dx\,\hat{i}) = -\alpha x^2\, dx$

$$W = \int_{x_1}^{x_2} (-\alpha x^2)\,dx = -\tfrac{\alpha}{3} x^3 \Big|_{x_1}^{x_2} = -\tfrac{\alpha}{3}(x_2^3 - x_1^3) = -\tfrac{\alpha}{3}\left((0.30\text{m})^3 - (0.10\text{m})^3\right)$$

$$W = -\tfrac{\alpha}{3}(0.026\,\text{m}^3)$$

c) $d\vec{l} = dx\,\hat{i}$ as in part (b), but now $x_1 = 0.30$ m and $x_2 = 0.10$ m

$$W = -\tfrac{\alpha}{3} x^3 \Big|_{x_1}^{x_2} = +\tfrac{\alpha}{3}(0.026\,\text{m}^3)$$

d) The total work for the displacement along the x-axis from 0.10 m to 0.30 m and then back to 0.10 m is the sum of the results of parts (b) and (c), which is zero.

The total work is zero when the starting and ending points are the same; the force is conservative.

$$W_{x_1 \to x_2} = -\tfrac{\alpha}{3}(x_2^3 - x_1^3) = \tfrac{\alpha}{3} x_1^3 - \tfrac{\alpha}{3} x_2^3$$

The definition of the potential energy function is $W_{x_1 \to x_2} = U_1 - U_2$.
Comparison of the two expressions for W gives $U = \tfrac{1}{3}\alpha x^3$.

7-31

$f = \mu_k mg = (0.30)(2.5\,\text{kg})(9.80\,\text{m/s}^2) = 7.35$ N; direction of $\vec{f}$ is opposite to the motion.

a) For the motion you to Beth the friction force is directed opposite to the displacement $\vec{s}$ and
$W_1 = -fs = -(7.35\,\text{N})(8.0\,\text{m}) = -58.8$ J

For the motion from Beth to Carlos the friction force is again directed opposite to the displacement and $W_2 = -58.8$ J

$$W_{tot} = W_1 + W_2 = -58.8\,\text{J} - 58.8\,\text{J} = \underline{-118\,\text{J}}$$

b) $s = \sqrt{2(8.0\,\text{m})^2} = 11.3$ m

$\vec{f}$ is opposite to $\vec{s}$, so $W = -fs = -(7.35\,\text{N})(11.3\,\text{m}) = \underline{-83\,\text{J}}$

7-31 (cont)

c)

you ←f— • Kim   For the motion you to Kim
       —s→ $W = -fs = -(7.35\,N)(8.0\,m) = -58.8\,J$

you ←—s— • Kim   For the motion Kim to you
       —f→ $W = -fs = -58.8\,J$

The total work for the round trip is $-58.8\,J - 58.8\,J = -118\,J$

d) Parts (a) and (b) show that for two different paths between you and Carlos, the work done by friction is different.

Part (c) shows that when the starting and ending points are the same, the total work is not zero.

Both these results show that the friction force is nonconservative.

7-33

Use coordinates where the origin is at one atom. The other atom then has coordinate x.

$$F_x = -\frac{dU}{dx} = -\frac{d}{dx}\left(-\frac{C_6}{x^6}\right) = +C_6 \frac{d}{dx}\left(\frac{1}{x^6}\right) = -\frac{6C_6}{x^7}$$

The minus sign means that F_x is directed in the $-x$-direction, toward the origin.

The force has magnitude $6C_6/x^7$ and is attractive.

7-35

$$U(x,y) = k(x^2 + y^2) + k'xy$$

$$F_x = -\frac{\partial U}{\partial x} = -2kx - k'y$$

$$F_y = -\frac{\partial U}{\partial y} = -2ky - k'x$$

$$\vec{F} = -(2kx + k'y)\hat{i} - (2ky + k'x)\hat{j}$$

7-37

a) $U = \frac{a}{r^{12}} - \frac{b}{r^6}$

$F = -\frac{dU}{dr} = -\frac{12a}{r^{13}} + \frac{6b}{r^7}$

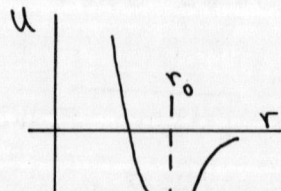
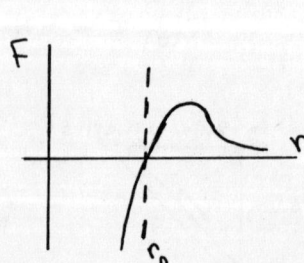

7-37 (cont)

b) At equilibrium $F=0 \Rightarrow \frac{dU}{dr}=0$

$F=0 \Rightarrow -\frac{12a}{r^{13}} + \frac{6b}{r^7} = 0$

$6br^6 = 12a \Rightarrow r = \left(\frac{2a}{b}\right)^{1/6}$

U is a minimum at this r; the equilibrium is stable.

c) At $r = \left(\frac{2a}{b}\right)^{1/6}$, $U = \frac{a}{r^{12}} - \frac{b}{r^6} = a\left(\frac{b}{2a}\right)^2 - b\left(\frac{b}{2a}\right) = \frac{b^2}{4a} - \frac{b^2}{2a} = -\frac{b^2}{4a}$

At $r \to \infty$, $U = 0$. The energy that must be added is $-\Delta U = \frac{b^2}{4a}$.

d) $r = \left(\frac{2a}{b}\right)^{1/6} = 1.21 \times 10^{-10}$ m $\Rightarrow \frac{2a}{b} = 3.138 \times 10^{-60}$ m^6 $\Rightarrow \frac{b}{4a} = 1.593 \times 10^{59}$ m^{-6}

$\frac{b^2}{4a} = b\left(\frac{b}{4a}\right) = 8.27 \times 10^{-19}$ J

$b(1.593 \times 10^{59}$ m$^{-6}) = 8.27 \times 10^{-19}$ J $\Rightarrow b = \underline{5.19 \times 10^{-78}}$ J·m^6

Then $\frac{2a}{b} = 3.138 \times 10^{-60}$ m^6 gives

$a = \frac{b}{2}(3.138 \times 10^{-60}$ m$^6) = \frac{1}{2}(5.19 \times 10^{-78}$ J·m$^6)(3.138 \times 10^{-60}$ m$^6) = \underline{8.14 \times 10^{-138}}$ J·m^{12}

Problems

7-41

Work is done on the block by the spring and by friction, so $W_{other} = W_f$ and $U = U_{el}$.

$K_1 + U_1 + W_{other} = K_2 + U_2$

$K_1 = K_2 = 0$

$U_1 = U_{1,el} = \frac{1}{2}kx_1^2 = \frac{1}{2}(100$ N/m$)(0.20$ m$)^2 = 2.00$ J

$U_2 = U_{2,el} = 0$, since after the block leaves the spring has given up all its stored energy

$W_{other} = W_f = (f_k \cos\phi) s = \mu_k mg(\cos\phi)s = -\mu_k mgs$, since $\phi = 180°$

(The friction force is directed opposite to the displacement and does negative work.)

Putting all this into $K_1 + U_1 + W_{other} = K_2 + U_2$ gives

$U_{1,el} + W_f = 0$

(The potential energy originally stored in the spring is taken out of the system by the negative work done by friction.)

$\mu_k mgs = U_{1,el}$

$\mu_k = \frac{U_{1,el}}{mgs} = \frac{2.00 \text{ J}}{(0.50 \text{ kg})(9.80 \text{ m/s}^2)(1.00 \text{ m})} = \underline{0.408}$

7-45

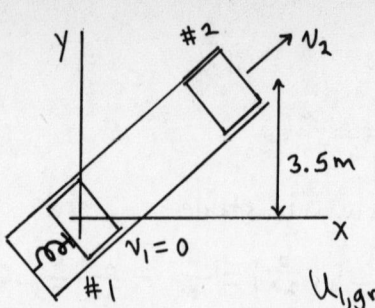

Take $y=0$ at his initial position.

$$K_1 + U_1 + W_{other} = K_2 + U_2$$

$K_1 = 0$, $K_2 = \frac{1}{2}mv_2^2$

$W_{other} = W_{fric} = -f_s = -(40N)(5.0m) = -200J$

$U_{1,grav} = 0$, $U_{1,el} = \frac{1}{2}kd^2$, where d is the distance the spring is initially compressed.

$F = kd$ so $d = \frac{F}{k} = \frac{5100N}{1100 N/m} = 4.636 m$

and $U_{1,el} = \frac{1}{2}(1100 \, N/m)(4.636m)^2 = 11,823 J$

$U_{2,grav} = mgy_2 = (80kg)(9.80m/s^2)(3.5m) = 2744 J$, $U_{2,el} = 0$

Then $K_1 + U_1 + W_{other} = K_2 + U_2$ gives

$11,823 J - 200 J = \frac{1}{2}mv_2^2 + 2744 J$

$\frac{1}{2}mv_2^2 = 8879 J$

$v_2 = \sqrt{\frac{2(8879 J)}{80 kg}} = \underline{15 m/s}$

7-47

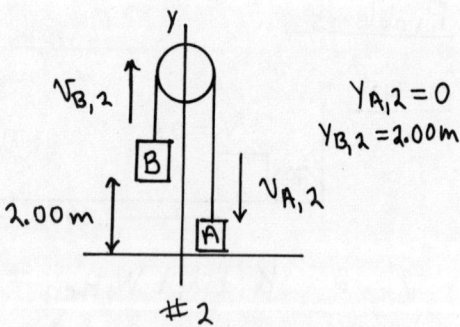

$y_{A,1} = 2.00m$
$y_{B,1} = 0$
$v_{A,1} = v_{B,1} = 0$

$v_{B,2}$
$y_{A,2} = 0$
$y_{B,2} = 2.00m$
$v_{A,2}$

#1   #2

The tension force does positive work on the 4.0kg block and an equal amount of negative work on the 12.0 kg block, so the net work done by the tension is zero.

Work is done on the system only by gravity, so $W_{other} = 0$ and $U = U_{grav}$

$$K_1 + U_1 + W_{other} = K_2 + U_2$$

$K_1 = 0$
$K_2 = \frac{1}{2}m_A v_{A,2}^2 + \frac{1}{2}m_B v_{B,2}^2$

But since the two blocks are connected by a rope they move together and have the same speed: $v_{A,2} = v_{B,2} = v_2$.

Thus $K_2 = \frac{1}{2}(m_A + m_B) v_2^2 = (8.00 kg) v_2^2$.

$U_1 = m_A g y_{A,1} = (12.0 kg)(9.80 m/s^2)(2.00m) = 235.2 J$
$U_2 = m_B g y_{B,2} = (4.0 kg)(9.80 m/s^2)(2.00m) = 78.4 J$

7-47 (cont)

Put all this into $K_1 + U_1 + W_{other} = K_2 + U_2$

$\Rightarrow U_1 = K_2 + U_2$

$235.2 \text{ J} = (8.00 \text{ kg}) v_2^2 + 78.4 \text{ J}$

$v_2 = \sqrt{\dfrac{235.2 \text{ J} - 78.4 \text{ J}}{8.00 \text{ kg}}} = \underline{4.43 \text{ m/s}}$

7-49

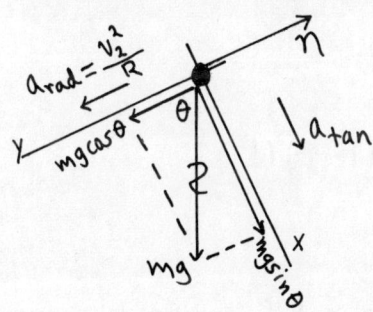

Let point #2 be where the skier loses contact with the snowball.

Loses contact $\Rightarrow n \to 0$

$y_1 = R$, $y_2 = R\cos\theta$

First, analyze the forces on the skier when she is at point #2. For this use coordinates that are in the tangential and radial directions. The skier moves in an arc of a circle, so her acceleration is $a_{rad} = v_2^2/R$, directed in towards the center of the snowball.

$\Sigma F_y = ma_y$

$mg\cos\theta - n^{\to 0} = m\dfrac{v_2^2}{R}$

$mg\cos\theta = m\dfrac{v_2^2}{R}$

$\boxed{v_2^2 = Rg\cos\theta}$

Now use conservation of energy to get another equation relating v_2 to θ:

$K_1 + U_1 + W_{other} = K_2 + U_2$

The only force that does work on the skier is gravity $\Rightarrow W_{other} = 0$.

$K_1 = 0$
$U_1 = mgy_1 = mgR$
$K_2 = \frac{1}{2}mv_2^2$
$U_2 = mgy_2 = mgR\cos\theta$

Then $mgR = \frac{1}{2}mv_2^2 + mgR\cos\theta$

$\boxed{v_2^2 = 2gR(1-\cos\theta)}$

Combine this with the $\Sigma F_y = ma_y$ equation

$Rg\cos\theta = 2gR(1-\cos\theta)$

$\cos\theta = 2 - 2\cos\theta$

$3\cos\theta = 2 \Rightarrow \cos\theta = \frac{2}{3} \Rightarrow \underline{\theta = 48.2°}$

7-51

a) Speed at ground if steps off platform of height h:
$$K_1 + U_1 + W_{other} = K_2 + U_2$$
$$mgh = \tfrac{1}{2}mv_2^2 \Rightarrow v_2^2 = 2gh$$

Motion from top to bottom of pole: (take $y=0$ at bottom)
$$K_1 + U_1 + W_{other} = K_2 + U_2$$
$$mgd - fd = \tfrac{1}{2}mv_2^2$$
Use $v_2^2 = 2gh \Rightarrow mgd - fd = mgh$
$$fd = mg(d-h)$$
$$f = mg\left(\tfrac{d-h}{d}\right) = mg\left(1 - \tfrac{h}{d}\right)$$

For $h=d$ this gives $f=0$ as it should (friction has no effect)

For $h=0$, $v_2 = 0$ (no motion). The equation for f gives $f = mg$ in this special case. When $f = mg$ the forces on him cancel and he doesn't accelerate down the pole, which agrees with $v_2 = 0$.

b) $f = mg\left(1 - \tfrac{h}{d}\right) = (80\,kg)(9.80\,m/s^2)\left(1 - \tfrac{1.0\,m}{3.5\,m}\right) = 560\,N$

c) Take $y = 0$ at bottom of pole, so $y_1 = d$ and $y_2 = y$.
$$K_1 + U_1 + W_{other} = K_2 + U_2$$
$$0 + mgd - f(d-y) = \tfrac{1}{2}mv^2 + mgy$$
$$\tfrac{1}{2}mv^2 = mg(d-y) - f(d-y)$$

Use $f = mg\left(1 - \tfrac{h}{d}\right) \Rightarrow \tfrac{1}{2}mv^2 = mg(d-y) - mg\left(1 - \tfrac{h}{d}\right)(d-y)$
$$\tfrac{1}{2}mv^2 = mg\tfrac{h}{d}(d-y)$$
$$v = \sqrt{2gh\left(1 - \tfrac{y}{d}\right)}$$

(Note that gives the expected results for $y=0$ and for $y=d$.)

7-53

a)

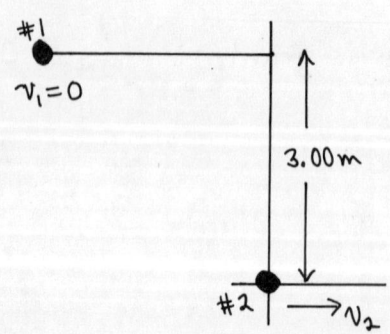

$y_1 = 3.00\,m$
$y_2 = 0$

The tension in the string is at all points in the motion perpendicular to the displacement, so $W_T = 0$.

The only force that does work on the ball is gravity, so $W_{other} = 0$.

$$K_1 + U_1 + W_{other}^{\,0} = K_2 + U_2$$
$$K_1 = 0, \quad K_2 = \tfrac{1}{2}mv_2^2$$

7-53 (cont)

$U_1 = mgy_1$, $U_2 = 0$

Thus $U_1 = K_2$
$mgy_1 = \frac{1}{2}mv_2^2$
$v_2 = \sqrt{2gy_1} = \sqrt{2(9.80 \text{ m/s}^2)(3.00 \text{ m})} = \underline{7.67 \text{ m/s}}$

b) Free-body diagram for the ball as it swings through its lowest point:

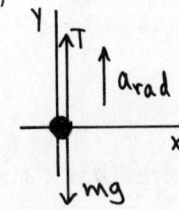

The acceleration $\vec{a}_{rad}$ is directed in toward the center of the circular path, so at this point it is upward.

$\Sigma F_y = ma_y$
$T - mg = ma_{rad}$
$T = m(g + a_{rad}) = m\left(g + \frac{v_2^2}{R}\right)$, where the radius R for the circular motion is the length L of the string.

It is instructive to use the algebraic expression for v_2 from part (a) rather than just putting in the numerical value:
$v_2 = \sqrt{2gy_1} = \sqrt{2gL}$, so $v_2^2 = 2gL$
Then $T = m\left(g + \frac{v_2^2}{L}\right) = m\left(g + \frac{2gL}{L}\right) = 3mg$; the tension at this point is three times the weight of the ball.
$T = 3mg = 3(0.200 \text{ kg})(9.80 \text{ m/s}^2) = \underline{5.88 \text{ N}}$

7-55

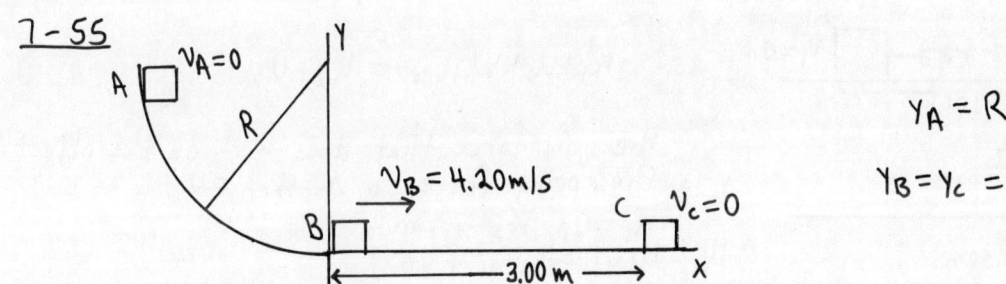

$y_A = R$

$y_B = y_C = 0$

a) Apply conservation of energy to the motion from B to C:
$K_B + U_B + W_{other} = K_C + U_C$

The only force that does work on the package during this part of the motion is friction $\Rightarrow W_{other} = W_f = f_k(\cos\phi)s = \mu_k mg(\cos 180°)s = -\mu_k mgs$
$K_B = \frac{1}{2}mv_B^2$, $K_C = 0$
$U_B = 0$, $U_C = 0$

Thus $K_B + W_f = 0$ (The negative friction work takes away all the kinetic energy.)
$\frac{1}{2}mv_B^2 - \mu_k mgs = 0$
$\mu_k = \frac{v_B^2}{2gs} = \frac{(4.20 \text{ m/s})^2}{2(9.80 \text{ m/s}^2)(3.00 \text{ m})} = \underline{0.300}$

7-55 (cont)

b) Apply conservation of energy to the motion from A to B:
$$K_A + U_A + W_{other} = K_B + U_B$$

Work is done by gravity and by friction, so $W_{other} = W_f$.
$K_A = 0$
$U_A = mgy_A = mgR = (0.200 \text{ kg})(9.80 \text{ m/s}^2)(1.60 \text{ m}) = 3.14 \text{ J}$
$K_B = \frac{1}{2} m v_B^2 = \frac{1}{2}(0.200 \text{ kg})(4.20 \text{ m/s})^2 = 1.76 \text{ J}$
$U_B = 0$

Thus $U_A + W_f = K_B$
$W_f = K_B - U_A = 1.76 \text{ J} - 3.14 \text{ J} = \underline{-1.38 \text{ J}}$
(W_f is negative as expected; the friction force does negative work since it is directed opposite to the displacement.)

7-61

$F_x = -\alpha x - \beta x^2$, $\alpha = 70.0 \text{ N/m}$ and $\beta = 12.0 \text{ N/m}^2$

a) $W_{F_x} = U_1 - U_2 = \int_{x_1}^{x_2} F_x(x) dx$

Let $x_1 = 0$ and $U_1 = 0$. Let x_2 be some arbitrary point x, so $U_2 = U(x)$
Then
$$U(x) = -\int_0^x F_x(x) dx = -\int_0^x (-\alpha x - \beta x^2) dx = \int_0^x (\alpha x + \beta x^2) dx = \frac{1}{2}\alpha x^2 + \frac{1}{3}\beta x^3$$

b)

$K_1 + U_1 + W_{other} = K_2 + U_2$

The only force that does work on the object is the spring force, so $W_{other} = 0$.
$K_1 = 0$, $K_2 = \frac{1}{2} m v_2^2$
$U_1 = U(x_1) = \frac{1}{2}\alpha x_1^2 + \frac{1}{3}\beta x_1^3$
$U_1 = \frac{1}{2}(70.0 \text{ N/m})(1.00 \text{ m})^2 + \frac{1}{3}(12.0 \text{ N/m}^2)(1.00 \text{ m})^3 = 39.0 \text{ J}$
$U_2 = U(x_2) = \frac{1}{2}\alpha x_2^2 + \frac{1}{3}\beta x_2^3$
$U_2 = \frac{1}{2}(70.0 \text{ N/m})(0.50 \text{ m})^2 + \frac{1}{3}(12.0 \text{ N/m}^2)(0.50 \text{ m})^3 = 9.25 \text{ J}$

Thus $39.0 \text{ J} = \frac{1}{2} m v_2^2 + 9.25 \text{ J}$
$v_2 = \sqrt{\dfrac{2(39.0 \text{ J} - 9.25 \text{ J})}{0.800 \text{ kg}}} = \underline{8.62 \text{ m/s}}$

7-63

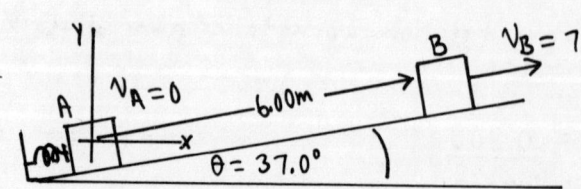

$v_B = 7.00 \text{ m/s}$

The normal force is $n = mg\cos\theta$, so $f_k = \mu_k n = \mu_k mg\cos\theta$.

$y_A = 0$; $y_B = (6.00 \text{ m})\sin 37.0° = 3.61 \text{ m}$

7-63 (cont)

Apply conservation of energy to the motion of the block from point A to point B: $K_A + U_A + W_{other} = K_B + U_B$

Work is done by gravity, by the spring force, and by friction, so $W_{other} = W_f$ and $U = U_{el} + U_{grav}$

$K_A = 0$
$K_B = \frac{1}{2}mv_B^2 = \frac{1}{2}(2.00\text{kg})(7.00\text{m/s})^2 = 49.0 \text{ J}$
$U_A = U_{el,A} + \cancel{U_{grav,A}} = U_{el,A}$
$U_B = U_{el,B} + U_{grav,B} = mgy_B = (2.00\text{kg})(9.80\text{m/s}^2)(3.61\text{m}) = 70.76 \text{ J}$
$W_{other} = W_f = (f_k \cos\phi)s = \mu_k mg \cos\theta (\cos 180°)s = -\mu_k mg \cos\theta \, s$
$W_{other} = -(0.50)(2.00\text{kg})(9.80\text{m/s}^2)(\cos 37.0°)(6.00\text{m}) = -46.96 \text{ J}$

Thus $U_{el,A} - 46.96 \text{ J} = 49.0 \text{ J} + 70.76 \text{ J}$
$U_{el,A} = 46.96 \text{ J} + 49.0 \text{ J} + 70.76 \text{ J} = \underline{167 \text{ J}}$

7-65

a) Apply $K_A + U_A + W_{other} = K_B + U_B$ to the motion from A to B.

$K_A = 0, \quad K_B = \frac{1}{2}mv_B^2$
$U_A = 0, \quad U_B = U_{B,el} = \frac{1}{2}Kx_B^2$, where $x_B = 0.25\text{m}$
$W_{other} = W_F = Fx_B$

Thus $Fx_B = \frac{1}{2}mv_B^2 + \frac{1}{2}Kx_B^2$.
(The work done by F goes partly to the potential energy of the stretched spring and partly to the kinetic energy of the block.)

$Fx_B = (20.0\text{N})(0.25\text{m}) = 5.0 \text{ J}$
$\frac{1}{2}Kx_B^2 = \frac{1}{2}(40.0\text{N/m})(0.25\text{m})^2 = 1.25 \text{ J}$
$\Rightarrow \quad 5.0 \text{ J} = \frac{1}{2}mv_B^2 + 1.25 \text{ J}$

$v_B = \sqrt{\frac{2(3.75 \text{ J})}{0.500 \text{kg}}} = \underline{3.87 \text{ m/s}}$

b) Let point C be where the block is closest to the wall. When the block is at point C the spring is compressed an amount $|x_c|$, so the block is $0.60\text{m} - |x_c|$ from the wall, and the distance between B and C is $x_B + |x_c|$.

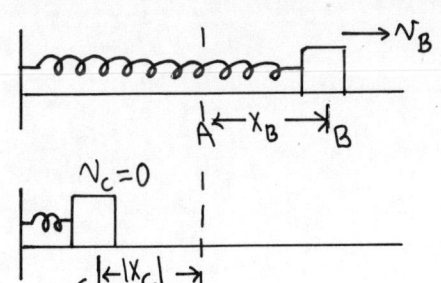

$K_B + U_B + W_{other} = K_C + U_C$

$W_{other} = 0$ (F is removed)
$K_B = \frac{1}{2}mv_B^2 = 5.0 \text{ J} - 1.25 \text{ J} = 3.75 \text{ J}$ (from part (a))
$U_B = \frac{1}{2}Kx_B^2 = 1.25 \text{ J}$
$K_C = 0$ (instantaneously at rest at point closest to wall)
$U_C = \frac{1}{2}k|x_c|^2$

105

7-65 (cont)

Thus $3.75\,\text{J} + 1.25\,\text{J} = \tfrac{1}{2}k|x_c|^2$

$|x_c| = \sqrt{\dfrac{2(5.0\,\text{J})}{40.0\,\text{N/m}}} = 0.50\,\text{m}$

The distance of the block from the wall is $0.60\,\text{m} - 0.50\,\text{m} = \underline{0.10\,\text{m}}$.

7-69

$\vec{F} = -\alpha x y^2 \hat{j}$, $\alpha = 1.50\,\text{N/m}^3$

a)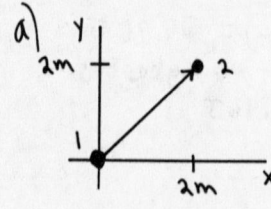

$d\vec{l} = dx\,\hat{i} + dy\,\hat{j}$
$\vec{F}\cdot d\vec{l} = -\alpha x y^2\,dy$
On the path, $x=y$, so $\vec{F}\cdot d\vec{l} = -\alpha y^3\,dy$

$W = \int_1^2 \vec{F}\cdot d\vec{l} = \int_{y_1}^{y_2}(-\alpha y^3\,dy) = -\tfrac{\alpha}{4} y^4\Big|_{y_1}^{y_2} = -\tfrac{\alpha}{4}(y_2^4 - y_1^4)$

$y_1 = 0$, $y_2 = 2.00\,\text{m}$, so $W = -\tfrac{1}{4}(1.50\,\text{N/m}^3)(2.00\,\text{m})^4 = \underline{-6.00\,\text{J}}$

b)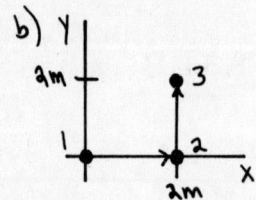

For the displacement from point 1 to point 2 $d\vec{l} = dx\,\hat{i}$, so $\vec{F}\cdot d\vec{l} = 0$ and $W=0$.
(The force is perpendicular to the displacement at each point along the path, so $W=0$.)

For the displacement from point 2 to point 3 $d\vec{l} = dy\,\hat{j}$, so $\vec{F}\cdot d\vec{l} = -\alpha x y^2\,dy$.
On this path, $x = 2.00\,\text{m}$, so $\vec{F}\cdot d\vec{l} = -(1.50\,\text{N/m}^3)(2.00\,\text{m})y^2\,dy = -(3.00\,\text{N/m}^2)y^2\,dy$

$W = \int_2^3 \vec{F}\cdot d\vec{l} = -(3.00\,\text{N/m}^2)\int_{y_2}^{y_3} y^2\,dy = -(3.00\,\text{N/m}^2)\tfrac{1}{3}(y_3^3 - y_2^3)$

$W = -(3.00\,\text{N/m}^2)\left(\tfrac{1}{3}\right)(2.00\,\text{m})^3 = \underline{-8.00\,\text{J}}$

c) For these two paths between the same starting and ending points the work done is different, so the force is nonconservative.

CHAPTER 8

Exercises 3, 7, 11, 13, 15, 21, 23, 25, 27, 31, 33, 35, 39, 41, 43, 45, 47, 53, 57

Problems 59, 65, 69, 71, 73, 77, 81, 85, 87, 89, 93

Exercises

8-3

$P_{1x} = mv_{1x} = (0.145 \text{ kg})(3.40 \text{ m/s}) = 0.493 \text{ kg·m/s}$
$P_{2x} = mv_{2x} = (0.057 \text{ kg})(-6.20 \text{ m/s}) = -0.353 \text{ kg·m/s}$

$P_x = P_{1x} + P_{2x} = 0.493 \text{ kg·m/s} - 0.353 \text{ kg·m/s} = \underline{+0.140 \text{ kg·m/s}}$
(plus sign means that is in the +x-direction)

8-7

Take the x-axis to be toward the right, so $v_{1x} = +3.00 \text{ m/s}$.

a) $J_x = P_{2x} - P_{1x}$
$J_x = F_x(t_2 - t_1) = (+5.00 \text{ N})(4.00 \text{ s}) = +20.0 \text{ kg·m/s}$
$\Rightarrow P_{2x} = J_x + P_{1x} = +20.0 \text{ kg·m/s} + (2.00 \text{ kg})(+3.00 \text{ m/s}) = +26.0 \text{ kg·m/s}$
$v_{2x} = \dfrac{P_{2x}}{m} = \dfrac{26.0 \text{ kg·m/s}}{2.00 \text{ kg}} = \underline{+13.0 \text{ m/s}}$ (to the right)

b) $J_x = F_x(t_2 - t_1) = (-7.00 \text{ N})(4.00 \text{ s}) = -28.0 \text{ kg·m/s}$ (negative since force is to left)
$P_{2x} = J_x + P_{1x} = -28.0 \text{ kg·m/s} + (2.00 \text{ kg})(+3.00 \text{ m/s}) = -22.0 \text{ kg·m/s}$
$v_{2x} = \dfrac{P_{2x}}{m} = \dfrac{-22.0 \text{ kg·m/s}}{2.00 \text{ kg}} = \underline{-11.0 \text{ m/s}}$ (to the left)

8-11

Take the x-axis to be toward the right, so $F_x = +(A + Bt^2)$

a) $J_x = \int_{t_1}^{t_2} F_x \, dt = \int_0^{t_2} (A + Bt^2) \, dt = (At + \tfrac{1}{3} Bt^3)\Big|_0^{t_2} = At_2 + \tfrac{1}{3} Bt_2^3$

The impulse has magnitude $J = At_2 + \tfrac{1}{3} Bt_2^3$ and is directed to the right.

b) $J_x = P_{2x} - P_{1x} \Rightarrow P_{2x} = J_x + P_{1x}$
Initially at rest $\Rightarrow P_{1x} = 0$.
$v_{2x} = \dfrac{P_{2x}}{m} = \dfrac{J_x}{m} = \dfrac{A}{m} t_2 + \dfrac{B}{3m} t_2^3$

Her speed is $v = \dfrac{A}{m} t_2 + \dfrac{B}{3m} t_2^3$ and she is moving to the right.

8-13

a) $\vec{J} = \int_{t_1}^{t_2} \vec{F}\, dt = \int_{t_1}^{t_2} [(1.80\times10^7 \text{ N/s})t - (9.00\times10^9 \text{ N/s}^2)t^2]\hat{i}\, dt$, where $t_1 = 0$ and $t_2 = 2.00\times10^{-3}$ s.

$\vec{J} = [(1.80\times10^7 \text{ N/s})(\frac{1}{2}t_2^2) - (9.00\times10^9 \text{ N/s}^2)(\frac{1}{3}t_2^3)]\hat{i}$

$\vec{J} = [\frac{1}{2}(1.80\times10^7 \text{ N/s})(2.00\times10^{-3}\text{s})^2 - \frac{1}{3}(9.00\times10^9 \text{ N/s}^2)(2.00\times10^{-3}\text{s})^3]\hat{i} = \underline{(12.0 \text{ kg·m/s})\hat{i}}$

b) $\vec{w} = -(mg)\hat{j}$ (constant)

$\vec{J} = \vec{w}(t_2 - t_1) = -mg\, t_2 \hat{j} = -(0.145 \text{ kg})(9.80 \text{ m/s}^2)(2.00\times10^{-3}\text{s})\hat{j}$
$\vec{J} = -(0.00284 \text{ kg·m/s})\hat{j}$

c) $F_{av} = \frac{J}{(t_2-t_1)} = \frac{12.0 \text{ kg·m/s}}{2.00\times10^{-3}\text{s}} = \underline{6000 \text{ N}}$

d) $\vec{J} = \vec{P_2} - \vec{P_1} \Rightarrow \vec{P_2} = \vec{P_1} + \vec{J}$

$\vec{J} = (12.0 \text{ kg·m/s})\hat{i} - (0.00284 \text{ kg·m/s})\hat{j}$
$\vec{P_1} = m\vec{v_1} = (0.145 \text{ kg})[(-40.0 \text{ m/s})\hat{i} - (5.0 \text{ m/s})\hat{j}] = (-5.80 \text{ kg·m/s})\hat{i} - (0.725 \text{ kg·m/s})\hat{j}$

Then
$\vec{P_2} = (-5.80 \text{ kg·m/s})\hat{i} - (0.725 \text{ kg·m/s})\hat{j} + (12.0 \text{ kg·m/s})\hat{i} - (0.00284 \text{ kg·m/s})\hat{j}$
$\vec{P_2} = (6.20 \text{ kg·m/s})\hat{i} - (0.728 \text{ kg·m/s})\hat{j}$

$\vec{v_2} = \frac{\vec{P_2}}{m} = (42.8 \text{ m/s})\hat{i} - (5.0 \text{ m/s})\hat{j}$

8-15

a) Let Gretzky be object A and the defender be object B. Let the $+x$-direction be the direction in which Gretzky is moving initially. No horizontal external forces, so P_x is constant.

before: A $v_{A1} = 13.0$ m/s →   B $v_{B1} = 5.00$ m/s ←

after: A $v_{A2} = 2.50$ m/s →   B $v_{B2} = ?$

P_x constant $\Rightarrow m_A v_{A1x} + m_B v_{B1x} = m_A v_{A2x} + m_B v_{B2x}$

If we multiply through by g we get an equation that uses the object's weight rather than its mass:

$w_A v_{A1x} + w_B v_{B1x} = w_A v_{A2x} + w_B v_{B2x}$

The components can be positive or negative depending on the directions of the velocities relative to our coordinate system.

Putting in the numbers,
$(756 \text{ N})(+13.0 \text{ m/s}) + (900 \text{ N})(-5.00 \text{ m/s}) = (756 \text{ N})(+2.50 \text{ m/s}) + (900 \text{ N})v_{B2x}$
$9828 \text{ N·m/s} - 4500 \text{ N·m/s} = 1890 \text{ N·m/s} + (900 \text{ N})v_{B2x}$

8-15 (cont)

$$v_{B2x} = \frac{9828\,N\cdot m/s - 4500\,N\cdot m/s - 1890\,N\cdot m/s}{900\,N} = 3.82\,m/s$$

After the collision the defender is moving at **3.82 m/s** in the same direction as Gretzky.

b) $K_1 = \frac{1}{2} m_A v_{A1}^2 + \frac{1}{2} m_B v_{B1}^2 = \frac{1}{2}\left(\frac{756\,N}{9.80\,m/s^2}\right)(13.0\,m/s)^2 + \frac{1}{2}\left(\frac{900\,N}{9.80\,m/s^2}\right)(5.00\,m/s)^2$

$K_1 = 6519\,J + 1148\,J = 7667\,J$

(Note that the kinetic energy of an object is always positive, and that it does not depend on the direction of the object's velocity.)

$K_2 = \frac{1}{2} m_A v_{A2}^2 + \frac{1}{2} m_B v_{B2}^2 = \frac{1}{2}\left(\frac{756\,N}{9.80\,m/s^2}\right)(2.50\,m/s)^2 + \frac{1}{2}\left(\frac{900\,N}{9.80\,m/s^2}\right)(3.82\,m/s)^2$

$K_2 = 241\,J + 670\,J = 911\,J$

$\Delta K = K_2 - K_1 = 911\,J - 7667\,J = -6756\,J$

The kinetic energy **decreases by 6760 J**.

8-21

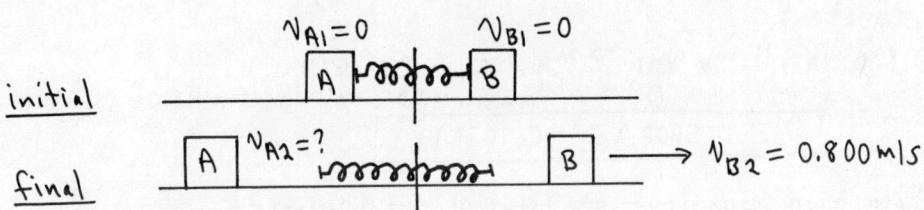

a) No horizontal force $\Rightarrow P_x$ is constant

$m_A \cancel{v_{A1x}^0} + m_B \cancel{v_{B1x}^0} = m_A v_{A2x} + m_B v_{B2x}$

$0 = m_A v_{A2x} + m_B v_{B2x}$

$v_{A2x} = -\left(\frac{m_B}{m_A}\right) v_{B2x} = -\left(\frac{3.00\,kg}{1.00\,kg}\right)(+0.800\,m/s) = -2.40\,m/s$

Block A has a final speed of 2.40 m/s, and moves off in the opposite direction to B.

b) Use energy conservation

$K_1 + U_1 + W_{other} = K_2 + U_2$

Only the spring force does work $\Rightarrow W_{other} = 0$ and $U = U_{el}$.
$K_1 = 0$ (the blocks initially are at rest)
$U_2 = 0$ (no potential energy is left stored in the spring)
$K_2 = \frac{1}{2} m_A v_{A2}^2 + \frac{1}{2} m_B v_{B2}^2 = \frac{1}{2}(1.00\,kg)(2.40\,m/s)^2 + \frac{1}{2}(3.00\,kg)(0.800\,m/s)^2 = 3.84\,J$
$U_1 = U_{1,el}$ the potential energy stored in the compressed spring.

Thus $U_{1,el} = K_2 = \underline{3.84\,J}$

8-23

Take the x-axis to lie along the initial velocity of A.

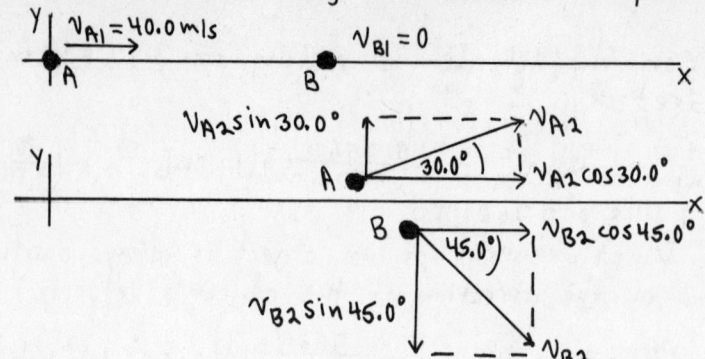

a) P_x is constant

$$\Rightarrow m_A v_{A1x} + m_B v_{B1x} = m_A v_{A2x} + m_B v_{B2x}$$
$$\cancel{m} v_{A1} = \cancel{m} v_{A2} \cos 30.0° + \cancel{m} v_{B2} \cos 45.0°$$

$$\boxed{40.0 \text{ m/s} = 0.8660 \, v_{A2} + 0.7071 \, v_{B2}}$$

P_y is constant

$$\Rightarrow m_A v_{A1y} + m_B v_{B1y} = m_A v_{A2y} + m_B v_{B2y}$$
$$0 = \cancel{m} v_{A2} \sin 30.0° - \cancel{m} v_{B2} \sin 45.0°$$

$$\boxed{0 = 0.5000 \, v_{A2} - 0.7071 \, v_{B2}}$$

Add these two equations $\Rightarrow 1.366 \, v_{A2} = 40.0 \text{ m/s}$
$$v_{A2} = \underline{29.3 \text{ m/s}}$$

Then $v_{B2} = \dfrac{0.5000}{0.7071} v_{A2} = \dfrac{0.5000}{0.7071} (29.3 \text{ m/s}) = \underline{20.7 \text{ m/s}}$.

b) $K_1 = K_{A1} = \tfrac{1}{2} m v_{A1}^2$
$K_2 = K_{A2} + K_{B2} = \tfrac{1}{2} m v_{A2}^2 + \tfrac{1}{2} m v_{B2}^2$
$\Delta K = K_2 - K_1$

The fraction of the original kinetic energy of puck A dissipated during the collision is

$$-\dfrac{\Delta K}{K_1} = -\dfrac{K_2 - K_1}{K_1} = -\left(\dfrac{\tfrac{1}{2} m v_{A2}^2 + \tfrac{1}{2} m v_{B2}^2 - \tfrac{1}{2} m v_{A1}^2}{\tfrac{1}{2} m v_{A1}^2} \right) = 1 - \left(\dfrac{v_{A2}}{v_{A1}}\right)^2 - \left(\dfrac{v_{B2}}{v_{A1}}\right)^2$$

$$-\dfrac{\Delta K}{K_1} = 1 - \left(\dfrac{29.3 \text{ m/s}}{40.0 \text{ m/s}}\right)^2 - \left(\dfrac{20.7 \text{ m/s}}{40.0 \text{ m/s}}\right)^2 = 1 - 0.5366 - 0.2679 = 0.1956 \, ; \, \underline{19.6\%}$$

8-25

Let Ken be object A and Kim be object B.

before: $v_{A1} = 3.00 \text{ m/s}$ [A][B] $\rightarrow v_{B1} = 3.00 \text{ m/s}$

8-25 (cont)

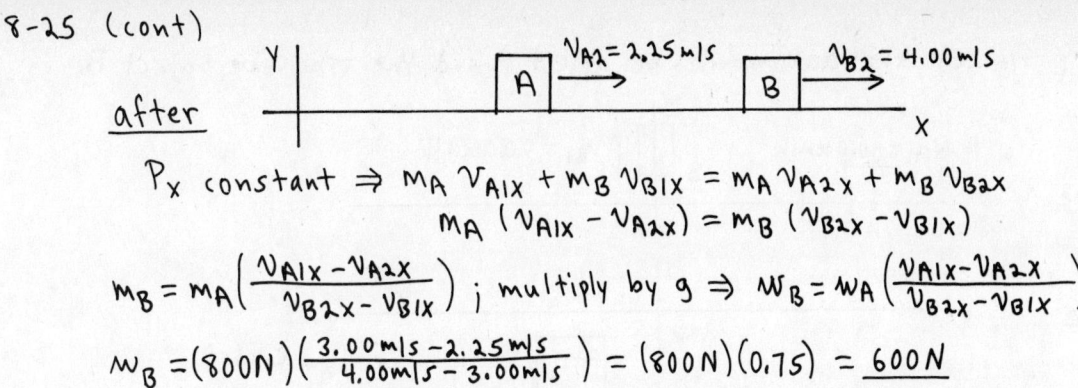
after

P_x constant $\Rightarrow m_A v_{A1x} + m_B v_{B1x} = m_A v_{A2x} + m_B v_{B2x}$
$m_A(v_{A1x} - v_{A2x}) = m_B(v_{B2x} - v_{B1x})$

$m_B = m_A \left(\dfrac{v_{A1x} - v_{A2x}}{v_{B2x} - v_{B1x}} \right)$; multiply by $g \Rightarrow w_B = w_A \left(\dfrac{v_{A1x} - v_{A2x}}{v_{B2x} - v_{B1x}} \right)$

$w_B = (800 N)\left(\dfrac{3.00 m/s - 2.25 m/s}{4.00 m/s - 3.00 m/s}\right) = (800 N)(0.75) = \underline{600 N}$

8-27
a) before

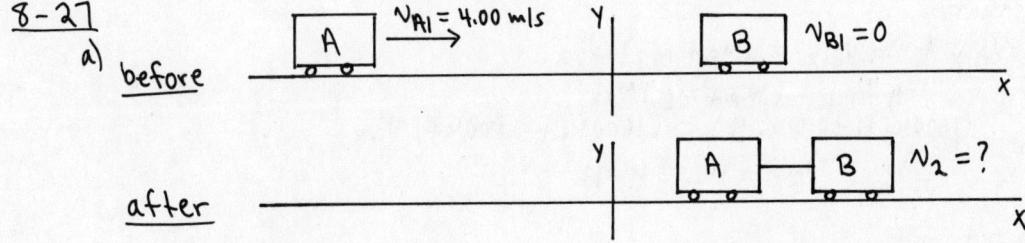
after

P_x is constant
$\Rightarrow m_A v_{A1x} + m_B v_{B1x} = (m_A + m_B) v_{2x}$
$m_A v_{A1} = (m_A + m_B) v_2$

$v_2 = \left(\dfrac{m_A}{m_A + m_B}\right) v_{A1} = \left(\dfrac{27,000 kg}{27,000 kg + 81,000 kg}\right) 4.00 m/s = \underline{1.00 m/s}$

b) $K_1 = \frac{1}{2} m_A v_{A1}^2 + \frac{1}{2} m_B v_{B1}^{2\,\to 0} = \frac{1}{2}(27,000 kg)(4.00 m/s)^2 = 2.16 \times 10^5 J$
$K_2 = \frac{1}{2}(m_A + m_B) v_2^2 = \frac{1}{2}(108,000 kg)(1.00 m/s)^2 = 5.40 \times 10^4 J$

$\Delta K = K_2 - K_1 = 5.40 \times 10^4 J - 2.16 \times 10^5 J = \underline{-1.62 \times 10^5 J}$ (decreases)

c) before

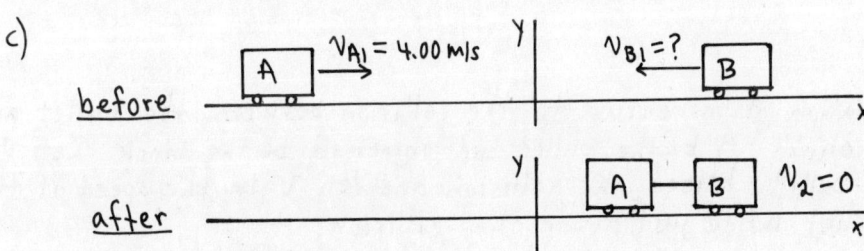
after

P_x is constant
$\Rightarrow m_A v_{A1x} + m_B v_{B1x} = (m_A + m_B) v_{2x}^{\to 0}$
$m_A v_{A1} - m_B v_{B1} = 0$

$v_{B1} = \left(\dfrac{m_A}{m_B}\right) v_{A1} = \left(\dfrac{27,000 kg}{81,000 kg}\right)(4.00 m/s) = \underline{1.33 m/s}$

8-31

Let the automobile be object A and the truck be object B.

before: $v_{A1} = 40.0$ km/h (A moving east), $v_{B1} = 20.0$ km/h (B moving south)

after: A+B moving at angle θ below +x-axis, with $v_{2x} = v_2 \cos\theta$, $v_{2y} = v_2 \sin\theta$

P_x is constant
$$\Rightarrow m_A v_{A1x} + m_B v_{B1x} = (m_A + m_B) v_{2x}$$
$$m_A v_{A1} = (m_A + m_B) v_{2x}$$
$$(1600 \text{ kg})(40.0 \text{ km/h}) = (1600 \text{ kg} + 2800 \text{ kg}) v_{2x}$$
$$v_{2x} = 14.55 \text{ km/h}$$

P_y is constant
$$\Rightarrow m_A v_{A1y} + m_B v_{B1y} = (m_A + m_B) v_{2y}$$
$$-m_B v_{B1} = (m_A + m_B) v_{2y}$$
$$-(2800 \text{ kg})(20.0 \text{ km/h}) = (1600 \text{ kg} + 2800 \text{ kg}) v_{2y}$$
$$v_{2y} = -12.73 \text{ km/h}$$

$$v_2 = \sqrt{v_{2x}^2 + v_{2y}^2} = \sqrt{(14.55 \text{ km/h})^2 + (-12.73 \text{ km/h})^2} = \underline{19.3 \text{ km/h}}$$

$$\tan\theta = \frac{v_{2y}}{v_{2x}} = \frac{-12.73 \text{ km/h}}{14.55 \text{ km/h}} = -0.875 \Rightarrow \theta = -41.2° \text{ (minus means clockwise from +x-axis)}$$

The wreckage is moving in the direction $\underline{41.2° \text{ S of E}}$.

8-33

Apply conservation of momentum to the collision between the bullet and the block. Let object A be the bullet and object B be the block. Let v_A be the speed of the bullet before the collision and let V be the speed of the block with the bullet inside just after the collision.

before: A $\rightarrow v_A$; B $v_{B1} = 0$

after: A+B $\rightarrow V$

P_x is constant $\Rightarrow \boxed{m_A v_A = (m_A + m_B) V}$

8-33 (cont)

Apply conservation of energy to the motion of the block after the collision.

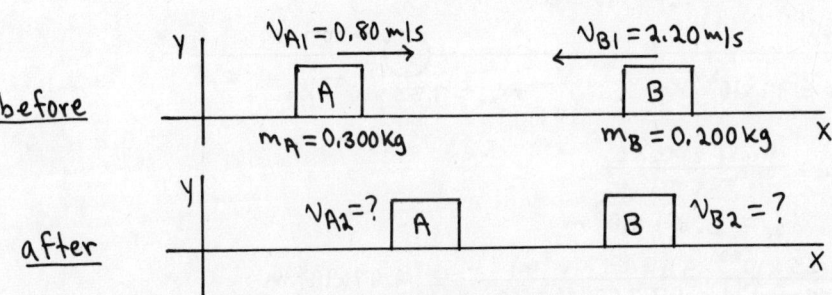

$K_1 + U_1 + W_{other} = K_2 + U_2$

Work is done by friction ⇒ $W_{other} = W_f = (f_k \cos\phi)s = -f_k s = -\mu_k mgs$
$U_1 = U_2 = 0$ (no work done by gravity)
$K_1 = \frac{1}{2}mV^2$
$K_2 = 0$ (block has come to rest)

Thus $\frac{1}{2}mV^2 - \mu_k mgs = 0$
$V = \sqrt{2\mu_k gs} = \sqrt{2(0.20)(9.80 \text{ m/s}^2)(0.250\text{m})} = 0.9899 \text{ m/s}$

Use this in the conservation of momentum equation
$V_A = \left(\frac{m_A + m_B}{m_A}\right)V = \left(\frac{5.00 \times 10^{-3}\text{ kg} + 1.50\text{ kg}}{5.00 \times 10^{-3}\text{ kg}}\right)(0.9899 \text{ m/s}) = \underline{298 \text{ m/s}}$

8-35

before: $v_{A1} = 0.80$ m/s, $m_A = 0.300$ kg; $v_{B1} = 2.20$ m/s, $m_B = 0.200$ kg
after: $v_{A2} = ?$, $v_{B2} = ?$

From conservation of x-component of momentum:
$m_A v_{A1x} + m_B v_{B1x} = m_A v_{A2x} + m_B v_{B2x}$
$m_A v_{A1} - m_B v_{B1} = m_A v_{A2x} + m_B v_{B2x}$

$(0.300\text{kg})(0.80\text{m/s}) - (0.200\text{kg})(2.20\text{m/s}) = (0.300\text{kg})v_{A2x} + (0.200\text{kg})v_{B2x}$
$0.240 \text{ kg·m/s} - 0.440 \text{ kg·m/s} = (0.300\text{kg})v_{A2x} + (0.200\text{kg})v_{B2x}$
$\boxed{-2.00 \text{ m/s} = 3v_{A2x} + 2v_{B2x}}$

From the relative velocity relation for an elastic collision (Eq. 8-27):
$v_{B2x} - v_{A2x} = -(v_{B1x} - v_{A1x}) = -(-2.20 \text{ m/s} - 0.80 \text{ m/s}) = +3.00 \text{ m/s}$
Multiply this equation by three ⇒ $\boxed{9.00 \text{ m/s} = 3v_{B2x} - 3v_{A2x}}$

Add the two equations ⇒ $7.00 \text{ m/s} = 5v_{B2x}$
$v_{B2x} = +1.40 \text{ m/s}$

and then $v_{A2x} = v_{B2x} - 3.00 \text{ m/s} = 1.40 \text{ m/s} - 3.00 \text{ m/s} = -1.60 \text{ m/s}$

The 0.300 kg block (block A) is moving to the left at $\underline{1.60 \text{ m/s}}$ and the 0.200 kg block (block B) is moving to the right at $\underline{1.40 \text{ m/s}}$.

8-39

a) Let A be the proton and B be the target nucleus. The collision is elastic, all velocities lie along a line, and B is at rest before the collision. Hence the results of Eqs. (8-24) and (8-25) apply.

Eq. (8-24): $m_B(v+v_A) = m_A(v-v_A)$, where v is the velocity component of A before the collision and v_A is the velocity component of A after the collision.

Here, $v = 2.00 \times 10^7$ m/s (take direction of incident beam to be positive)
$v_A = -1.50 \times 10^7$ m/s (negative since traveling in direction opposite to incident beam)

$$m_B = m_A\left(\frac{v-v_A}{v+v_A}\right) = m\left(\frac{2.00 \times 10^7 \text{ m/s} + 1.50 \times 10^7 \text{ m/s}}{2.00 \times 10^7 \text{ m/s} - 1.50 \times 10^7 \text{ m/s}}\right) = m\left(\frac{3.50}{0.50}\right) = \underline{7.0\,m}$$

b) Eq. (8-25): $v_B = \left(\frac{2m_A}{m_A+m_B}\right)v = \left(\frac{2m}{m+7m}\right)(2.00 \times 10^7 \text{ m/s}) = \underline{5.00 \times 10^6 \text{ m/s}}$

8-41

Apply Eq. (8-28) with the earth as mass #1 and the moon as mass #2. Take the origin at the earth and let the moon lie on the positive x-axis.

[Diagram: earth ($m_E = 5.97 \times 10^{24}$ kg) at origin and moon ($m_M = 7.35 \times 10^{22}$ kg) at 3.84×10^8 m on positive x-axis]

$$X_{cm} = \frac{m_1 x_1 + m_2 x_2}{m_1 + m_2}$$

$x_1 = 0$ and $x_2 = 3.84 \times 10^8$ m

$$X_{cm} = \frac{(7.35 \times 10^{22} \text{ kg})(3.84 \times 10^8 \text{ m})}{5.97 \times 10^{24} \text{ kg} + 7.35 \times 10^{22} \text{ kg}} = 4.67 \times 10^6 \text{ m}$$

The center of mass is 4.67×10^6 m from the center of the earth and is on the line connecting the centers of the earth and of the moon.

8-43

$X_{cm} = 3.0$ m, $Y_{cm} = 0$; $\vec{v}_{cm} = (6.0 \text{ m/s})\hat{j}$

a) $X_{cm} = \frac{m_1 x_1 + m_2 x_2}{m_1 + m_2} = \frac{m_1(0) + (0.10 \text{ kg})(12.0 \text{ m})}{m_1 + 0.10 \text{ kg}} = \frac{1.20 \text{ kg} \cdot \text{m}}{m_1 + 0.10 \text{ kg}}$

$X_{cm} = 3.0$ m $\Rightarrow$ 3.0 m $= \frac{1.20 \text{ kg} \cdot \text{m}}{m_1 + 0.10 \text{ kg}}$

$m_1 + 0.10$ kg $= \frac{1.20 \text{ kg} \cdot \text{m}}{3 \text{ m}} = 0.40$ kg

$\underline{m_1 = 0.30 \text{ kg}}$

b) $\vec{P} = M\vec{v}_{cm} = (m_1 + m_2)\vec{v}_{cm} = (0.10 \text{ kg} + 0.30 \text{ kg})(6.0 \text{ m/s})\hat{j} = \underline{(2.4 \text{ kg} \cdot \text{m/s})\hat{j}}$

c) $\vec{v}_{cm} = \frac{m_1 \vec{v}_1 + m_2 \vec{v}_2}{m_1 + m_2}$

8-43 (cont)

particle 2 at rest $\Rightarrow v_2 = 0$

Then $\vec{v}_1 = \left(\frac{m_1+m_2}{m_1}\right)\vec{v}_{cm} = \left(\frac{0.30\text{kg}+0.10\text{kg}}{0.30\text{kg}}\right)(6.0\text{m/s})\hat{j} = \underline{(8.0\text{m/s})\hat{j}}$

8-45

a) $y_{cm} = \frac{m_1 y_1 + m_2 y_2}{m_1 + m_2}$

$m_1 + m_2 = \frac{m_1 y_1 + m_2 y_2}{y_{cm}} = \frac{m_1(0) + (0.60\text{kg})(80\text{m})}{24\text{m}} = \underline{2.0\text{kg}}$

b) $\vec{a}_{cm} = \frac{d\vec{v}_{cm}}{dt} = \underline{(12.0\text{m/s}^3)t\,\hat{j}}$

c) $\sum \vec{F}_{ext} = M\vec{a}_{cm} = (2.0\text{kg})(12.0\text{m/s}^3)t\,\hat{j}$

At $t = 3.0\text{s}$, $\sum \vec{F}_{ext} = (2.0\text{kg})(12.0\text{m/s}^3)(3.0\text{s})\hat{j} = \underline{(72.0\text{N})\hat{j}}$

8-47

Apply Eq. (8-39): $a = -\frac{V_{ex}}{m}\frac{dm}{dt}$

$\frac{dm}{dt} = -\frac{ma}{V_{ex}} = -\frac{(7000\text{kg})(25.0\text{m/s}^2)}{2000\text{m/s}} = \underline{-87.5\text{kg/s}}$

So in 1s the rocket must eject 87.5 kg of gas.

8-53

Use Eq. (8-40): $v - v_0 = V_{ex}\ln\left(\frac{m_0}{m}\right)$

$v_0 = 0$ ("fired from rest"), so $\frac{v}{V_{ex}} = \ln\left(\frac{m_0}{m}\right)$

$\Rightarrow \frac{m_0}{m} = e^{v/V_{ex}}$, or $\frac{m}{m_0} = e^{-v/V_{ex}}$

If v is the final speed then m is the mass left when all the fuel has been expended; $\frac{m}{m_0}$ is the fraction of the initial mass that is not fuel.

a) $v = 1.00\times 10^{-3}c = 2.998\times 10^5 \text{m/s} \Rightarrow \frac{m}{m_0} = e^{-(2.998\times 10^5 \text{m/s}/2400\text{m/s})} = 6\times 10^{-55}$

this is clearly not feasible, for so little of the initial mass to not be fuel.

b) $v = 3000\text{m/s} \Rightarrow \frac{m}{m_0} = e^{-(3000\text{m/s}/2400\text{m/s})} = \underline{0.287}$

28.7% of the total initial mass not fuel, so 71.3% is fuel; this is possible.

8-57

Let $\vec{p}_e$, $\vec{p}_{\bar{\nu}}$, and $\vec{p}_N$ be the final momenta of the electron, the antineutrino, and the recoiling ^{210}Po nucleus.

8-57 (cont)

before / **after** diagrams:
- $v_1 = 0$, ^{210}Bi
- $P_e = 3.60 \times 10^{-22}$ kg·m/s (e⁻)
- ^{210}B with P_{Ny}, P_N, P_{Nx}
- $P_{\bar{y}} = 5.20 \times 10^{-22}$ kg·m/s

a) P_x is conserved and initially $P_x = 0$

$$\Rightarrow 0 = P_{Nx} - P_e$$

$$P_{Nx} = P_e = 3.60 \times 10^{-22} \text{ kg·m/s}$$

P_y is conserved and initially $P_y = 0$

$$0 = P_{Ny} - P_{\bar{y}}$$

$$P_{Ny} = P_{\bar{y}} = 5.20 \times 10^{-22} \text{ kg·m/s}$$

Then $P_N = \sqrt{P_{Nx}^2 + P_{Ny}^2} = \sqrt{(3.60 \times 10^{-22})^2 + (5.20 \times 10^{-22})^2}$ kg·m/s $= \underline{6.32 \times 10^{-22} \text{ kg·m/s}}$

b) $K_N = \frac{1}{2} m_N v_N^2$ and $P_N = m_N v_N \Rightarrow K_N = \frac{P_N^2}{2 m_N}$

$$K_N = \frac{(6.32 \times 10^{-22} \text{ kg·m/s})^2}{2(3.50 \times 10^{-25} \text{ kg})} = \underline{5.71 \times 10^{-19} \text{ J}}$$

Problems

8-59

$$\vec{F} = (\alpha t^2)\hat{i} + (\beta + \gamma t)\hat{j}; \quad \alpha = 30.0 \text{ N/s}^2, \beta = 40.0 \text{ N}, \gamma = 5.0 \text{ N/s}$$

$$J_x = \int_{t_1}^{t_2} F_x(t)\, dt = \int_0^{t_2} (\alpha t^2)\, dt = \frac{1}{3}\alpha t_2^3 = \frac{1}{3}(30.0 \text{ N/s}^2)(0.500s)^3 = 1.25 \text{ N·s}$$

$$J_y = \int_{t_1}^{t_2} F_y(t)\, dt = \int_0^{t_2} (\beta + \gamma t)\, dt = (\beta t_2 + \frac{1}{2}\gamma t_2^2)$$

$$J_y = (40.0 \text{ N})(0.500s) + \frac{1}{2}(5.00 \text{ N/s})(0.500s)^2 = 20.6 \text{ N·s}$$

$$J_x = P_{2x} - P_{1x} = m(v_{2x} - v_{1x}^{\,0})$$

$$v_{2x} = \frac{J_x}{m} = \frac{1.25 \text{ N·s}}{2.00 \text{ kg}} = 0.625 \text{ m/s}$$

$$J_y = P_{2y} - P_{1y} = m(v_{2y} - v_{1y}^{\,0})$$

$$v_{2y} = \frac{J_y}{m} = \frac{20.6 \text{ N·s}}{2.00 \text{ kg}} = 10.3 \text{ m/s}$$

Thus $\vec{v} = (0.625 \text{ m/s})\hat{i} + (10.3 \text{ m/s})\hat{j}$.

8-65

Use a coordinate system attached to the ground. Take the +x-axis to be east (along the tracks) and the y-axis to be north (parallel to the ground and perpendicular to the tracks). Then P_x is conserved but P_y is **not** conserved due to the sideways force exerted by the tracks, the force that keeps the handcar on the tracks.

a) Let A be the 30.0 kg mass and B be the car (mass 170 kg). After the mass is thrown sideways relative to the car, it still has the same eastward component of velocity, 5.00 m/s, as it had before it was thrown.

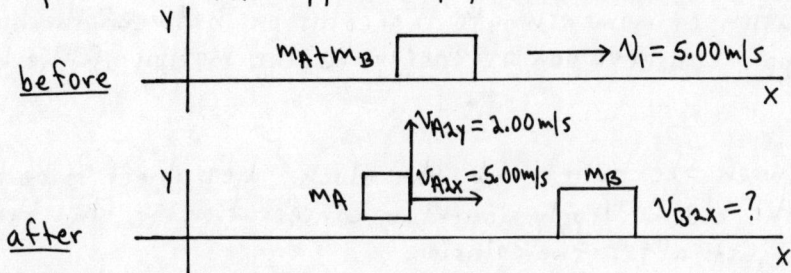

P_x is conserved
$$\Rightarrow (m_A + m_B) V_1 = m_A V_{A2x} + m_B V_{B2x}$$
$$(200 \text{ kg})(5.00 \text{ m/s}) = (30.0 \text{ kg})(5.00 \text{ m/s}) + (170 \text{ kg}) V_{B2x}$$

$$V_{B2x} = \frac{1000 \text{ kg·m/s} - 150 \text{ kg·m/s}}{170 \text{ kg}} = 5.00 \text{ m/s}$$

The final velocity of the car is 5.00 m/s, east (unchanged).

b) We need the final velocity of A relative to the ground
$$\vec{V}_{A/E} = \vec{V}_{A/B} + \vec{V}_{B/E}$$

$V_{B/E} = +5.00$ m/s
$V_{A/B} = -5.00$ m/s (minus since the mass is moving west relative to the car)
This gives $V_{A/E} = 0$; the mass is at rest relative to the earth after it is thrown backwards from the car.

As in part (a), $(m_A + m_B) V_1 = m_A V_{A2x} + m_B V_{B2x}$
Now $V_{A2x} = 0$, so $(m_A + m_B) V_1 = m_B V_{B2x}$

$$V_{B2x} = \left(\frac{m_A + m_B}{m_B}\right) V_1 = \left(\frac{200 \text{ kg}}{170 \text{ kg}}\right)(5.00 \text{ m/s}) = 5.88 \text{ m/s}$$

The final velocity of the car is **5.88 m/s, east**.

c) Let A be the 30.0 kg mass and B be the car (mass $m_B = 200$ kg).

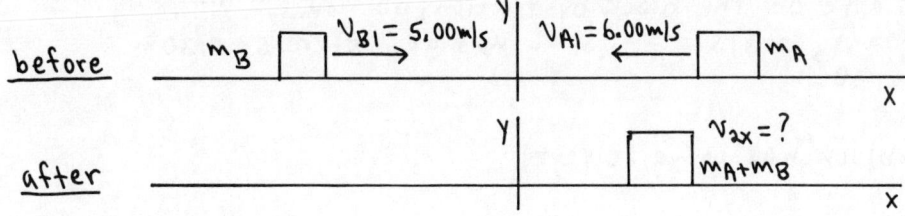

117

8-65 (cont)

P_x is conserved

$\Rightarrow m_A v_{A1x} + m_B v_{B1x} = (m_A + m_B) v_{2x}$

$- m_A v_{A1} + m_B v_{B1} = (m_A + m_B) v_{2x}$

$v_{2x} = \dfrac{m_B v_{B1} - m_A v_{A1}}{m_A + m_B} = \dfrac{(200\,kg)(5.00\,m/s) - (30.0\,kg)(6.00\,m/s)}{200\,kg + 30.0\,kg} = 3.57\,m/s$

The final velocity of the car is $\underline{3.57\,m/s,\,east}$.

8-69

Apply conservation of momentum to the collision between the bullet and the block and apply conservation of energy to the motion of the block after the collision.

Collision between the bullet and the block: Let object A be the bullet and object B be the block. Apply momentum conservation to find the speed v_{B2} of the block just after the collision.

before: $A \Rightarrow v_{A1} = 500\,m/s$; B , $v_{B1} = 0$

after: B, $v_{B2} = ?$; $\Rightarrow v_{A2} = 100\,m/s$

P_x is conserved, so

$m_A v_{A1x} + m_B \cancel{v_{B1x}}^{0} = m_A v_{A2x} + m_B v_{B2x}$

$m_A v_{A1} = m_A v_{A2} + m_B v_{B2x}$

$v_{B2x} = \dfrac{m_A (v_{A1} - v_{A2})}{m_B} = \dfrac{(4.00 \times 10^{-3}\,kg)(500\,m/s - 100\,m/s)}{1.00\,kg} = 1.60\,m/s$

Motion of the block after the collision:

Let point 1 in the motion be just after the collision, where the block has the speed 1.60 m/s calculated above, and let point 2 be where the block has come to rest.

#1 $v_1 = 1.60\,m/s$ #2 $v_2 = 0$; 0.30 m

$K_1 + U_1 + W_{other} = K_2 + U_2$

Work is done on the block by friction, so $W_{other} = W_f$.

$W_{other} = W_f = (f_k \cos\phi) s = -f_k s = -\mu_k m g s$, where $s = 0.30\,m$

$U_1 = 0,\ U_2 = 0$

$K_1 = \tfrac{1}{2} m v_1^2$

$K_2 = 0$ (block has come to rest)

118

8-69 (cont)

Thus $\frac{1}{2}mv_1^2 - \mu_K mgs = 0$.

$$\mu_K = \frac{v_1^2}{2gs} = \frac{(1.60 \text{ m/s})^2}{2(9.80 \text{ m/s}^2)(0.30\text{m})} = \underline{0.435}$$

b) For the bullet,
$K_1 = \frac{1}{2}mv_1^2 = \frac{1}{2}(4.00 \times 10^{-3} \text{kg})(500 \text{ m/s})^2 = 500 \text{ J}$
$K_2 = \frac{1}{2}mv_2^2 = \frac{1}{2}(4.00 \times 10^{-3} \text{kg})(100 \text{ m/s})^2 = 20 \text{ J}$
$\Delta K = K_2 - K_1 = 20 \text{ J} - 500 \text{ J} = -480 \text{ J}$

The kinetic energy of the bullet decreases by $\underline{480 \text{ J}}$.

c) Immediately after the collision the speed of the block is 1.60 m/s so its kinetic energy is $K = \frac{1}{2}mv^2 = \frac{1}{2}(1.00 \text{ kg})(1.60 \text{ m/s})^2 = \underline{1.28 \text{ J}}$.

Note that the collision is highly inelastic. The bullet loses 480 J of kinetic energy, but only 1.28 J is gained by the block. But momentum is conserved in the collision. All the momentum lost by the bullet is gained by the block.

8-71

Use the free-body diagram for the frame when it hangs at rest on the end of the spring to find the force constant K of the spring. Let s be the amount the spring is stretched.

$a=0$, ks (the spring force) upward, mg downward

$\sum F_y = ma_y$
$-mg + ks = 0$
$k = \frac{mg}{s} = \frac{(0.100 \text{ kg})(9.80 \text{ m/s}^2)}{0.050 \text{ m}} = 19.6 \text{ N/m}$

Next find the speed of the putty when it reaches the frame. The putty falls with acceleration $a=g$, downward.

$v_0 = 0$
$y - y_0 = 0.300 \text{ m}$
$v = ?$
$a = +9.80 \text{ m/s}^2$

$v^2 = v_0^2 + 2a(y-y_0)$
$v = \sqrt{2a(y-y_0)}$
$v = \sqrt{2(9.80 \text{ m/s}^2)(0.300 \text{ m})} = 2.425 \text{ m/s}$

Apply conservation of momentum to the collision between the putty (A) and the frame (B):

before: v_{A1} downward, $v_{B1} = 0$
after: $v_2 = ?$ downward

8-71 (cont)

P_y is conserved, so $-m_A v_{A1} = -(m_A + m_B) v_2$

$v_2 = \left(\dfrac{m_A}{m_A + m_B}\right) v_{A1} = \left(\dfrac{0.200 \text{ kg}}{0.300 \text{ kg}}\right)(2.425 \text{ m/s}) = 1.617 \text{ m/s}$

Apply conservation of energy to the motion of the frame on the end of the spring after the collision. Let point #1 be just after the putty strikes and point #2 be when the frame has its maximum downward displacement. Let d be the amount the frame moves downward.

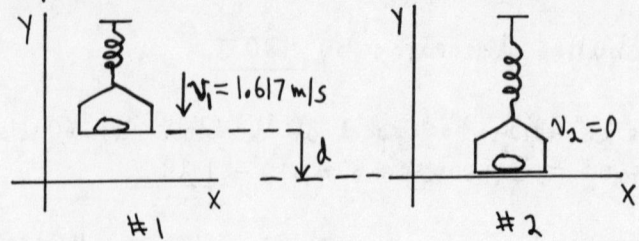

#1   #2

When the frame is at position #1 the spring is stretched a distance $x_1 = 0.050$ m. When the frame is at position #2 the spring is stretched a distance $x_2 = 0.050$ m $+ d$. Use coordinates with the $+y$-direction upward and $y = 0$ at the lowest point reached by the frame, so that $y_1 = d$ and $y_2 = 0$. Work is done on the frame by gravity and by the spring force, so $W_{other} = 0$, and $U = U_{el} + U_{gravity}$.

$$K_1 + U_1 + W_{other} = K_2 + U_2$$

$W_{other} = 0$
$K_1 = \tfrac{1}{2} m v_1^2 = \tfrac{1}{2}(0.300 \text{ kg})(1.617 \text{ m/s})^2 = 0.3920 \text{ J}$, $K_2 = 0$
$U_1 = U_{1,el} + U_{1,grav} = \tfrac{1}{2} k x_1^2 + mgy_1 = \tfrac{1}{2}(19.6 \text{ N/m})(0.050 \text{ m})^2 + (0.300 \text{ kg})(9.80 \text{ m/s}^2) d$
$U_1 = 0.0245 \text{ J} + (2.94 \text{ N}) d$
$U_2 = U_{2,el} + U_{2,grav} = \tfrac{1}{2} k x_2^2 + \cancel{mgy_2}^0 = \tfrac{1}{2}(19.6 \text{ N/m})(0.050 \text{ m} + d)^2$
$U_2 = 0.0245 \text{ J} + (0.98 \text{ N}) d + (9.8 \text{ N/m}) d^2$

Thus $0.3920 \text{ J} + 0.0245 \text{ J} + (2.94 \text{ N}) d = 0.0245 \text{ J} + (0.98 \text{ N}) d + (9.8 \text{ N/m}) d^2$
$(9.8 \text{ N/m}) d^2 - (1.96 \text{ N}) d - 0.3920 \text{ J} = 0$

$d = \dfrac{1}{2(9.8)}\left[1.96 \pm \sqrt{(1.96)^2 - 4(9.8)(-0.3920)}\right] \text{ m} = 0.100 \text{ m} \pm 0.2236 \text{ m}$

The solution we want is a positive (downward) distance, so
$d = 0.100 \text{ m} + 0.2236 \text{ m} = \underline{0.324 \text{ m}}$

8-73

Let A be the bullet and B be the stone.

a) $v_{A1} = 450$ m/s $v_{B1} = 0$

120

8-73 (cont)

[Sketch: y-axis with v_{B2y} pointing up, v_{B2} at angle θ above x-axis, v_{B2x} along x-axis; below origin $v_{A2} = 300$ m/s pointing down]

P_x is conserved, so

$$m_A v_{A1x} + m_B \cancel{v_{B1x}}^0 = m_A \cancel{v_{A2x}}^0 + m_B v_{B2x}$$

$$m_A v_{A1} = m_B v_{B2x}$$

$$v_{B2x} = \left(\frac{m_A}{m_B}\right) v_{A1} = \left(\frac{4.00 \times 10^{-3} \text{ kg}}{0.100 \text{ kg}}\right)(450 \text{ m/s}) = 18.0 \text{ m/s}$$

P_y is conserved, so

$$m_A \cancel{v_{A1y}}^0 + m_B \cancel{v_{B1y}}^0 = m_A v_{A2y} + m_B v_{B2y}$$

$$0 = -m_A v_{A2} + m_B v_{B2y}$$

$$v_{B2y} = \left(\frac{m_A}{m_B}\right) v_{A2} = \left(\frac{4.00 \times 10^{-3} \text{ kg}}{0.100 \text{ kg}}\right)(300 \text{ m/s}) = 12.0 \text{ m/s}$$

$$v_{B2} = \sqrt{v_{B2x}^2 + v_{B2y}^2} = \sqrt{(18.0 \text{ m/s})^2 + (12.0 \text{ m/s})^2} = \underline{21.6 \text{ m/s}}$$

$$\tan\theta = \frac{v_{B2y}}{v_{B2x}} = \frac{12.0 \text{ m/s}}{18.0 \text{ m/s}} = 0.667 \Rightarrow \theta = 33.7° \text{ (defined in the sketch)}$$

b) To answer this question compare K_1 and K_2 for the system:

$$K_1 = \tfrac{1}{2} m_A v_{A1}^2 + \tfrac{1}{2} m_B v_{B1}^2 = \tfrac{1}{2}(4.00 \times 10^{-3} \text{ kg})(450 \text{ m/s})^2 = 405 \text{ J}$$

$$K_2 = \tfrac{1}{2} m_A v_{A2}^2 + \tfrac{1}{2} m_B v_{B2}^2 = \tfrac{1}{2}(4.00 \times 10^{-3} \text{ kg})(300 \text{ m/s})^2 + \tfrac{1}{2}(0.100 \text{ kg})(21.6 \text{ m/s})^2 = 203 \text{ J}$$

$\Delta K = K_2 - K_1 = 203 \text{ J} - 405 \text{ J} = -202 \text{ J}$

The kinetic energy of the system decreases by 202 J as a result of the collision; the collision is _not_ elastic.

8-77

a) $K = \tfrac{1}{2} m_A v_A^2 + \tfrac{1}{2} m_B v_B^2$

Note that $\vec{v}_A'$ and $\vec{v}_B'$ as defined in the problem are the velocities of A and B in coordinates moving with the center of mass. Note also that $m_A \vec{v}_A' + m_B \vec{v}_B' = M \vec{v}_{cm}'$ where $\vec{v}_{cm}'$ is the velocity of the cm in these coordinates. But that's zero, so $m_A \vec{v}_A' + m_B \vec{v}_B' = 0$; we can use this in the proof.

$\vec{v}_A = \vec{v}_A' + \vec{v}_{cm} \Rightarrow v_A^2 = v_A'^2 + v_{cm}^2 + 2 \vec{v}_A' \cdot \vec{v}_{cm}$ (This uses that for a
$\vec{v}_B = \vec{v}_B' + \vec{v}_{cm} \Rightarrow v_B^2 = v_B'^2 + v_{cm}^2 + 2 \vec{v}_B' \cdot \vec{v}_{cm}$ vector $\vec{A}$, $A^2 = \vec{A} \cdot \vec{A}$.)

Thus $K = \tfrac{1}{2} m_A v_A'^2 + \tfrac{1}{2} m_A v_{cm}^2 + m_A \vec{v}_A' \cdot \vec{v}_{cm} + \tfrac{1}{2} m_B v_B'^2 + \tfrac{1}{2} m_B v_{cm}^2 + m_B \vec{v}_B' \cdot \vec{v}_{cm}$

$K = \tfrac{1}{2}(m_A + m_B) v_{cm}^2 + \tfrac{1}{2}(m_A v_A'^2 + m_B v_B'^2) + (m_A \vec{v}_A' + m_B \vec{v}_B') \cdot \vec{v}_{cm}$

But $m_A + m_B = M$ and as noted earlier $m_A \vec{v}_A' + m_B \vec{v}_B' = 0$, so

8-77 (cont)
$$K = \tfrac{1}{2} M V_{cm}^2 + \tfrac{1}{2}(m_A V_A'^2 + m_B V_B'^2)$$
This is the result the problem asked us to derive.

b) In the collision $\vec{P} = M \vec{V}_{cm}$ is constant, so $\tfrac{1}{2} M V_{cm}^2$ stays constant. The asteroids can lose all their relative kinetic energy but $\tfrac{1}{2} M V_{cm}^2$ must remain.

8-81

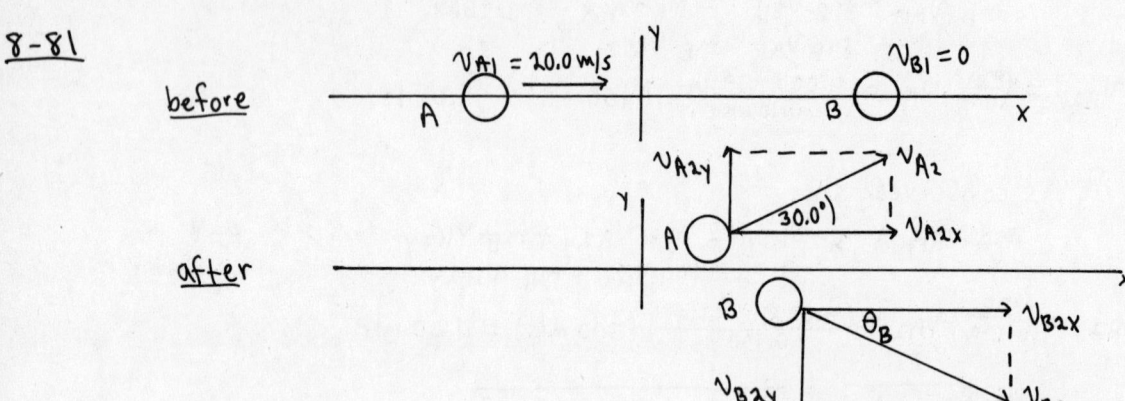

The result of Problem 8-80 part (d) applies here and says that $30.0° + \theta_B = 90.0°$, so that $\theta_B = \underline{60.0°}$. (A and B move off in perpendicular directions.)

P_x is conserved
$\Rightarrow m_A V_{A1x} + m_B \cancel{V_{B1x}}^0 = m_A V_{A2x} + m_B V_{B2x}$

But $m_A = m_B$, so $\boxed{V_{A1} = V_{A2} \cos 30.0° + V_{B2} \cos 60.0°}$

P_y is conserved
$\Rightarrow \cancel{m_A V_{A1y}}^0 + \cancel{m_B V_{B1y}}^0 = m_A V_{A2y} + m_B V_{B2y}$
$0 = V_{A2y} + V_{B2y}$
$0 = V_{A2} \sin 30.0° - V_{B2} \sin 60.0°$
$\boxed{V_{B2} = \left(\dfrac{\sin 30.0°}{\sin 60.0°}\right) V_{A2}}$

Use this result in the first equation
$\Rightarrow V_{A1} = V_{A2} \cos 30.0° + \left(\dfrac{\sin 30.0° \cos 60.0°}{\sin 60.0°}\right) V_{A2}$

$V_{A1} = 1.155 V_{A2}$

$V_{A2} = \dfrac{V_{A1}}{1.115} = \dfrac{20.0 \text{ m/s}}{1.115} = \underline{17.3 \text{ m/s}}$

And then $V_{B2} = \left(\dfrac{\sin 30.0°}{\sin 60.0°}\right)(17.3 \text{ m/s}) = \underline{10.0 \text{ m/s}}$

8-85

Use coordinates fixed to the ice, with the direction you walk the +x-direction. $\vec{v}_{cm}$ is constant and initially $v_{cm} = 0$.

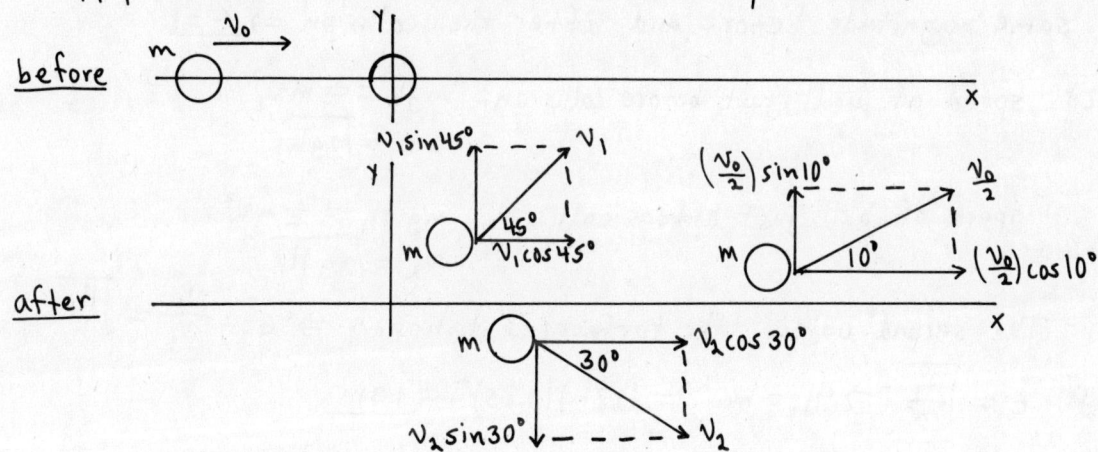

$$v_{cm} = \frac{m_p \vec{v}_p + m_s \vec{v}_s}{m_p + m_s} = 0$$

$$\Rightarrow m_p \vec{v}_p + m_s \vec{v}_s = 0$$

$$m_p v_{px} + m_s v_{sx} = 0$$

$$v_{sx} = -\left(\frac{m_p}{m_s}\right) v_{px} = -\left(\frac{m_p}{4m_p}\right) 3.00 \text{ m/s} = -0.750 \text{ m/s}$$

The slab moves at **0.750 m/s**, in the direction opposite to the direction you are walking.

8-87

Apply momentum conservation in the x and y direction:

P_x is conserved
$$\Rightarrow \cancel{m} v_0 = \cancel{m} \left(\frac{v_0}{2}\right) \cos 10° + \cancel{m} v_1 \cos 45° + \cancel{m} v_2 \cos 30°$$
$$v_0 = 0.4924 v_0 + 0.7071 v_1 + 0.8660 v_2$$
$$\boxed{0.5076 v_0 = 0.7071 v_1 + 0.8660 v_2}$$

P_y is conserved
$$\Rightarrow 0 = m \left(\frac{v_0}{2} \sin 10°\right) + m v_1 \sin 45° - m v_2 \sin 30°$$
$$\boxed{0.0868 v_0 = -0.7071 v_1 + 0.500 v_2}$$

Add these two equations $\Rightarrow 0.5944 v_0 = 1.366 v_2$
$$v_2 = 0.4351 v_0 = 0.4351 (5.0 \times 10^6 \text{ m/s}) = \mathbf{2.18 \times 10^6 \text{ m/s}}$$

8-87 (cont)

Then $v_1 = \dfrac{0.5076\, v_0 - 0.8660\, v_2}{0.7071} = \dfrac{(0.5076)(5.0 \times 10^6 \text{ m/s}) - (0.8660)(2.18 \times 10^6 \text{ m/s})}{0.7071}$

$v_1 = 9.2 \times 10^5 \text{ m/s}$

The two emitted neutrons have speeds of $\underline{2.2 \times 10^6 \text{ m/s}}$ and $\underline{9.2 \times 10^5 \text{ m/s}}$.

The speeds of the Ba and Kr nuclei are related by P_z conservation.
P_z is conserved $\Rightarrow 0 = m_{Ba}\, v_{Ba} - m_{Kr}\, v_{Kr}$

$v_{Kr} = \left(\dfrac{m_{Ba}}{m_{Kr}}\right) v_{Ba} = \left(\dfrac{2.3 \times 10^{-25} \text{ kg}}{1.5 \times 10^{-25} \text{ kg}}\right) v_{Ba} = 1.5\, v_{Ba}$; $\boxed{v_{Kr} = 1.5\, v_{Ba}}$

(We can't say what these speeds are, but they must satisfy this relation. The value of v_{Ba} depends on energy considerations.)

8-89

a) Objects stick together $\Rightarrow$ relative speed after the collision $= 0 \Rightarrow \underline{\epsilon = 0}$.

b) In an elastic collision the relative velocity of the two bodies has the same magnitude before and after the collision $\Rightarrow \underline{\epsilon = 1}$.

c) speed of ball just before collision: $mgh = \tfrac{1}{2} m v_1^2$

$v_1 = \sqrt{2gh}$

speed of ball just after collision: $mg H_1 = \tfrac{1}{2} m v_2^2$

$v_2 = \sqrt{2g H_1}$

The second object (the surface) is stationary $\Rightarrow \epsilon = \dfrac{v_2}{v_1} = \sqrt{\dfrac{H_1}{h}}$

d) $\epsilon = \sqrt{\dfrac{H_1}{h}} \Rightarrow H_1 = h\epsilon^2 = (1.8 \text{ m})(0.85)^2 = \underline{1.3 \text{ m}}$

e) $H_1 = h\epsilon^2$
$H_2 = H_1 \epsilon^2 = (h\epsilon^2)\epsilon^2 = h\epsilon^4$
$H_3 = H_2 \epsilon^2 = (h\epsilon^4)\epsilon^2 = h\epsilon^6$
$\vdots$
$H_n = H_{n-1}\epsilon^2 = h\epsilon^{2(n-1)} \epsilon^2 = h\epsilon^{2n}$

f) 10th bounce $\Rightarrow n = 10$

$H_{10} = h\epsilon^{20} = 1.8 \text{ m} (0.85)^{20} = 0.070 \text{ m} = \underline{7.0 \text{ cm}}$

8-93

a) Eq. (8-40) $\Rightarrow v - v_0^{\,0} = v_{ex} \ln\left(\frac{m_0}{m}\right)$

$$v = v_{ex} \ln\left(\frac{m_0}{m}\right)$$

The total initial mass of the rocket is $m_0 = 12{,}000 \text{ kg} + 1000 \text{ kg} = 13{,}000 \text{ kg}$. Of this, $8000 \text{ kg} + 600 \text{ kg} = 8600 \text{ kg}$ is fuel, so the mass m left after all the fuel is burned is $13{,}000 \text{ kg} - 8600 \text{ kg} = 4400 \text{ kg}$.

$$v = v_{ex} \ln\left(\frac{13{,}000 \text{ kg}}{4400 \text{ kg}}\right) = \underline{1.08 \, v_{ex}}$$

b) First stage: $v = v_{ex} \ln\left(\frac{m_0}{m}\right)$

$m_0 = 13{,}000 \text{ kg}$

The first stage has 8000 kg of fuel, so the mass left after the first stage fuel has burned is $13{,}000 \text{ kg} - 8000 \text{ kg} = 5000 \text{ kg}$.

$$v = v_{ex} \ln\left(\frac{13{,}000 \text{ kg}}{5000 \text{ kg}}\right) = \underline{0.956 \, v_{ex}}$$

c) Second stage:

$m_0 = 1000 \text{ kg}$

$m = 1000 \text{ kg} - 600 \text{ kg} = 400 \text{ kg}$

$$v = v_0 + v_{ex} \ln\left(\frac{m_0}{m}\right) = 0.956 \, v_{ex} + v_{ex} \ln\left(\frac{1000 \text{ kg}}{400 \text{ kg}}\right) = \underline{1.87 \, v_{ex}}$$

d) $v = 7.00 \text{ km/s}$

$$v_{ex} = \frac{v}{1.87} = \frac{7.00 \text{ km/s}}{1.87} = \underline{3.74 \text{ km/s}}$$

CHAPTER 9

Exercises 5, 7, 9, 11, 17, 19, 21, 25, 27, 31, 33, 35, 37, 43, 45, 47, 49, 51, 55

Problems 59, 61, 63, 65, 67, 71, 73, 75, 83

Exercises

9-5

$\theta = \gamma t + \beta t^3$; $\gamma = 0.800$ rad/s, $\beta = 0.0160$ rad/s^3

a) $\omega = \frac{d\theta}{dt} = \gamma + 3\beta t^2$

b) $t = 0 \Rightarrow \omega = \gamma = \underline{0.800 \text{ rad/s}}$

c) $t = 5.00\text{s} \Rightarrow \omega = 0.800 \text{ rad/s} + 3(0.0160 \text{ rad/s}^3)(5.00\text{s})^2 = \underline{2.00 \text{ rad/s}}$

$\omega_{av} = \frac{\Delta \theta}{\Delta t} = \frac{\theta_2 - \theta_1}{t_2 - t_1}$

$t_1 = 0 \Rightarrow \theta_1 = 0$

$t_2 = 5.00\text{s} \Rightarrow \theta_2 = (0.800 \text{ rad/s})(5.00\text{s}) + (0.0160 \text{ rad/s}^3)(5.00\text{s})^3 = 6.00 \text{ rad}$

So $\omega_{av} = \frac{6.00 \text{ rad} - 0}{5.00 \text{ s} - 0} = \underline{1.20 \text{ rad/s}}$

ω at 5.00 s is larger than ω_{av} for the time interval 0 to 5.00 s. The angular velocity is increasing in time so its value at the end of the interval is larger than its average value during the interval.

9-7

$\theta = a + bt^2 + ct^3$

$\omega = \frac{d\theta}{dt} = 2bt + 3ct^2$

$\alpha = \frac{d\omega}{dt} = 2b + 6ct$

9-9

a) $\alpha = 0.450 \text{ rad/s}^2$
$\omega_0 = 0$ (starts from rest)
$\omega = 8.00 \text{ rad/s}$
$t = ?$

$\omega = \omega_0 + \alpha t$

$t = \frac{\omega - \omega_0}{\alpha} = \frac{8.00 \text{ rad/s} - 0}{0.450 \text{ rad/s}^2} = \underline{17.8 \text{ s}}$

b) $\theta - \theta_0 = ?$

$\theta - \theta_0 = \omega_0 t + \frac{1}{2}\alpha t^2 = 0 + \frac{1}{2}(0.450 \text{ rad/s}^2)(17.8\text{s})^2 = 71.1 \text{ rad}$

$\theta - \theta_0 = 71.1 \text{ rad} \left(\frac{1 \text{ rev}}{2\pi \text{ rad}}\right) = \underline{11.3 \text{ rev}}$

126

9-11

a) $\omega_0 = (900 \text{ rev/min})\left(\frac{1 \text{ min}}{60 \text{ s}}\right) = 15.0 \text{ rev/s}$ 　　　$\omega = \omega_0 + \alpha t$

$\omega = (400 \text{ rev/min})\left(\frac{1 \text{ min}}{60 \text{ s}}\right) = 6.67 \text{ rev/s}$ 　　$\alpha = \frac{\omega - \omega_0}{t} = \frac{6.67 \text{ rev/s} - 15.0 \text{ rev/s}}{6.00 \text{ s}}$

$t = 6.00 \text{ s}$

$\alpha = ?$ 　　　　　　　　　　　　　　　　$\alpha = -1.39 \text{ rev/s}^2$

$\theta - \theta_0 = ?$

$\theta - \theta_0 = \omega_0 t + \tfrac{1}{2}\alpha t^2 = (15.0 \text{ rev/s})(6.00 \text{ s}) + \tfrac{1}{2}(-1.39 \text{ rev/s}^2)(6.00 \text{ s})^2 = \underline{65.0 \text{ rev}}$

b) $\omega = 0$ (comes to rest) 　　　　　$\omega = \omega_0 + \alpha t$

$\omega_0 = 6.67 \text{ rev/s}$

$\alpha = -1.39 \text{ rev/s}^2$ 　　　　　$t = \frac{\omega - \omega_0}{\alpha} = \frac{0 - 6.67 \text{ rev/s}}{-1.39 \text{ rev/s}^2} = \underline{4.80 \text{ s}}$

$t = ?$

9-17

a) Consider the motion from $t = 0$ to $t = 2.00 \text{ s}$:

$\theta - \theta_0 = ?$ 　　　　　$\theta - \theta_0 = \omega_0 t + \tfrac{1}{2}\alpha t^2$

$\omega_0 = 24.0 \text{ rad/s}$ 　　　$\theta - \theta_0 = (24.0 \text{ rad/s})(2.00 \text{ s}) + \tfrac{1}{2}(60.0 \text{ rad/s}^2)(2.00 \text{ s})^2$

$\alpha = 60.0 \text{ rad/s}^2$ 　　$\theta - \theta_0 = 48.0 \text{ rad} + 120.0 \text{ rad} = 168.0 \text{ rad}$

$t = 2.00 \text{ s}$ 　　　　　total: $168 \text{ rad} + 432 \text{ rad} = \underline{600 \text{ rad}}$

Note: At $t = 2.00 \text{ s}$, $\omega = \omega_0 + \alpha t = 24.0 \text{ rad/s} + (60.0 \text{ rad/s}^2)(2.00 \text{ s}) = 144 \text{ rad/s}$, when the breaker trips.

b) Consider the motion from when the circuit breaker trips until stops. For this calculation let $t = 0$ when the breaker trips.

$t = ?$ 　　　　　　　　　　　　$\theta - \theta_0 = \left(\frac{\omega_0 + \omega}{2}\right) t$

$\theta - \theta_0 = 432 \text{ rad}$

$\omega = 0$ 　　　　　　　　　　$t = \frac{2(\theta - \theta_0)}{\omega_0 + \omega} = \frac{2(432 \text{ rad})}{144 \text{ rad/s} + 0} = \underline{6.00 \text{ s}}$

$\omega_0 = 144 \text{ rad/s}$ (from part (a))

The wheel stops 6.00 s after the breaker trips so $2.00 \text{ s} + 6.00 \text{ s} = 8.00 \text{ s}$ from the beginning.

c) $\alpha = ?$; consider the same motion as in part (b)

$\omega = \omega_0 + \alpha t$

$\alpha = \frac{\omega - \omega_0}{t} = \frac{0 - 144 \text{ rad/s}}{6.00 \text{ s}} = \underline{-24.0 \text{ rad/s}^2}$

9-19

a) $v = r\omega$, but in this equation ω must be in rad/s

$\omega = (500 \text{ rev/min})\left(\frac{2\pi \text{ rad}}{1 \text{ rev}}\right)\left(\frac{1 \text{ min}}{60 \text{ s}}\right) = 52.36 \text{ rad/s}$

9-19 (cont)

$r = 0.075 \text{ m}$
$v = r\omega = (0.075 \text{ m})(52.36 \text{ rad/s}) = \underline{3.93 \text{ m/s}}$

b) $v = r\omega \Rightarrow \omega = \dfrac{v}{r} = \dfrac{0.60 \text{ m/s}}{0.035 \text{ m}} = 17.14 \text{ rad/s}$

$\omega = 17.14 \text{ rad/s} \left(\dfrac{1 \text{ rev}}{2\pi \text{ rad}}\right)\left(\dfrac{60 \text{ s}}{1 \text{ min}}\right) = \underline{164 \text{ rev/min}}$

9-21

The tangential speed of a blade tip is $v = r\omega$.

$\omega = (150 \text{ rev/min})\left(\dfrac{2\pi \text{ rad}}{1 \text{ rev}}\right)\left(\dfrac{1 \text{ min}}{60 \text{ s}}\right) = 15.71 \text{ rad/s}$

$v = r\omega = (5.0 \text{ m})(15.71 \text{ rad/s}) = 78.54 \text{ m/s}$

The upward velocity of the entire blade has magnitude 8.0 m/s. The tangential velocity and the upward velocity of a blade tip are perpendicular, so their resultant has magnitude

$v_{res} = \sqrt{(78.54 \text{ m/s})^2 + (8.0 \text{ m/s})^2} = \underline{78.9 \text{ m/s}}$

9-25

a) at the start
$t = 0$
flywheel starts from rest $\Rightarrow \omega = \omega_0 = 0$
$a_{tan} = r\alpha = (0.200 \text{ m})(0.600 \text{ rad/s}^2) = \underline{0.120 \text{ m/s}^2}$
$a_{rad} = r\omega^2 = 0$

$a = \sqrt{a_{rad}^2 + a_{tan}^2} = \underline{0.120 \text{ m/s}^2}$

b) $\theta - \theta_0 = 120°$
$a_{tan} = r\alpha = \underline{0.120 \text{ m/s}^2}$

Calculate ω:
$\theta - \theta_0 = 120°\left(\dfrac{\pi \text{ rad}}{180°}\right) = 2.094 \text{ rad}$
$\omega_0 = 0$
$\alpha = 0.600 \text{ rad/s}^2$
$\omega = ?$

$\omega^2 = \omega_0^2 + 2\alpha(\theta - \theta_0)$
$\omega = \sqrt{2\alpha(\theta-\theta_0)} = \sqrt{2(0.600 \text{ rad/s}^2)(2.094 \text{ rad})}$
$\omega = 1.585 \text{ rad/s}$

Then $a_{rad} = r\omega^2 = (0.200 \text{ m})(1.585 \text{ rad/s})^2 = \underline{0.502 \text{ m/s}^2}$.

$a = \sqrt{a_{rad}^2 + a_{tan}^2} = \sqrt{(0.502 \text{ m/s}^2)^2 + (0.120 \text{ m/s}^2)^2} = \underline{0.517 \text{ m/s}^2}$

c) $\underline{\theta - \theta_0 = 240°}$

$a_{tan} = r\alpha = \underline{0.120 \text{ m/s}^2}$

9-25 (cont)

Calculate ω:

$\theta - \theta_0 = 240° \left(\dfrac{\pi \text{ rad}}{180°}\right) = 4.189 \text{ rad}$

$\omega_0 = 0$

$\alpha = 0.600 \text{ rad/s}^2$

$\omega = ?$

$\omega^2 = \omega_0^{2\,0} + 2\alpha(\theta - \theta_0)$

$\omega = \sqrt{2\alpha(\theta - \theta_0)} = \sqrt{2(0.600 \text{ rad/s}^2)(4.189 \text{ rad})}$

$\omega = 2.242 \text{ rad/s}$

Then $a_{rad} = r\omega^2 = (0.200 \text{ m})(2.242 \text{ rad/s})^2 = 1.01 \text{ m/s}^2$.

$a = \sqrt{a_{rad}^2 + a_{tan}^2} = \sqrt{(1.01 \text{ m/s}^2)^2 + (0.120 \text{ m/s}^2)^2} = 1.02 \text{ m/s}^2$

9-27

$a_{rad} = r\omega^2 \Rightarrow r = \dfrac{a_{rad}}{\omega^2}$, where ω must be in rad/s

$a_{rad} = 2000g = 2000(9.80 \text{ m/s}^2) = 19,600 \text{ m/s}^2$

$\omega = (5000 \text{ rev/min})\left(\dfrac{1 \text{ min}}{60 \text{ s}}\right)\left(\dfrac{2\pi \text{ rad}}{1 \text{ rev}}\right) = 523.6 \text{ rad/s}$

$r = \dfrac{a_{rad}}{\omega^2} = \dfrac{19,600 \text{ m/s}^2}{(523.6 \text{ rad/s})^2} = 0.0715 \text{ m}$

(The diameter is then 0.143 m, which is larger than 0.127 m.)

9-31

a) $a_{rad} = r\omega^2$

$F_{rad} = ma_{rad} = mr\omega^2$

$\dfrac{F_{rad,2}}{F_{rad,1}} = \left(\dfrac{\omega_2}{\omega_1}\right)^2 = \left(\dfrac{640 \text{ rev/min}}{423 \text{ rev/min}}\right)^2 = 2.29$

(Note: Since a ratio is used the units cancel and there is no need to convert ω to rad/s.)

b) $v = r\omega$

$\dfrac{v_2}{v_1} = \dfrac{\omega_2}{\omega_1} = \dfrac{640 \text{ rev/min}}{423 \text{ rev/min}} = 1.51$

c) $v = r\omega$

$\omega = (640 \text{ rev/min})\left(\dfrac{2\pi \text{ rad}}{1 \text{ rev}}\right)\left(\dfrac{1 \text{ min}}{60 \text{ s}}\right) = 67.0 \text{ rad/s}$

$v = r\omega = (0.235 \text{ m})(67.0 \text{ rad/s}) = 15.7 \text{ m/s}$

$a = r\omega^2 = (0.235 \text{ m})(67.0 \text{ rad/s})^2 = 1055 \text{ m/s}^2$

$\dfrac{a}{g} = \dfrac{1055 \text{ m/s}^2}{9.80 \text{ m/s}^2} = 108 \Rightarrow a = 108g$

9-33

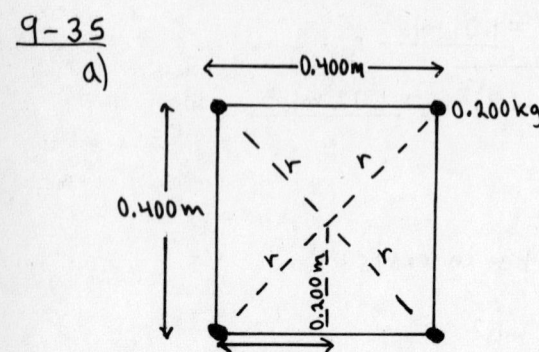

$I = I_{rod} + 2 I_{cap}$

$I = \frac{1}{12} M L^2 + 2(m)\left(\frac{L}{2}\right)^2 = \left(\frac{1}{12}M + \frac{1}{2}m\right) L^2$

9-35
a)

$r = \sqrt{(0.200\,m)^2 + (0.200\,m)^2} = 0.2828\,m$

$I = \sum_i m_i r_i^2 = 4(0.200\,kg)(0.2828\,m)^2$

$I = \underline{0.0640\,kg \cdot m^2}$

b)

$r = 0.200\,m$

$I = \sum_i m_i r_i^2 = 4(0.200\,kg)(0.200\,m)^2$

$I = \underline{0.0320\,kg \cdot m^2}$

9-37

$I = \sum_i m_i r_i^2 \Rightarrow I = I_{rim} + I_{spokes}$

$I_{rim} = MR^2 = (1.60\,kg)(0.300\,m)^2 = 0.144\,kg \cdot m^2$

Each spoke can be treated as a slender rod with the axis through one end $\Rightarrow I_{spokes} = 8\left(\frac{1}{3} ML^2\right) = \frac{8}{3}(0.320\,kg)(0.300\,m)^2 = 0.0768\,kg \cdot m^2$

$I = I_{rim} + I_{spokes} = 0.144\,kg \cdot m^2 + 0.0768\,kg \cdot m^2 = \underline{0.221\,kg \cdot m^2}$

9-43

$K = \frac{1}{2} I \omega^2$

$a_{rad} = R\omega^2 \Rightarrow \omega = \sqrt{\frac{a_{rad}}{R}} = \sqrt{\frac{4000\,m/s^2}{1.20\,m}} = 57.74\,rad/s$

Disk $\Rightarrow I = \frac{1}{2} MR^2 = \frac{1}{2}(80.0\,kg)(1.20\,m)^2 = 57.6\,kg \cdot m^2$

Thus $K = \frac{1}{2} I \omega^2 = \frac{1}{2}(57.6\,kg \cdot m^2)(57.74\,rad/s)^2 = \underline{9.60 \times 10^4\,J}$

9-45

a) $K = \frac{1}{2}I\omega^2$
   $\omega = \frac{2\pi}{T}$ $\Rightarrow K = \frac{1}{2}I\left(\frac{2\pi}{T}\right)^2 = \frac{2\pi^2 I}{T^2}$

b) $\frac{dK}{dt} = 2I\pi^2 \left(\frac{dT^{-2}}{dt}\right) = 2I\pi^2\left(-\frac{2}{T^3}\right)\frac{dT}{dt} = -\frac{4\pi^2 I}{T^3}\frac{dT}{dt}$

c) $K = \frac{2\pi^2 I}{T^2} = \frac{2\pi^2 (8.0 \text{ kg}\cdot\text{m}^2)}{(2.0\text{s})^2} = \underline{39.5 \text{ J}}$

d) $\frac{dK}{dt} = -\frac{4\pi^2 I}{T^3}\frac{dT}{dt} = -\frac{4\pi^2 (8.0 \text{ kg}\cdot\text{m}^2)}{(2.0\text{s})^3}(0.0050) = \underline{-0.197 \text{ J/s}}$

9-47

Eq. (9-18) says $U = Mgy_{cm}$

The positive work done by the wrestler must equal in magnitude the negative work done by gravity
$\Rightarrow W = -W_{grav} = U_2 - U_1 = Mg(\Delta y_{cm})$
$W = (120 \text{ kg})(9.80 \text{ m/s}^2)(0.600 \text{ m}) = \underline{706 \text{ J}}$

9-49

thin-walled hollow sphere, axis along a diameter $\Rightarrow I = \frac{2}{3}MR^2$.

For solid sphere with mass M and radius R, $I_{cm} = \frac{2}{5}MR^2$, for axis along a diameter.

Find d so that $I_p = I_{cm} + Md^2$ with $I_p = \frac{2}{3}MR^2$:
$\frac{2}{3}MR^2 = \frac{2}{5}MR^2 + Md^2$
$\left(\frac{2}{3} - \frac{2}{5}\right)R^2 = d^2$
$d = \sqrt{\frac{10-6}{15}} R = \frac{2R}{\sqrt{15}} = \underline{0.516 R}$

Axis is parallel to a diameter and is 0.516R from the center.

9-51

From part (c) of Fig. 9-2, $I_{cm} = \frac{1}{12}M(a^2 + a^2) = \frac{1}{6}Ma^2$.

$I_p = I_{cm} + Md^2$

The distance d of P from the cm is
$d = \sqrt{\left(\frac{a}{2}\right)^2 + \left(\frac{a}{2}\right)^2} = \frac{a}{\sqrt{2}}$.

$I_p = I_{cm} + Md^2 = \frac{1}{6}Ma^2 + M\left(\frac{a}{\sqrt{2}}\right)^2 = Ma^2\left(\frac{1}{6} + \frac{1}{2}\right) = \frac{2Ma^2}{3}$.

9-55

Eq. (9-20): $I = \int r^2 dm$

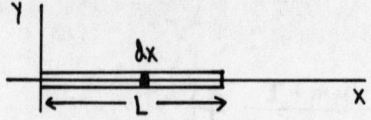

Take the x-axis to lie along the rod, with the origin at the left end. Consider a thin slice at coordinate x and width dx. The mass per unit length for this rod is $\frac{M}{L}$, so the mass of this slice is $dm = (\frac{M}{L})dx$.

$$I = \int_0^L x^2 (\frac{M}{L}) dx = \frac{M}{L} \int_0^L x^2 dx = \frac{M}{L} (\frac{1}{3}L^3) = \frac{1}{3} ML^2$$

This result agrees with Table 9-2.

Problems

9-59

$\theta(t) = \gamma t^2 - \beta t^3$; $\gamma = 3.20$ rad/s^2, $\beta = 0.400$ rad/s^3

a) $\omega(t) = \frac{d\theta}{dt} = \frac{d}{dt}(\gamma t^2 - \beta t^3) = 2\gamma t - 3\beta t^2$

b) $\alpha(t) = \frac{d\omega}{dt} = \frac{d}{dt}(2\gamma t - 3\beta t^2) = 2\gamma - 6\beta t$

c) The maximum angular velocity occurs when $\alpha = 0$.

$2\gamma - 6\beta t = 0 \Rightarrow t = \frac{2\gamma}{6\beta} = \frac{\gamma}{3\beta} = \frac{3.20 \text{ rad/s}^2}{3(0.400 \text{ rad/s}^3)} = 2.667 \text{ s}$

At this t, $\omega = 2\gamma t - 3\beta t^2 = 2(3.20 \text{ rad/s}^2)(2.667 \text{s}) - 3(0.400 \text{ rad/s}^3)(2.667 \text{s})^2 = 8.53$ rad/s

The maximum positive angular velocity is $\underline{8.53 \text{ rad/s}}$ and it occurs at $\underline{2.67 \text{ s}}$.

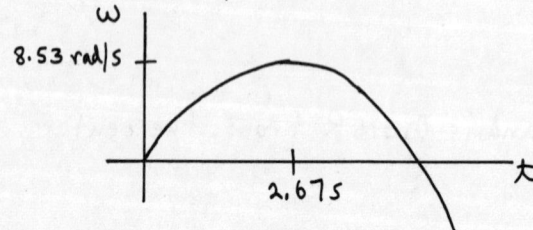

9-61

a) $v_1 = r_1 \omega_1$

$\omega_1 = (60.0 \text{ rev/s})(\frac{2\pi \text{ rad}}{1 \text{ rev}}) = 377$ rad/s

$v_1 = r_1 \omega_1 = (0.35 \times 10^{-2} \text{ m})(377 \text{ rad/s}) = \underline{1.3 \text{ m/s}}$

9-61 (cont)
b) $v_1 = v_2$
$r_1 \omega_1 = r_2 \omega_2$
$\omega_2 = \frac{r_1}{r_2} \omega_1 = \left(\frac{0.35 \text{ cm}}{2.00 \text{ cm}}\right)(377 \text{ rad/s}) = \underline{66 \text{ rad/s}}$

9-63
a) $\frac{v_{toy}}{v_{scale}} = \frac{L_{toy}}{L_{real}} \Rightarrow v_{toy} = v_{scale}\left(\frac{L_{toy}}{L_{real}}\right) = (800 \text{ km/h})\left(\frac{0.150 \text{ m}}{3.0 \text{ m}}\right) = 40.0 \text{ km/h}$
$v_{toy} = (40.0 \text{ km/h})\left(\frac{1000 \text{ m}}{1 \text{ km}}\right)\left(\frac{1 \text{ h}}{3600 \text{ s}}\right) = \underline{11.1 \text{ m/s}}$

b) $K = \frac{1}{2}mv^2 = \frac{1}{2}(0.120 \text{ kg})(11.1 \text{ m/s})^2 = \underline{7.41 \text{ J}}$

c) $K = \frac{1}{2}I\omega^2 \Rightarrow \omega = \sqrt{\frac{2K}{I}} = \sqrt{\frac{2(7.41 \text{ J})}{4.00 \times 10^{-5} \text{ kg} \cdot \text{m}^2}} = \underline{609 \text{ rad/s}}$

9-65
a) $W_{tot} = K_2 - K_1 \Rightarrow K_2 = K_1 + W_{tot}$

$W_{tot} = -5000 \text{ J}$ (the amount of energy given up by the flywheel)

$K = \frac{1}{2}I\omega_1^2$, but ω_1 must be in rad/s
$\omega_1 = (300 \text{ rev/min})\left(\frac{2\pi \text{ rad}}{1 \text{ rev}}\right)\left(\frac{1 \text{ min}}{60 \text{ s}}\right) = 31.42 \text{ rad/s}$
$K_1 = \frac{1}{2}(16.0 \text{ kg} \cdot \text{m}^2)(31.42 \text{ rad/s})^2 = 7898 \text{ J}$

Then $K_2 = K_1 + W_{tot} = 7898 \text{ J} - 5000 \text{ J} = 2898 \text{ J}$
$K_2 = \frac{1}{2}I\omega_2^2 \Rightarrow \omega_2 = \sqrt{\frac{2K_2}{I}} = \sqrt{\frac{2(2898 \text{ J})}{16.0 \text{ kg} \cdot \text{m}^2}} = 19.03 \text{ rad/s}$
$\omega_2 = 19.03 \text{ rad/s}\left(\frac{1 \text{ rev}}{2\pi \text{ rad}}\right)\left(\frac{60 \text{ s}}{1 \text{ min}}\right) = \underline{182 \text{ rev/min}}$

b) The 5000 J of energy must be restored to the flywheel, so
$P_{av} = \frac{\Delta W}{\Delta t} = \frac{5000 \text{ J}}{5.00 \text{ s}} = \underline{1000 \text{ W}}$

9-67
a) $a_{rad} = r\omega^2$
$a_{rad,1} = r\omega_1^2, \quad a_{rad,2} = r\omega_2^2$
$\Delta a_{rad} = a_{rad,2} - a_{rad,1} = r(\omega_2^2 - \omega_1^2)$

One of the constant angular acceleration equations can be written
$\omega_2^2 = \omega_1^2 + 2\alpha(\theta_2 - \theta_1) \Rightarrow \omega_2^2 - \omega_1^2 = 2\alpha(\theta_2 - \theta_1)$

9-67 (cont)

Thus $\Delta a_{rad} = r\, 2\alpha(\theta_2-\theta_1) = 2r\alpha(\theta_2-\theta_1)$, as was to be shown.

b) $\alpha = \dfrac{\Delta a_{rad}}{2r(\theta_2-\theta_1)} = \dfrac{85.0\,m/s^2 - 25.0\,m/s^2}{2(0.300\,m)(15.0\,rad)} = 6.67\,rad/s^2$

$a_{tan} = r\alpha = (0.300\,m)(6.67\,rad/s^2) = \underline{2.00\,m/s^2}$

c) $K = \tfrac{1}{2}I\omega^2$

$K_2 = \tfrac{1}{2}I\omega_2^2$, $K_1 = \tfrac{1}{2}I\omega_1^2$

$\Delta K = K_2 - K_1 = \tfrac{1}{2}I(\omega_2^2 - \omega_1^2) = \tfrac{1}{2}I(2\alpha(\theta_2-\theta_1)) = I\alpha(\theta_2-\theta_1)$, as was to be shown

d) $I = \dfrac{\Delta K}{\alpha(\theta_2-\theta_1)} = \dfrac{90.0\,J - 40.0\,J}{(6.67\,rad/s^2)(15.0\,rad)} = \underline{0.500\,kg\cdot m^2}$

9-71

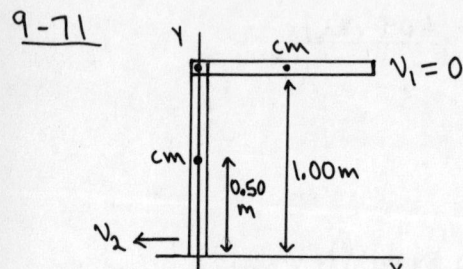

Take the origin of coordinates at the lowest point reached by the stick and take the positive y-direction to be upward.

a) Use Eq. (9-18): $U = Mgy_{cm}$

$\Delta U = U_2 - U_1 = Mg(y_{cm2} - y_{cm1})$

The center of mass of the meter stick is at its geometrical center, so $y_{cm1} = 1.00\,m$ and $y_{cm2} = 0.50\,m$.

Then $\Delta U = (0.060\,kg)(9.80\,m/s^2)(0.50\,m - 1.00\,m) = \underline{-0.294\,J}$

b) Use conservation of energy: $K_1 + U_1 + W_{other} = K_2 + U_2$

Gravity is the only force that does work on the meter stick, so $W_{other} = 0$.

$K_1 = 0$

Thus $K_2 = U_1 - U_2 = -\Delta U$, where ΔU was calculated in part (a).

$K_2 = \tfrac{1}{2}I\omega_2^2 \Rightarrow \tfrac{1}{2}I\omega_2^2 = -\Delta U$ and $\omega_2 = \sqrt{\dfrac{2(-\Delta U)}{I}}$

Stick pivoted about one end $\Rightarrow I = \tfrac{1}{3}ML^2$ where $L = 1.00\,m$, so

$\omega_2 = \sqrt{\dfrac{6(-\Delta U)}{ML^2}} = \sqrt{\dfrac{6(0.294\,J)}{(0.060\,kg)(1.00\,m)^2}} = \underline{5.42\,rad/s}$

c) $v = r\omega = (1.00\,m)(5.42\,rad/s) = \underline{5.42\,m/s}$

d) $v_{0y} = 0$, $v = ?$

$y - y_0 = -1.00\,m$

$a_y = -9.80\,m/s^2$

$v^2 = v_{0y}^2 + 2a_y(y-y_0)$

$v = -\sqrt{2a_y(y-y_0)} = -\sqrt{2(-9.80\,m/s^2)(-1.00\,m)} = \underline{-4.43\,m/s}$

The magnitude of the answer in part (c) is larger.

9-73

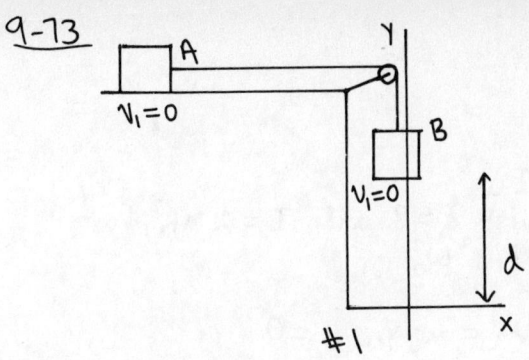 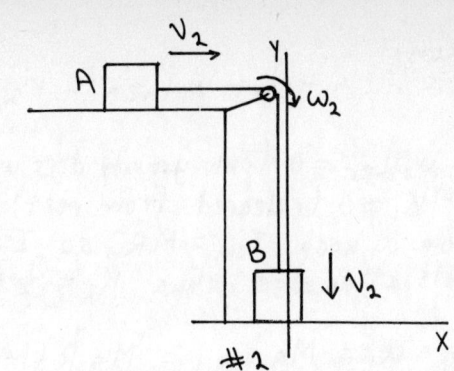

Use the work-energy relation $K_1 + U_1 + W_{other} = K_2 + U_2$.
Use coordinates where $+y$ is upward and where the origin is at the position of block B after it has descended.

The tension in the rope does positive work on block A and negative work of the same magnitude on block B, so the net work done by the tension in the rope is zero.
Gravity does work on block B and kinetic friction does work on block A.
Thus $W_{other} = W_f = -\mu_K m_A g d$.
$K_1 = 0$ (system is released from rest)
$U_1 = m_B g y_{B1} = m_B g d$
$U_2 = m_B g y_{B2} = 0$

$K_2 = \frac{1}{2} m_A v_2^2 + \frac{1}{2} m_B v_2^2 + \frac{1}{2} I \omega_2^2$.
But $v(\text{blocks}) = R\omega$ (pulley), so $\omega_2 = \frac{v_2}{R}$ and
$K_2 = \frac{1}{2}(m_A + m_B) v_2^2 + \frac{1}{2} I \left(\frac{v_2}{R}\right)^2 = \frac{1}{2}\left(m_A + m_B + \frac{I}{R^2}\right) v_2^2$

Putting all this into the work-energy relation gives
$m_B g d - \mu_K m_A g d = \frac{1}{2}\left(m_A + m_B + \frac{I}{R^2}\right) v_2^2$

$\left(m_A + m_B + \frac{I}{R^2}\right) v_2^2 = 2gd(m_B - \mu_K m_A)$

$$v_2 = \sqrt{\frac{2gd(m_B - \mu_K m_A)}{m_A + m_B + I/R^2}}$$

9-75

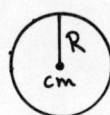 The center of mass of the hoop is at its geometrical center.

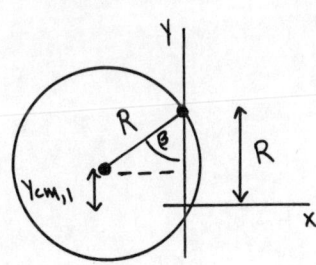 Take the origin to be at the original location of the center of the hoop, before it is rotated to one side.

$y_{cm,1} = R - R\cos\beta = R(1 - \cos\beta)$
$y_{cm,2} = 0$ (at equilibrium position hoop is at original position)

9-75 (cont)

$$K_1 + U_1 + W_{other} = K_2 + U_2$$

$W_{other} = 0$ (only gravity does work)
$K_1 = 0$ (released from rest), $K_2 = \frac{1}{2}I\omega_2^2$

For a hoop, $I_{cm} = MR^2$, so $I = Md^2 + MR^2$ with $d = R$ and $I = 2MR^2$, for axis at an edge. Thus $K_2 = \frac{1}{2}(2MR^2)\omega_2^2 = MR^2\omega_2^2$.

$$U_1 = Mg\, y_{cm,1} = MgR(1-\cos\beta), \quad U_2 = mg\, y_{cm,2} = 0$$

Thus $K_1 + U_1 + W_{other} = K_2 + U_2$ gives
$$MgR(1-\cos\beta) = MR^2\omega_2^2$$
$$\omega_2 = \sqrt{\frac{g(1-\cos\beta)}{R}}$$

9-83

Let L be the length of the cylinder. Divide the cylinder into thin cylindrical shells of inner radius r and outer radius $r+dr$. An end view is

$\rho = \alpha r$

The mass of the thin cylindrical shell is
$$dm = \rho\, dV = \rho(2\pi r\, dr)L = 2\pi\alpha L r^2 dr$$

$$I = \int r^2 dm = 2\pi\alpha L \int_0^R r^4 dr = 2\pi\alpha L\left(\frac{1}{5}R^5\right) = \frac{2}{5}\pi\alpha L R^5$$

Relate M to α:
$$M = \int dm = 2\pi\alpha L \int_0^R r^2 dr = 2\pi\alpha L\left(\frac{1}{3}R^3\right) = \frac{2}{3}\pi\alpha L R^3, \text{ so } \pi\alpha L R^3 = \frac{3M}{2}.$$

Use this in the above result for I
$$\Rightarrow I = \frac{2}{5}\left(\frac{3M}{2}\right)R^2 = \underline{\frac{3}{5}MR^2}$$

CHAPTER 10

Exercises 1, 3, 5, 9, 13, 17, 19, 23, 25, 29, 31, 33, 37, 39, 43

Problems 49, 51, 55, 57, 59, 61, 63, 67, 69, 73, 75, 79, 81

Exercises

10-1

a)

$\tau = F\ell$
$\ell = r\sin\phi = (4.00\text{ m})\sin 90° = 4.00\text{ m}$
$\tau = (15.0\text{ N})(4.00\text{ m}) = \underline{60.0\text{ N·m}}$

The force tends to produce a counterclockwise (↺) rotation about the axis; by the right-hand rule the vector $\vec{\tau}$ is directed out of the plane of the figure.

b)

$\tau = F\ell$
$\ell = r\sin\phi = (4.00\text{ m})\sin 120° = 3.464\text{ m}$
$\tau = (15.0\text{ N})(3.464\text{ m}) = \underline{52.0\text{ N·m}}$

The force tends to produce a counterclockwise (↺) rotation about the axis; by the right-hand rule the vector $\vec{\tau}$ is directed out of the plane of the figure.

c)

$\tau = F\ell$
$\ell = r\sin\phi = (4.00\text{ m})\sin 30° = 2.00\text{ m}$
$\tau = (15.0\text{ N})(2.00\text{ m}) = \underline{30.0\text{ N·m}}$

The force tends to produce a counterclockwise (↺) rotation about the axis; by the right-hand rule the vector $\vec{\tau}$ is directed out of the plane of the figure.

d)

$\tau = F\ell$
$\ell = r\sin\phi = (2.00\text{ m})\sin 60° = 1.732\text{ m}$
$\tau = (15.0\text{ N})(1.732\text{ m}) = \underline{26.0\text{ N·m}}$

The force tends to produce a clockwise (↻) rotation about the axis; by the right-hand rule the vector $\vec{\tau}$ is directed into the plane of the figure.

e)

$\tau = F\ell$
$r = 0$ so $\ell = 0 \Rightarrow \underline{\tau = 0}$

f)

$\tau = F\ell$
$\ell = r\sin\phi, \quad \phi = 180° \Rightarrow \ell = 0 \Rightarrow \underline{\tau = 0}$

10-3

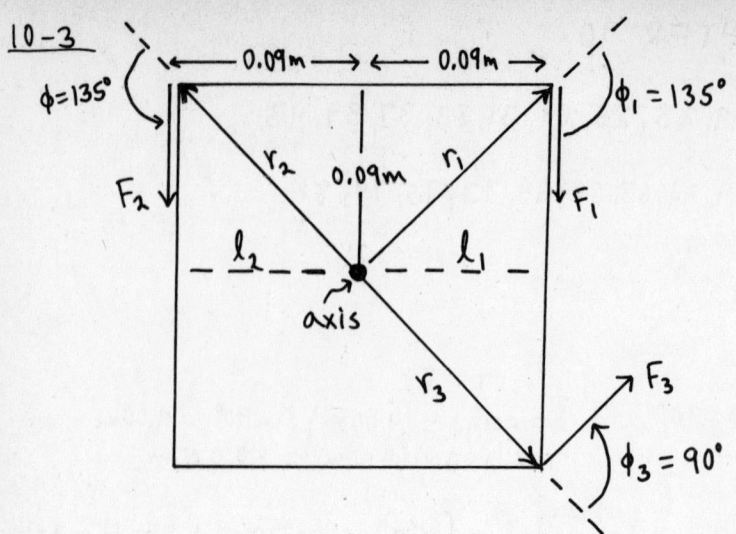

Let $(+)$ be the positive sense of rotation.

$r_1 = r_2 = r_3 = \sqrt{(0.090m)^2 + (0.090m)^2}$
$r_1 = r_2 = r_3 = 0.1273\,m$

$T_1 = -F_1 \ell_1$
$\ell_1 = r_1 \sin\phi_1 = (0.1273\,m)\sin 135°$
$\ell_1 = 0.0900\,m$
$T_1 = -(28.0N)(0.0900m) = -2.52\,N\cdot m$
($\vec{T}_1$ is directed into paper)

$T_2 = +F_2 \ell_2$
$\ell_2 = r_2 \sin\phi_2 = (0.1273m)\sin 135° = 0.0900\,m$
$T_2 = +(16.0N)(0.0900m) = +1.44\,N\cdot m$
($\vec{T}_2$ is directed out of paper)

$T_3 = +F_3 \ell_3$
$\ell_3 = r_3 \sin\phi_3 = (0.1273m)\sin 90° = 0.1273m$
$T_3 = +(18.0N)(0.1273m) = +2.29\,N\cdot m$ ($\vec{T}_3$ is directed out of paper)

$\sum T = T_1 + T_2 + T_3 = -2.52\,N\cdot m + 1.44\,N\cdot m + 2.29\,N\cdot m = \underline{+1.21\,N\cdot m}$

$+ \Rightarrow$ the net torque tends to produce a counterclockwise ($\circlearrowleft$) rotation; the net vector torque is directed out of the plane of the paper.

10-5

$\vec{r} = (-0.300m)\hat{i} + (0.600m)\hat{j}$
$\vec{F} = (4.00N)\hat{i} - (5.00N)\hat{j}$

$\vec{T} = \vec{r} \times \vec{F} = [(-0.300m)\hat{i} + (0.600m)\hat{j}] \times [(4.00N)\hat{i} - (5.00N)\hat{j}]$
$\vec{T} = -(1.20\,N\cdot m)(\hat{i}\times\hat{i}) + (1.50\,N\cdot m)\hat{i}\times\hat{j} + (2.40\,N\cdot m)\hat{j}\times\hat{i} - (3.00\,N\cdot m)\hat{j}\times\hat{j}$
$\hat{i}\times\hat{i} = \hat{j}\times\hat{j} = 0$
$\hat{i}\times\hat{j} = \hat{k},\ \hat{j}\times\hat{i} = -\hat{k}$
Thus $\vec{T} = (1.50\,N\cdot m)\hat{k} + (2.40\,N\cdot m)(-\hat{k}) = \underline{(-0.90\,N\cdot m)\hat{k}}$.

Note: The calculation gives that $\vec{T}$ is in the $-z$-direction. This agrees with what one gets from the right-hand rule.

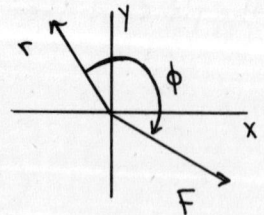

When the fingers of your right hand curl from the direction of $\vec{r}$ into the direction of $\vec{F}$ (through the smaller of the two angles, angle ϕ), your thumb points into the paper (the direction of $\vec{T}$, the $-z$-direction).

10-9

a) From Example 10-3, $T = \dfrac{mg}{1+2m/M} = \dfrac{mMg}{M+2m}$

$\Sigma F_y = ma_y$
$n - T - Mg = 0$
$n = T + Mg = \dfrac{mMg}{M+2m} + Mg = \dfrac{mMg + M(M+2m)g}{M+2m} = \dfrac{M(M+3m)g}{M+2m}$

$n = \left(\dfrac{M+3m}{M+2m}\right)Mg = \left(\dfrac{M+3m}{1+2m/M}\right)g$

b) Re-write the expression for n:

$n = \dfrac{mMg}{M+2m} + Mg = Mg + mg\left(\dfrac{M}{M+2m}\right) = Mg + mg + mg\left(\dfrac{M}{M+2m} - 1\right)$

$n = (M+m)g - mg\left(\dfrac{M+2m-M}{M+2m}\right) = (M+m)g - \dfrac{2m}{2m+M}$; n is _less than_ $(M+m)g$.

The block is accelerating downward, so the tension is less than its weight. Or, you could say that part of the system is accelerating downward, so the total upward force n on the system must be less than the total downward force $(M+m)g$ on the system.

c) The force diagrams in Example 10-3 are unchanged, so this has no effect on T and n.

10-13

Use the kinematic information to solve for the angular acceleration of the grindstone. Assume that the grindstone is rotating counterclockwise and let that be the positive sense of rotation, ↺ +.

$\omega_0 = 1100 \text{ rev/min}\left(\dfrac{2\pi \text{ rad}}{1 \text{ rev}}\right)\left(\dfrac{1 \text{ min}}{60 \text{ s}}\right) = 115.2 \text{ rad/s}$

$t = 10.0 \text{ s}$
$\omega = 0$ (comes to rest)
$\alpha = ?$

$\omega = \omega_0 + \alpha t$
$\alpha = \dfrac{\omega - \omega_0}{t}$

$\alpha = \dfrac{0 - 115.2 \text{ rad/s}}{10.0 \text{ s}} = -11.5 \text{ rad/s}^2$

Now consider the torques on the grindstone and apply $\Sigma \tau = I\alpha$, with ↺ +

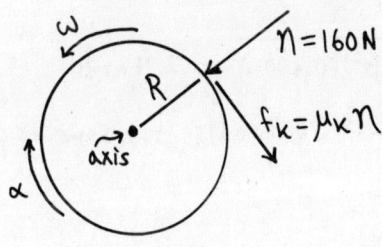

The normal force has zero moment arm for rotation about an axis at the center of the grindstone, and therefore zero torque. The only torque on the grindstone is that due to the friction force f_k exerted by the ax; for this force the moment arm is $l = R$ and the torque is negative.

$\Sigma \tau = -f_k R = -\mu_k n R$; $I = \tfrac{1}{2}MR^2$ (solid disk, axis through center)
Thus $\Sigma \tau = I\alpha \Rightarrow -\mu_k n R = (\tfrac{1}{2}MR^2)\alpha$

$\mu_k = -\dfrac{MR\alpha}{2n} = -\dfrac{(50.0 \text{ kg})(0.300 \text{ m})(-11.5 \text{ rad/s}^2)}{2(160 \text{ N})} = \underline{+0.539}$

10-17

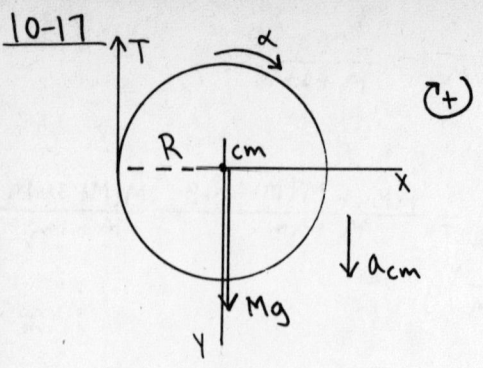

For translational motion of the center of mass of the hoop take the origin of coordinates at the cm and take +y to be downward. Then $\Sigma F_y = ma_y$ gives $\boxed{Mg - T = Ma_{cm}}$.

Apply $\Sigma \tau = I\alpha$ for rotation about the center of mass with the clockwise sense of rotation taken to be positive. The weight Mg has zero moment arm and therefore zero torque.

Thus $TR = I_{cm}\alpha$
For a hoop $I_{cm} = MR^2 \Rightarrow TR = MR^2 \alpha$
$T = MR\alpha$
$a_{cm} = R\alpha \Rightarrow \boxed{T = Ma_{cm}}$

Combine these two equations to eliminate $T \Rightarrow Mg - Ma_{cm} = Ma_{cm}$
$2a_{cm} = g \Rightarrow a_{cm} = g/2$
Then $T = M(\frac{g}{2}) = \frac{1}{2}Mg = \frac{1}{2}(0.120 \text{ kg})(9.80 \text{ m/s}^2) = \underline{0.588 \text{ N}}$

b) Apply the constant acceleration kinematic equations to the motion of the center of mass:
$v_{oy} = 0$
$y - y_0 = 0.600 \text{ m}$
$a_y = \frac{1}{2}g = 4.90 \text{ m/s}^2$
$t = ?$

$y - y_0 = v_{oy}t + \frac{1}{2}a_y t^2$
$t = \sqrt{\frac{2(y-y_0)}{a_y}} = \sqrt{\frac{2(0.600 \text{ m})}{4.90 \text{ m/s}^2}} = \underline{0.495 \text{ s}}$

c) We can use the constant angular acceleration equations for rotational motion:
$t = 0.495 \text{ s}$
$\omega_0 = 0$
$\alpha = \frac{a_{cm}}{R} = \frac{\frac{1}{2}(9.80 \text{ m/s}^2)}{0.0800 \text{ m}} = 61.25 \text{ rad/s}^2$
$\omega = ?$

$\omega = \omega_0 + \alpha t$
$\omega = 0 + (61.25 \text{ rad/s}^2)(0.495 \text{ s})$
$\omega = \underline{30.3 \text{ rad/s}}$

Alternatively, we can find v_{cm} at this point and then use $\omega = \frac{v_{cm}}{R}$:
$v_{oy} = 0$
$v_y = ?$
$a_y = 4.90 \text{ m/s}^2$
$y - y_0 = 0.600 \text{ m}$

$v_y^2 = v_{oy}^2 + 2a_y(y - y_0)$
$v_y = \sqrt{2a_y(y-y_0)} = \sqrt{2(4.90 \text{ m/s}^2)(0.600 \text{ m})} = 2.42 \text{ m/s}$

Then $\omega = \frac{v_{cm}}{R} = \frac{2.42 \text{ m/s}}{0.0800 \text{ m}} = 30.3 \text{ rad/s}$, the same as before.

10-19

$\omega_1 = 50.0 \text{ rad/s}$, #1
$\omega_2 = 0$, #2

Take $y = 0$ at the center of the wheel when it is at the bottom of the hill.

$K_1 + U_1 + W_{other} = K_2 + U_2$

10-19 (cont)

$W_{other} = W_{fric} = -3000 \text{ J}$ (the friction work is negative)

$K_1 = \frac{1}{2} I \omega_1^2$, $K_2 = 0$

$U_1 = 0$, $U_2 = Mgh$

Thus $\frac{1}{2} I \omega_1^2 + W_{fric} = Mgh$

$I = 0.800 MR^2 \Rightarrow \frac{1}{2}(0.800) MR^2 \omega_1^2 + W_{fric} = Mgh$

$h = \dfrac{0.400 MR^2 \omega_1^2 + W_{fric}}{Mg}$

$M = \dfrac{W}{g} = \dfrac{392 \text{ N}}{9.80 \text{ m/s}^2} = 40.0 \text{ kg}$

$h = \dfrac{(0.400)(40.0 \text{ kg})(0.600 \text{ m})^2 (50.0 \text{ rad/s})^2 - 3000 \text{ J}}{(40.0 \text{ kg})(9.80 \text{ m/s}^2)} = \underline{29.1 \text{ m}}$

10-23

a)

Apply $\sum \tau = I\alpha$ to find the angular acceleration α:

$FR = I\alpha$

$\alpha = \dfrac{FR}{I} = \dfrac{(25.0 \text{ N})(4.40 \text{ m})}{3200 \text{ kg} \cdot \text{m}^2} = 0.03438 \text{ rad/s}^2$

Use the constant α kinematic equations to find ω:

$\omega = ?$
$\omega_0 = 0$ (initially at rest)
$\alpha = 0.03438 \text{ rad/s}^2$
$t = 20.0 \text{ s}$

$\omega = \omega_0 + \alpha t$
$\omega = 0 + (0.03438 \text{ rad/s}^2)(20.0 \text{ s}) = 0.6876 \text{ rad/s}$
So $\omega = \underline{0.688 \text{ rad/s}}$.

b) This question can be answered either of two ways:
(1) $W = \tau \Delta\theta$ (Eq. (10-24))

$\Delta\theta = \theta - \theta_0 = \omega_0 t + \frac{1}{2}\alpha t^2 = 0 + \frac{1}{2}(0.03438 \text{ rad/s}^2)(20.0 \text{ s})^2 = 6.876 \text{ rad}$

$\tau = FR = (25.0 \text{ N})(4.40 \text{ m}) = 110 \text{ N·m}$

Then $W = \tau \Delta\theta = (110 \text{ N·m})(6.876 \text{ rad}) = \underline{756 \text{ J}}$.

or

(2) $W_{tot} = K_2 - K_1$ (the work-energy relation from Chapter 6)

$W_{tot} = W$, the work done by the child
$K_1 = 0$
$K_2 = \frac{1}{2} I \omega^2 = \frac{1}{2}(3200 \text{ kg·m}^2)(0.6876 \text{ rad/s})^2 = 756 \text{ J}$

Thus $W = 756 \text{ J}$, the same as before.

c) $P_{av} = \dfrac{\Delta W}{\Delta t} = \dfrac{756 \text{ J}}{20.0 \text{ s}} = \underline{37.8 \text{ W}}$

10-25

a) Use Eq.(10-26): $P = \tau\omega$, where ω must be in rad/s

$\omega = (4000 \text{ rev/min})\left(\frac{2\pi \text{ rad}}{1 \text{ rev}}\right)\left(\frac{1 \text{ min}}{60 \text{ s}}\right) = 418.9 \text{ rad/s}$

$\tau = \frac{P}{\omega} = \frac{1.80 \times 10^5 \text{ W}}{418.9 \text{ rad/s}} = \underline{4.30 \times 10^2 \text{ N·m}}$

b)

v constant $\Rightarrow a = 0 \Rightarrow T = w$

$\tau = TR \Rightarrow T = \frac{\tau}{R} = \frac{4.30 \times 10^2 \text{ N·m}}{0.250 \text{ m}} = 1.72 \times 10^3 \text{ N}$

Thus a weight $w = \underline{1.72 \times 10^3 \text{ N}}$ can be lifted.

c) $v = R\omega$ and the drum has $\omega = 418.9$ rad/s
$\Rightarrow v = (0.250 \text{ m})(418.9 \text{ rad/s}) = \underline{105 \text{ m/s}}$

10-29

Use $L = I\omega$.

The second hand makes 1 revolution in 1 min
$\Rightarrow \omega = (1.00 \text{ rev/min})\left(\frac{2\pi \text{ rad}}{1 \text{ rev}}\right)\left(\frac{1 \text{ min}}{60 \text{ s}}\right) = 0.1047 \text{ rad/s}$

Slender rod, axis about one end $\Rightarrow I = \frac{1}{3}ML^2 = \frac{1}{3}(15.0 \times 10^{-3} \text{ kg})(0.250 \text{ m})^2$
$I = 3.125 \times 10^{-4} \text{ kg·m}^2$

Then $L = I\omega = (3.125 \times 10^{-4} \text{ kg·m}^2)(0.1047 \text{ rad/s}) = \underline{3.27 \times 10^{-5} \text{ kg·m}^2/\text{s}}$.

10-31

Use $L = mvr \sin\phi$ (Eq. (10-28)):

$\phi = 143.1°$

$\ell = r\sin 36.9° = r\sin\phi$

$L = mvr \sin\phi$
$L = (0.300 \text{ kg})(12.0 \text{ m/s})(8.00 \text{ m})\sin 143.1°$
$L = \underline{17.3 \text{ kg·m}^2/\text{s}}$

10-33

a) $L_1 = L_2 \Rightarrow I_1\omega_1 = I_2\omega_2$

Block treated as a point mass $\Rightarrow I = mr^2$, where r is the distance of the block from the hole.

10-33 (cont)

$$\cancel{m}r_1^2 \omega_1 = \cancel{m}r_2^2 \omega_2$$
$$\omega_2 = \left(\frac{r_1}{r_2}\right)^2 \omega_1 = \left(\frac{0.200\,m}{0.100\,m}\right)^2 (1.75\,rad/s) = \underline{7.00\,rad/s}$$

b) $K_1 = \frac{1}{2} I_1 \omega_1^2 = \frac{1}{2} m r_1^2 \omega_1^2 = \frac{1}{2} m v_1^2$
$v_1 = r_1 \omega_1 = (0.200\,m)(1.75\,rad/s) = 0.350\,m/s$
$K_1 = \frac{1}{2} m v_1^2 = \frac{1}{2}(0.0300\,kg)(0.350\,m/s)^2 = 1.84 \times 10^{-3}\,J$

$K_2 = \frac{1}{2} m v_2^2$
$v_2 = r_2 \omega_2 = (0.100\,m)(7.00\,rad/s) = 0.700\,m/s$
$K_2 = \frac{1}{2} m v_2^2 = \frac{1}{2}(0.0300\,kg)(0.700\,m/s)^2 = 7.35 \times 10^{-3}\,J$

$\Delta K = K_2 - K_1 = 7.35 \times 10^{-3}\,J - 1.84 \times 10^{-3}\,J = \underline{5.51 \times 10^{-3}\,J}$

c) $W_{tot} = \Delta K$
But $W_{tot} = W_1$ the work done by the tension in the cord $\Rightarrow W = \underline{5.51 \times 10^{-3}\,J}$.

10-37

$L_1 = L_2$, with the axis at the hinge
(top view of door)

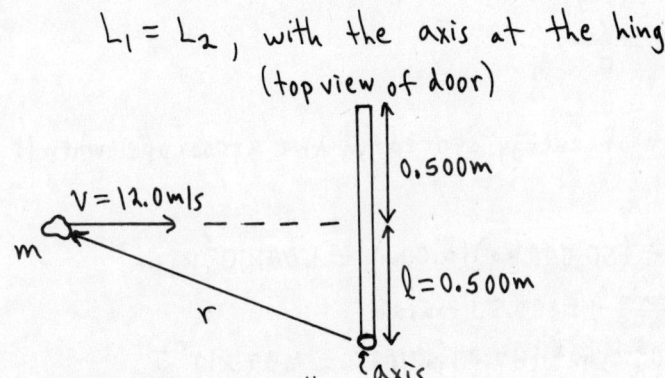 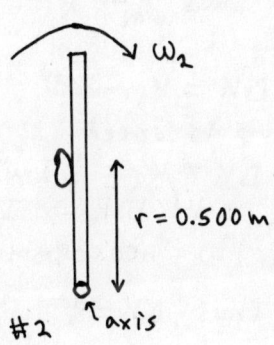

Before impact, $L_1 = L_{mud} + L_{door}$, and $L_{door} = 0$.
$L_{mud} = m v r \sin\phi = m v \ell$ (Eq. 10-28)
$L_{mud} = (0.500\,kg)(12.0\,m/s)(0.500\,m) = 3.00\,kg \cdot m^2/s$
$\Rightarrow L_1 = 3.00\,kg \cdot m^2/s$

After impact the mud sticks to the door and both objects rotate with the same angular velocity ω_2.
$I_2 = I_{door} + I_{mud}$
$I_{door} = \frac{1}{3} M L^2 = \frac{1}{3}(50.0\,kg)(1.00\,m)^2 = 16.67\,kg \cdot m^2$
Treat the mud as a point mass, so $I_{mud} = m r^2 = (0.500\,kg)(0.500\,m)^2 = 0.125\,kg \cdot m^2$
Thus $I_2 = 16.67\,kg \cdot m^2 + 0.125\,kg \cdot m^2 = 16.80\,kg \cdot m^2$.

Then $L_1 = L_2 \Rightarrow I_2 \omega_2 = 3.00\,kg \cdot m^2/s$
and $\omega_2 = \frac{3.00\,kg \cdot m^2/s}{16.80\,kg \cdot m^2} = \underline{0.179\,rad/s}$

10-39

a) $L_1 = L_2$

$I_1 \omega_1 = I_2 \omega_2 \Rightarrow \omega_2 = \left(\frac{I_1}{I_2}\right)\omega_1$

$I_1 = I_{tt} = \frac{1}{2}MR^2 = \frac{1}{2}(120\,kg)(2.00\,m)^2 = 240\,kg \cdot m^2$

$I_2 = I_{tt} + I_{bag} = 240\,kg \cdot m^2 + mR^2 = 240\,kg \cdot m^2 + (100\,kg)(2.00\,m)^2 = 640\,kg \cdot m^2$

$\omega_2 = \left(\frac{I_1}{I_2}\right)\omega_1 = \left(\frac{240\,kg \cdot m^2}{640\,kg \cdot m^2}\right)(3.00\,rad/s) = \underline{1.12\,rad/s}$

b) $K_1 = \frac{1}{2}I_1 \omega_1^2 = \frac{1}{2}(240\,kg \cdot m^2)(3.00\,rad/s)^2 = 1080\,J$

$K_2 = \frac{1}{2}I_2 \omega_2^2 = \frac{1}{2}(640\,kg \cdot m^2)(1.125\,rad/s)^2 = 405\,J$

$\Delta K = K_2 - K_1 = 405\,J - 1080\,J = \underline{-675\,J}$

The kinetic energy decreases because of the negative work done on the turntable and the bag by the friction force between these two objects.

10-43

a) By the work-energy relation $W = \Delta K$ the work done on the gyroscope equals its increase in kinetic energy.

Thus $P_{av} = \frac{W}{t} = \frac{\Delta K}{t} \Rightarrow t = \frac{\Delta K}{P}$

$\Delta K = K_2 - K_1^0$, the amount of energy stored in the gyroscope when it is up to speed

$\Delta K = K_2 = \frac{1}{2}I \omega^2$

solid disk $\Rightarrow I = \frac{1}{2}MR^2 = \frac{1}{2}(50,000\,kg)(2.00\,m)^2 = 1.00 \times 10^5\,kg \cdot m^2$

$\omega = 600\,rev/min \left(\frac{2\pi\,rad}{1\,rev}\right)\left(\frac{1\,min}{60\,s}\right) = 62.83\,rad/s$

Thus $\Delta K = \frac{1}{2}I \omega^2 = \frac{1}{2}(1.00 \times 10^5\,kg \cdot m^2)(62.83\,rad/s)^2 = 1.97 \times 10^8\,J$

$t = \frac{\Delta K}{P} = \frac{1.97 \times 10^8\,J}{7.46 \times 10^4\,W} = 2646\,s = \underline{44.1\,min}$

b) Eq. (10-36) says $\Omega = \frac{\tau}{L} = \frac{\tau}{I\omega}$, so $\tau = I\omega\Omega$

$\Omega = 1.00°/s \left(\frac{\pi\,rad}{180°}\right) = 0.01745\,rad/s$

$\tau = I\omega\Omega = (1.00 \times 10^5\,kg \cdot m^2)(62.83\,rad/s)(0.01745\,rad/s) = \underline{1.10 \times 10^5\,N \cdot m}$

Problems

10-49

a) $P = \tau\omega$

α constant $\Rightarrow \omega = \omega_0 + \alpha t = \alpha t$

$\tau = I\alpha \Rightarrow \alpha = \frac{\tau}{I}$ and $\omega = \frac{\tau t}{I}$

10-49 (cont)

Thus $P = \tau \left(\frac{\tau}{I}\right) = \frac{\tau^2}{I}$.

b) $\frac{P}{\tau^2} = \frac{1}{I} = $ constant, so $\frac{P_1}{\tau_1^2} = \frac{P_2}{\tau_2^2}$

$P_2 = P_1 \left(\frac{\tau_2}{\tau_1}\right)^2 = 900 \text{ W} \left(\frac{80.0 \text{ N·m}}{20.0 \text{ N·m}}\right)^2 = \underline{14.4 \text{ kW}}$

c) $P = \tau \omega$

$\omega^2 = \omega_0^2 + 2\alpha(\theta-\theta_0) = 2\left(\frac{\tau}{I}\right)(\theta-\theta_0)$

$\omega = \sqrt{\frac{2\tau}{I}(\theta-\theta_0)}$

Thus $P = \tau \omega = \tau \sqrt{\frac{2\tau}{I}(\theta-\theta_0)} = \sqrt{\frac{2(\theta-\theta_0)}{I}} \, \tau^{3/2}$.

d) $\frac{P}{\tau^{3/2}} = \sqrt{\frac{2(\theta-\theta_0)}{I}} = $ constant, so $\frac{P_1}{\tau_1^{3/2}} = \frac{P_2}{\tau_2^{3/2}}$

$P_2 = P_1 \left(\frac{\tau_2}{\tau_1}\right)^{3/2} = 900 \text{ W} \left(\frac{80.0 \text{ N·m}}{20.0 \text{ N·m}}\right)^{3/2} = \underline{7.20 \text{ kW}}$

e) There is no contradiction. When the net torque is varied, keeping α constant is different from keeping $\theta - \theta_0$ constant.

10-51

Use $\Sigma \tau = I\alpha$ to find α, and then use the constant α kinematic equations to solve for t.

$\Sigma \tau = F\ell = (220 \text{ N})(1.00 \text{ m}) = 220 \text{ N·m}$

From Table 9-2(d), $I = \frac{1}{3} M\ell^2 = \frac{1}{3} \left(\frac{700 \text{ N}}{9.80 \text{ m/s}^2}\right)(1.00 \text{ m})^2$

$I = 23.81 \text{ kg·m}^2$

$\Sigma \tau = I\alpha \Rightarrow \alpha = \frac{\Sigma \tau}{I} = \frac{220 \text{ N·m}}{23.81 \text{ kg·m}^2} = 9.24 \text{ rad/s}^2$

$\alpha = 9.24 \text{ rad/s}^2$
$\theta - \theta_0 = 90° \left(\frac{\pi \text{ rad}}{180°}\right) = \frac{\pi}{2} \text{ rad}$
$\omega_0 = 0$ (door initially at rest)
$t = ?$

$\theta - \theta_0 = \omega_0 t + \frac{1}{2}\alpha t^2$

$t = \sqrt{\frac{2(\theta-\theta_0)}{\alpha}} = \sqrt{\frac{2(\frac{\pi}{2} \text{ rad})}{9.24 \text{ rad/s}^2}} = \underline{0.583 \text{ s}}$

10-55

Force diagram for the crate:

$\Sigma F_y = ma_y$

$T - mg = ma$

$T = m(g+a) = 50 \text{ kg}(9.80 \text{ m/s}^2 + 0.80 \text{ m/s}^2) = \underline{530 \text{ N}}$

10-55 (cont)

Force diagram for the cylinder:

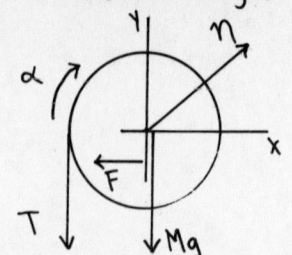

$\Sigma \tau = I\alpha$

$F\ell - TR = I\alpha$, where $\ell = 0.12\,m$ and $R = 0.25\,m$

$a = R\alpha \Rightarrow \alpha = \dfrac{a}{R}$

$F\ell = TR + \dfrac{Ia}{R}$

$F = T\dfrac{R}{\ell} + \dfrac{Ia}{R\ell} = 530\,N \left(\dfrac{0.25\,m}{0.12\,m}\right) + \dfrac{(0.92\,kg \cdot m^2)(0.80\,m/s^2)}{(0.25\,m)(0.12\,m)} = \underline{1130\,N}$

10-57

a)

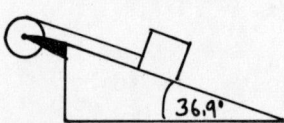

Apply $\Sigma \tau = I\alpha$ to the rotation of the flywheel about the axis:

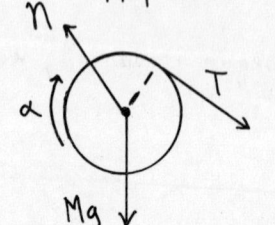

The forces n and Mg act at the axis so have zero torque $\Rightarrow \Sigma \tau = TR$

$\boxed{TR = I\alpha}$

Apply $\Sigma \vec{F} = m\vec{a}$ to the translational motion of the block:

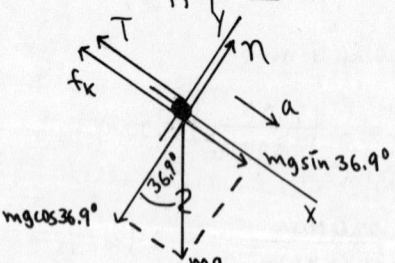

$\Sigma F_y = ma_y$
$n - mg\cos 36.9° = 0$
$n = mg\cos 36.9°$
$f_k = \mu_k n = \mu_k mg\cos 36.9°$

$\Sigma F_x = ma_x$
$mg\sin 36.9° - T - f_k = ma$
$mg\sin 36.9° - T - \mu_k mg\cos 36.9° = ma$
$\boxed{mg(\sin 36.9° - \mu_k \cos 36.9°) - T = ma}$

But we also know that $a_{block} = R\alpha_{wheel} \Rightarrow \alpha = \dfrac{a}{R}$

Use this in the first equation $\Rightarrow TR = I\dfrac{a}{R}$

$T = \left(\dfrac{I}{R^2}\right)a$

Use this to replace T in the second equation
$\Rightarrow mg(\sin 36.9° - \mu_k \cos 36.9°) - \left(\dfrac{I}{R^2}\right)a = ma$

$a = \dfrac{mg(\sin 36.9° - \mu_k \cos 36.9°)}{m + I/R^2} = \dfrac{(5.00\,kg)(9.80\,m/s^2)(\sin 36.9° - (0.25)\cos 36.9°)}{5.00\,kg + 0.400\,kg \cdot m^2/(0.200\,m)^2}$

$a = \underline{1.31\,m/s^2}$

b) $T = \dfrac{0.400\,kg \cdot m^2}{(0.200\,m)^2}(1.31\,m/s^2) = \underline{13.1\,N}$

10-59

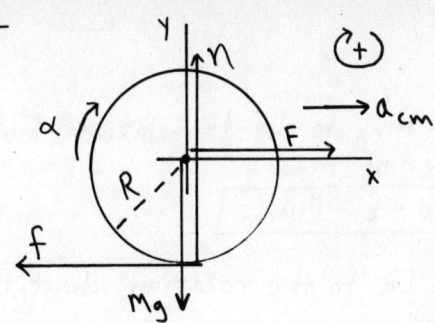

Apply $\sum \vec{F} = m\vec{a}$ to the translational motion of the center of mass:
$$\sum F_x = ma_x$$
$$\boxed{F - f = Ma_{cm}}$$

Apply $\sum \tau = I\alpha$ to the rotation about the center of mass;
$$\sum \tau = fR$$
thin-walled hollow cylinder $\Rightarrow I = MR^2$

Then $\sum \tau = I\alpha \Rightarrow fR = MR^2\alpha$
But $a_{cm} = R\alpha$, so $\boxed{f = Ma_{cm}}$

Use this in the first equation $\Rightarrow F - Ma_{cm} = Ma_{cm}$
$$a_{cm} = \frac{F}{2M}$$
And then $f = Ma_{cm} = M\left(\frac{F}{2M}\right) = \frac{F}{2}$.

10-61

This problem can be done either of two ways. We will do it both ways.

(1) **Conservation of energy**: $K_1 + U_1 + W_{other} = K_2 + U_2$

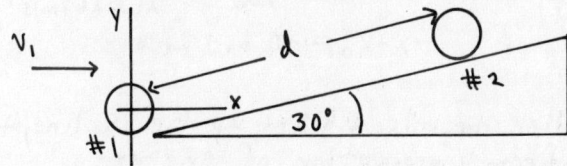

Take position #1 to be the location of the disk at the base of the ramp and #2 to be where the disk momentarily stops before rolling back down.

Take the origin of coordinates at the center of the disk at position #1 and take $+y$ to be upward. Then $y_1 = 0$ and $y_2 = d\sin 30°$, where d is the distance that the disk rolls up the ramp.

"rolls without slipping" and neglect rolling friction $\Rightarrow W_f = 0$ and only gravity does work on the disk, so $W_{other} = 0$.

$U_1 = Mgy_1 = 0$
$K_1 = \frac{1}{2}Mv_1^2 + \frac{1}{2}I_{cm}\omega_1^2$ (Eq. 10-11)
But $\omega_1 = v_1/R$ and $I_{cm} = \frac{1}{2}MR^2 \Rightarrow \frac{1}{2}I_{cm}\omega_1^2 = \frac{1}{2}\left(\frac{1}{2}MR^2\right)\left(\frac{v_1}{R}\right)^2 = \frac{1}{4}Mv_1^2$.
Thus $K_1 = \frac{1}{2}Mv_1^2 + \frac{1}{4}Mv_1^2 = \frac{3}{4}Mv_1^2$.
$U_2 = Mgy_2 = Mgd\sin 30°$
$K_2 = 0$ (disk is at rest here)

Thus $\frac{3}{4}Mv_1^2 = Mgd\sin 30°$
$$d = \frac{3v_1^2}{4g\sin 30°} = \frac{3(2.00 \text{ m/s})^2}{4(9.80 \text{ m/s}^2)\sin 30°} = \underline{0.612 \text{ m}}$$

10-61 (cont)
(2) Force and acceleration:

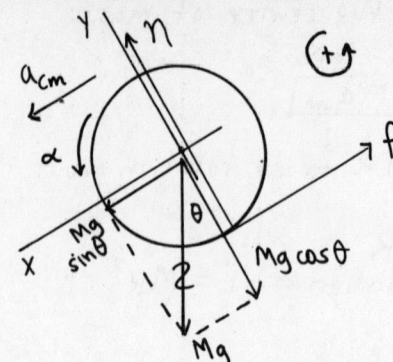

Apply $\Sigma F_x = ma_x$ to the translational motion of the center of mass:
$$\boxed{Mg\sin\theta - f = Ma_{cm}}$$

Apply $\Sigma \tau = I\alpha$ to the rotation about the center of mass:
$$fR = (\tfrac{1}{2}MR^2)\alpha$$
$$\boxed{f = \tfrac{1}{2}MR\alpha}$$

But $a_{cm} = R\alpha \Rightarrow f = \tfrac{1}{2}Ma_{cm}$

Use this in the first equation to eliminate f:
$$Mg\sin\theta - \tfrac{1}{2}Ma_{cm} = Ma_{cm}$$
$$\tfrac{3}{2}a_{cm} = g\sin\theta$$
$$a_{cm} = \tfrac{2}{3}g\sin\theta = \tfrac{2}{3}(9.80 \text{ m/s}^2)\sin 30° = 3.267 \text{ m/s}^2$$

Apply the constant acceleration equations to the motion of the center of mass. Note that in our coordinates the positive x-direction is down the incline.

$v_{0x} = -2.00$ m/s (directed up the incline) $v_x^2 = v_{0x}^2 + 2a_x(x-x_0)$
$a_x = +3.267$ m/s^2 $x - x_0 = -\dfrac{v_{0x}^2}{2a_x} = -\dfrac{(-2.00 \text{ m/s})^2}{2(3.267 \text{ m/s}^2)}$
$v_x = 0$ (momentarily comes to rest)
$x - x_0 = ?$ $x - x_0 = -0.612$ m

The calculation says that the disk travels 0.612 m up the incline, the same result that we obtained from conservation of energy.

10-63

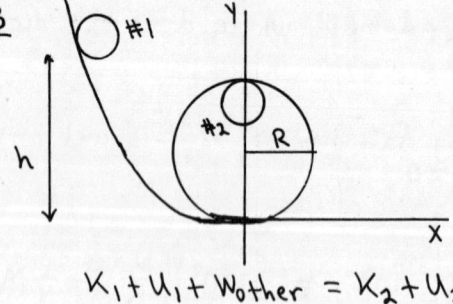

Take the origin at the lowest point in the track.
$y_{cm,1} = h$
$y_{cm,2} = 2R - r$
$I_{cm} = \tfrac{2}{5}mr^2$

$K_1 + U_1 + W_{other} = K_2 + U_2$

$W_{other} = 0$
$K_1 = 0$
$K_2 = \tfrac{1}{2}mv^2 + \tfrac{1}{2}I_{cm}\omega^2 = \tfrac{1}{2}mv^2 + \tfrac{1}{2}(\tfrac{2}{5}mr^2)\omega^2$
rolls without slipping $\Rightarrow v = r\omega \Rightarrow \omega = \dfrac{v}{r}$
Thus $K_2 = \tfrac{1}{2}mv^2 + \tfrac{1}{2}(\tfrac{2}{5}mr^2)(\dfrac{v^2}{r^2}) = mv^2(\tfrac{1}{2} + \tfrac{1}{5}) = \tfrac{7}{10}mv^2$
$U_1 = mg\, y_{cm,1} = mgh$

10-63 (cont)

$$U_2 = mg\, y_{cm,2} = mg(2R-r)$$

Thus $K_1 + U_1 + W_{other} = K_2 + U_2$ gives
$$mgh = \tfrac{7}{10} mv^2 + mg(2R-r)$$
$$\boxed{\tfrac{7}{10} v^2 = g(h - 2R + r)}$$

Force diagram for the marble at point #2:

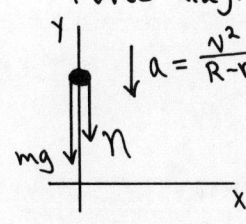

The radius of the circle in which the cm travels is $R-r$. When the marble just starts to leave the track (minimum h) $n \to 0$.
Then $\Sigma F_y = ma_y$ gives
$$mg = m\tfrac{v^2}{R-r}$$
$$v^2 = g(R-r)$$

Use this in the above equation $\Rightarrow \tfrac{7}{10} g(R-r) = g(h - 2R + r)$
$$h = \tfrac{7}{10} R + 2R - \tfrac{7}{10} r - r = \underline{(27R - 17r)/10}$$

b) No friction $\Rightarrow$ marble slides without rolling $\Rightarrow K_1 = \tfrac{1}{2} mv^2$; no $\tfrac{1}{2} I_{cm} \omega^2$ term.
This gives $\tfrac{1}{2} v^2 = g(h - 2R + r)$.
Combine with $v^2 = g(R-r) \Rightarrow \tfrac{1}{2} g(R-r) = g(h - 2R + r)$
$$\tfrac{1}{2} R - \tfrac{1}{2} r = h - 2R + r$$
$$h = \tfrac{5}{2} R - \tfrac{3}{2} r = \underline{(5R - 3r)/2}$$

Note that when $r = 0$ this result is the same as for Problem 7-46 part (a).

10-67

$$U_{el} = \tfrac{1}{2} kx^2 = \tfrac{1}{2}(200\,N/m)(0.15m)^2 = 2.25\,J$$
$$K_1 = 0.800\, U_{el} = 1.80\,J$$
$$U_2 = 0.900\, K_1 = 1.62\,J$$

a) $K_1 = \tfrac{1}{2} mv^2 + \tfrac{1}{2} I_{cm} \omega^2$
rolling without slipping $\Rightarrow \omega = \tfrac{v}{R}$
$I_{cm} = \tfrac{2}{5} mR^2$
$$K_1 = \tfrac{1}{2} mv^2 + \tfrac{1}{2}(\tfrac{2}{5} mR^2)(\tfrac{v}{R})^2 = mv^2(\tfrac{1}{2} + \tfrac{1}{5}) = \tfrac{7}{10} mv^2$$
$$v = \sqrt{\tfrac{10 K_1}{7m}} = \sqrt{\tfrac{10(1.80\,J)}{7(0.0590\,kg)}} = \underline{6.60\,m/s}$$

b) $v = 2 v_{cm}$

From Fig. (10-13), at the top of the ball
$$v = 2 v_{cm} = \underline{13.2\,m/s}$$

149

10-67 (cont)

c)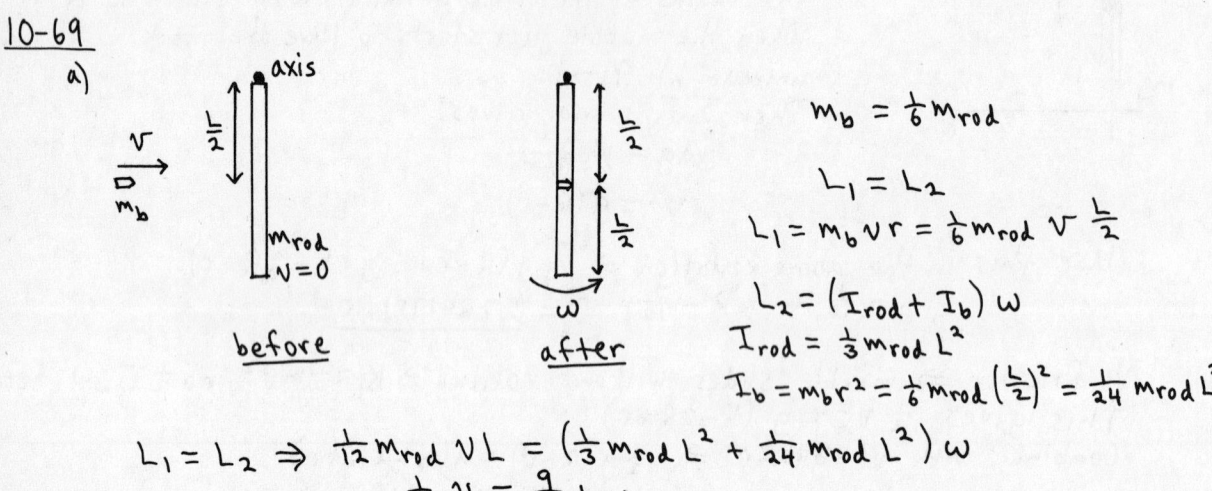

From Fig. (10-13), $v=0$ at the bottom of the ball.

d) $U_2 = mgh = 1.62 \text{ J}$

$h = \dfrac{1.62 \text{ J}}{mg} = \dfrac{1.62 \text{ J}}{(0.0590 \text{ kg})(9.80 \text{ m/s}^2)} = \underline{2.80 \text{ m}}$

10-69

a)

before / after

$m_b = \tfrac{1}{6} m_{rod}$

$L_1 = L_2$

$L_1 = m_b v r = \tfrac{1}{6} m_{rod} v \tfrac{L}{2}$

$L_2 = (I_{rod} + I_b) \omega$

$I_{rod} = \tfrac{1}{3} m_{rod} L^2$

$I_b = m_b r^2 = \tfrac{1}{6} m_{rod} \left(\tfrac{L}{2}\right)^2 = \tfrac{1}{24} m_{rod} L^2$

$L_1 = L_2 \Rightarrow \tfrac{1}{12} m_{rod} v L = \left(\tfrac{1}{3} m_{rod} L^2 + \tfrac{1}{24} m_{rod} L^2\right) \omega$

$\tfrac{1}{12} v = \tfrac{9}{24} L \omega$

$\omega = \dfrac{2v}{9L}$

b) $K_1 = \tfrac{1}{2} m v^2 = \tfrac{1}{12} m_{rod} v^2$

$K_2 = \tfrac{1}{2} I \omega^2 = \tfrac{1}{2}(I_{rod} + I_b)\omega^2 = \tfrac{1}{2}\left(\tfrac{1}{3} m_{rod} L^2 + \tfrac{1}{24} m_{rod} L^2\right)\left(\dfrac{2v}{9L}\right)^2$

$K_2 = \tfrac{1}{2}\left(\tfrac{9}{24}\right)\left(\tfrac{4}{81}\right) m_{rod} v^2 = \tfrac{1}{108} m_{rod} v^2$

$\dfrac{K_2}{K_1} = \dfrac{\tfrac{1}{108} m_{rod} v^2}{\tfrac{1}{12} m_{rod} v^2} = \dfrac{12}{108} = \tfrac{1}{9}$

10-73

$I_A = \tfrac{1}{2} I_B \Rightarrow I_B = 2 I_A$

There is no external torque on the system consisting of the two disks, so we can apply conservation of angular momentum, $L_1 = L_2$.

L_1 is the initial angular momentum of disk A, $L_1 = I_A \omega_0$.
L_2 is the final angular momentum of the two disks after they are connected and reach a common angular velocity ω, $L_2 = (I_A + I_B)\omega = 3 I_A \omega$

Then $L_1 = L_2 \Rightarrow I_A \omega_0 = 3 I_A \omega$

$\omega = \omega_0 / 3$

10-73 (cont)

$$W_{tot} = K_2 - K_1$$

$W_{tot} = W_f = -5000 J$
$K_1 = \frac{1}{2} I_A \omega_0^2$
$K_2 = \frac{1}{2}(I_A + I_B)\omega^2 = \frac{1}{2}(3I_A)\left(\frac{\omega_0}{3}\right)^2 = \frac{1}{3}\left(\frac{1}{2} I_A \omega_0^2\right) = \frac{1}{3} K_1$

Thus $-5000 J = \frac{1}{3} K_1 - K_1$
$\frac{2}{3} K_1 = 5000 J \Rightarrow K_1 = \frac{3}{2}(5000 J) = \underline{7500 J}$

10-75

a) Apply conservation of angular momentum to the collision between the bullet and the board:

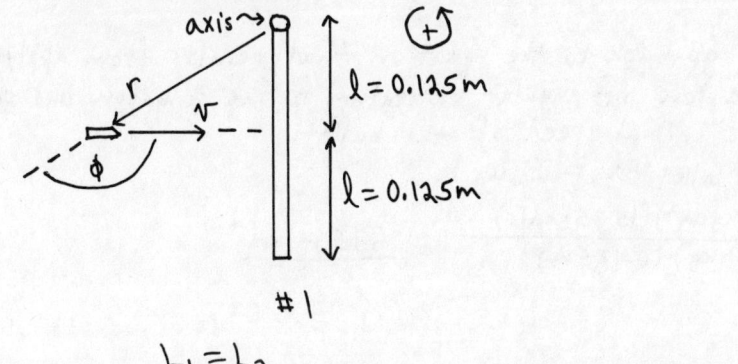
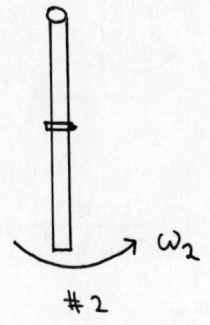

$L_1 = L_2$

$L_1 = mvr\sin\phi = mvl = (1.90 \times 10^{-3} kg)(360 m/s)(0.125 m) = 0.0855 kg \cdot m^2/s$
$L_2 = I_2 \omega_2$
$I_2 = I_{board} + I_{bullet} = \frac{1}{3} ML^2 + mr^2 = \frac{1}{3}(0.500 kg)(0.250 m)^2 + (1.90 \times 10^{-3} kg)(0.100 m)^2$
$I_2 = 0.01044 kg \cdot m^2$

Then $L_1 = L_2 \Rightarrow \omega_2 = \frac{L_1}{I_2} = \frac{0.0855 kg \cdot m^2/s}{0.01044 kg \cdot m^2} = \underline{8.19 rad/s}$

b) Apply conservation of energy to the motion of the board after the collision. Take the origin of coordinates at the center of the board and +y to be upward, so $y_{cm,1} = 0$ and $y_{cm,2} = h$, the height being asked for.

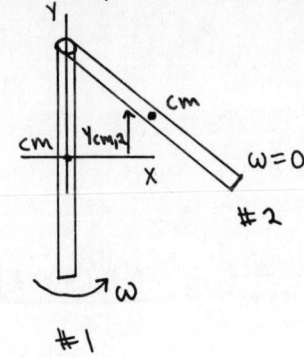

$K_1 + U_1 + W_{other} = K_2 + U_2$

Only gravity does work, so $W_{other} = 0$.
$K_1 = \frac{1}{2} I \omega^2$
$U_1 = mg y_{cm,1} = 0$
$K_2 = 0$
$U_2 = mg y_{cm,2} = mgh$

Thus $\frac{1}{2} I \omega^2 = mgh$.

151

10-75 (cont)

$$h = \frac{I\omega^2}{2mg} = \frac{(0.01044 \text{ kg} \cdot \text{m}^2)(8.19 \text{ rad/s})^2}{2(0.500 \text{ kg} + 1.90 \times 10^{-3} \text{ kg})(9.80 \text{ m/s}^2)} = 0.0712 \text{ m} = \underline{7.12 \text{ cm}}$$

c)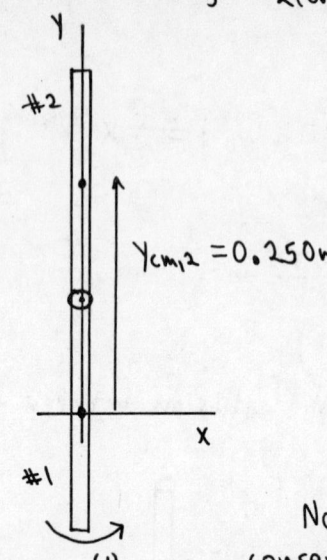

Apply conservation of energy as in part (b), except now we want $y_{cm,2} = h = 0.250$ m. Solve for the ω after the collision that is required for this to happen.

$$\tfrac{1}{2}I\omega^2 = mgh$$

$$\omega = \sqrt{\frac{2mgh}{I}} = \sqrt{\frac{2(0.500 \text{ kg} + 1.90 \times 10^{-3} \text{ kg})(9.80 \text{ m/s}^2)(0.250 \text{ m})}{0.01044 \text{ kg} \cdot \text{m}^2}}$$

$$\omega = 15.35 \text{ rad/s}$$

Now go back to the equation that results from applying conservation of angular momentum to the collision and solve for the initial speed of the bullet.

$$L_1 = L_2 \Rightarrow m_{bullet} v \ell = I_2 \omega_2$$

$$v = \frac{I_2 \omega_2}{m_{bullet} \ell} = \frac{(0.01044 \text{ kg} \cdot \text{m}^2)(15.35 \text{ rad/s})}{(1.90 \times 10^{-3} \text{ kg})(0.125 \text{ m})} = \underline{675 \text{ m/s}}$$

10-79

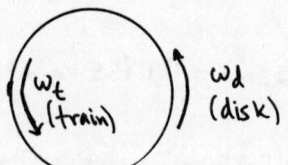

$L_1 = L_2$ ⊕↑

$L_1 = 0$ (before you switch on the train's engine; both the train and the platform are at rest)

$L_2 = L_{train} + L_{disk}$

The train is $\tfrac{1}{2}(0.95 \text{ m}) = 0.475$ m from the axis of rotation, so for it

$$I_t = m_t R_t^2 = (1.20 \text{ kg})(0.475 \text{ m})^2 = 0.2708 \text{ kg} \cdot \text{m}^2$$

$$\omega_{rel} = \frac{v_{rel}}{R_t} = \frac{0.600 \text{ m/s}}{0.475 \text{ m}} = 1.263 \text{ rad/s}$$

This is the angular velocity of the train relative to the disk. Relative to the earth $\omega_t = \omega_{rel} + \omega_d$.

Thus $L_{train} = I_t \omega_t = I_t (\omega_{rel} + \omega_d)$.

$L_2 = L_1 = 0 \Rightarrow L_{disk} = -L_{train}$

$L_{disk} = I_d \omega_d$, where $I_d = \tfrac{1}{2} m_d R_d^2$

So $L_{disk} = L_{train} \Rightarrow \tfrac{1}{2} m_d R_d^2 \omega_d = -I_t (\omega_{rel} + \omega_d)$

$$\omega_d = -\frac{I_t \omega_{rel}}{\tfrac{1}{2} m_d R_d^2 + I_t} = -\frac{(0.2708 \text{ kg} \cdot \text{m}^2)(1.263 \text{ rad/s})}{\tfrac{1}{2}(8.00 \text{ kg})(0.500 \text{ m})^2 + 0.2708 \text{ kg} \cdot \text{m}^2} = \underline{-0.269 \text{ rad/s}}$$

The minus sign tells us that the disk is rotating clockwise relative to the earth.

10-81

Eq. (10-36): $\Omega = \dfrac{\tau}{I\omega}$

The problem is asking for Ω.

The capsizing torque is $\tau = wr = (60.0\,\text{kg})(9.80\,\text{m/s}^2)(0.040\,\text{m}) = 23.52\,\text{N·m}$

$I = 0.085\,\text{kg·m}^2$

$v_{cm} = 5.00\,\text{m/s}$, so the angular velocity of the wheel about its axle is
$\omega = \dfrac{v_{cm}}{r} = \dfrac{5.00\,\text{m/s}}{0.33\,\text{m}} = 15.15\,\text{rad/s}$

Thus $\Omega = \dfrac{\tau}{I\omega} = \dfrac{23.52\,\text{N·m}}{(0.085\,\text{kg·m}^2)(15.15\,\text{rad/s})} = \underline{18.3\,\text{rad/s}}$.

CHAPTER 11

Exercises 1, 3, 5, 9, 13, 15, 19, 23, 27, 29, 33, 35, 37

Problems 41, 45, 49, 51, 61, 63, 65, 67, 69, 71, 73, 75, 77, 79

Exercises

11-1

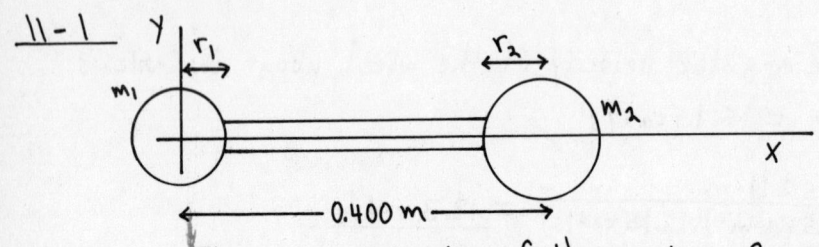

Use coordinates with the origin at the center of the 1.00 kg ball and the +x-axis along the rod.

The x-coordinate of the center of gravity is given by

$$X_{cm} = \frac{m_1 x_1 + m_2 x_2}{m_1 + m_2}$$

The center of gravity of ball #1 is at $x_1 = 0$ and the center of gravity of ball #2 is at $x_2 = 0.400 \text{m} + r_1 + r_2 = 0.400 \text{m} + 0.060 \text{m} + 0.080 \text{m} = 0.540 \text{m}$.

Then $X_{cm} = \frac{0 + (3.00 \text{kg})(0.540 \text{m})}{1.00 \text{kg} + 3.00 \text{kg}} = 0.405 \text{m}$.

The center of gravity is 0.405 m to the right of the center of the 1.00 kg ball.

11-3

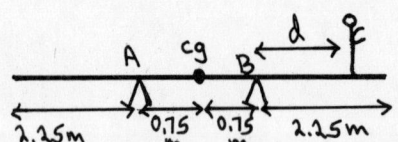

Let cg be the center of gravity of the plank, at the geometrical center of the plank. Let the child be a distance d to the right of the right-hand sawhorse. Let M be the mass of the plank and m the mass of the child. Let the sawhorses touch the plank at points A and B.

Free-body diagram for the plank:

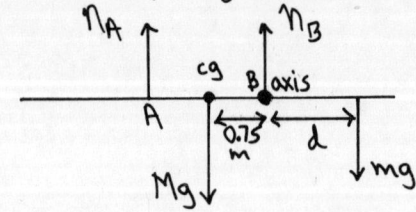

When the child is at the point where the plank just starts to tip, the plank starts to lose contact with the sawhorse at A, so $n_A \to 0$.

Use $\Sigma \tau = 0$, with axis at B and $\circlearrowleft +$

$+Mg(0.75\text{m}) - mg(d) = 0$

$d = (0.75\text{m})\frac{M}{m} = (0.75\text{m})\left(\frac{90\text{kg}}{60\text{kg}}\right) = 1.125 \text{m}$; this distance from B is halfway to the end of the plank.

154

11-5

Free-body diagram for the board:
(Let w_m be the weight of the motor. Take the origin of coordinates at the end where the 700 N force is applied (point A).

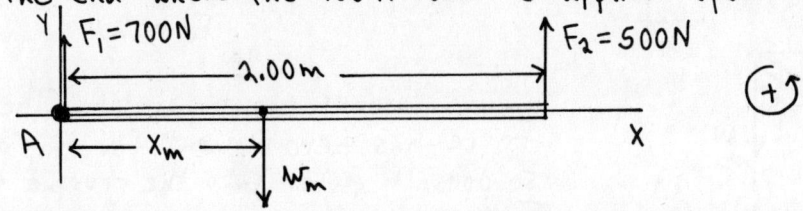

$\sum F_y = ma_y \Rightarrow F_1 + F_2 - w_m = 0$
$w_m = F_1 + F_2 = 700N + 500N = \underline{1200N}$

$\sum \tau_A = 0 \Rightarrow +F_2(2.00m) - w_m x_m = 0$
$x_m = 2.00m\left(\frac{F_2}{w_m}\right) = 2.00m\left(\frac{500N}{1200N}\right) = \underline{0.833\,m}$

The center of gravity of the motor is 0.833 m from the end where the 700 N force is applied.

11-9

a) Free-body diagram for the diving board:
Take the origin of coordinates at the left-hand end of the board (point A).

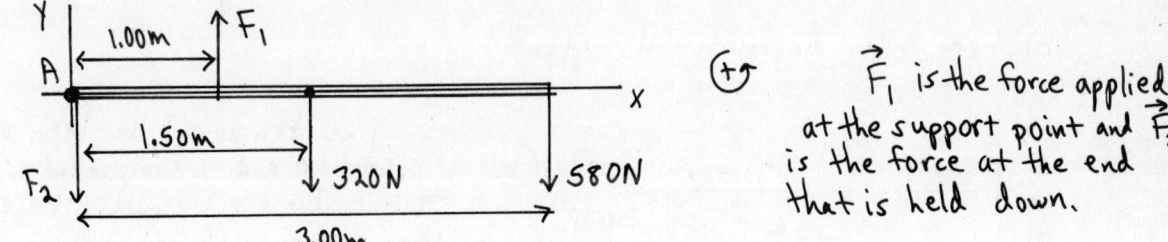

$\vec{F_1}$ is the force applied at the support point and $\vec{F_2}$ is the force at the end that is held down.

$\sum \tau_A = 0 \Rightarrow +F_1(1.00m) - (580N)(3.00m) - (320N)(1.50m) = 0$
$F_1 = \frac{(580N)(3.00m) + (320N)(1.50m)}{1.00m} = \underline{2220\,N}$

b) $\sum F_y = ma_y$
$F_1 - F_2 - 320N - 580N = 0$
$F_2 = F_1 - 320N - 580N = 2220N - 320N - 580N = \underline{1320\,N}$

11-13

(a) Free-body diagram for the strut. Take the origin of coordinates at the hinge (point A) and +y upward. Let F_h and F_v be the horizontal and vertical components of the force $\vec{F}$ exerted on the strut by the pivot. The tension in the vertical cable is the weight w of the suspended object. The weight w of the strut can be taken to act at the center of the

155

11-13 (cont)
strut. Let L be the length of the strut.

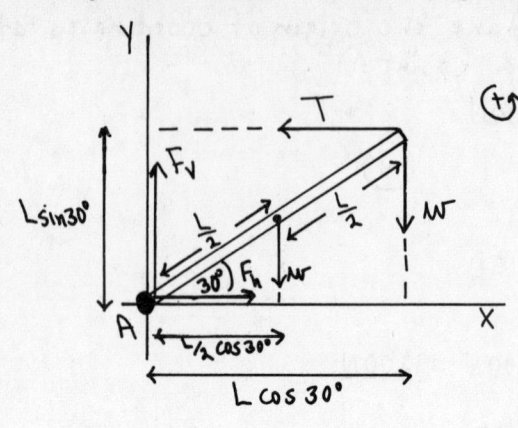

$\sum F_y = ma_y \Rightarrow F_v - w - w = 0$
$F_v = 2w$

Sum torques about point A. The pivot force has zero moment for this axis and so doesn't enter into the torque equation.
$\sum \tau_A = 0$
$TL\sin 30.0° - w(\frac{L}{2}\cos 30.0°) - w(L\cos 30.0°) = 0$
$T\sin 30.0° - \frac{3w}{2}\cos 30.0° = 0$
$T = \frac{3w \cos 30.0°}{2\sin 30.0°} = 2.60w$

Then $\sum F_x = ma_x \Rightarrow T - F_h = 0$
$F_h = 2.60w$

We have the components of $\vec{F}$ so can find its magnitude and direction:

$F = \sqrt{F_h^2 + F_v^2} = \sqrt{(2.60w)^2 + (2.00w)^2} = \underline{3.28w}$

$\tan\theta = \frac{F_v}{F_h} = \frac{2.00w}{2.60w} = 0.7692 \Rightarrow \theta = \underline{37.6°}$

(b) Free-body diagram for the strut:

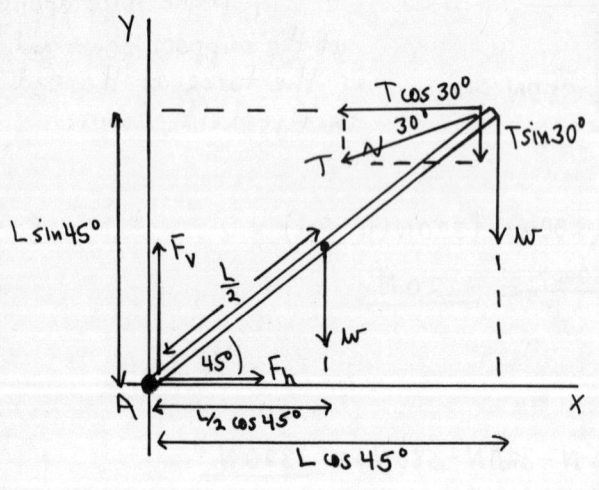

The tension T has been replaced by its x and y components. The torque due to T equals the sum of the torques of its components, and the latter are easier to calculate.

$\sum \tau_A = 0$
$+(T\cos 30.0°)(L\sin 45.0°) - (T\sin 30.0°)(L\cos 45.0°)$
$-w(\frac{L}{2}\cos 45.0°) - w(L\cos 45.0°) = 0$

The length L divides out of the equation. The equation can also be simplified by noting that $\sin 45.0° = \cos 45.0°$.
Then $T(\cos 30.0° - \sin 30.0°) = \frac{3w}{2}$.

$T = \frac{3w}{2(\cos 30.0° - \sin 30.0°)} = \underline{4.10w}$

$\sum F_x = ma_x$
$F_h - T\cos 30.0° = 0$
$F_h = T\cos 30.0° = (4.10w)(\cos 30.0°) = 3.55w$

11-13 (cont)

$$\Sigma F_y = ma_y$$
$$F_V - w - w - T\sin 30.0° = 0$$
$$F_V = 2w + (4.10w)\sin 30.0° = 4.05w$$

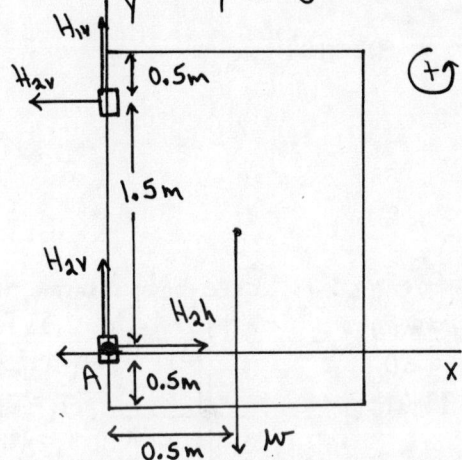

$$F = \sqrt{F_h^2 + F_v^2} = \sqrt{(3.55w)^2 + (4.05w)^2} = \underline{5.39w}$$

$$\tan\theta = \frac{F_V}{F_h} = \frac{4.05w}{3.65w} = 1.14 \Rightarrow \theta = \underline{48.8°}$$

11-15

Free-body diagram for the door:

Let $\vec{H}_1$ and $\vec{H}_2$ be the forces exerted by the upper and lower hinges. Take the origin of coordinates at the bottom hinge (point A) and +y upward.

We are given that $H_{1V} = H_{2V} = \frac{w}{2} = 165N$.

$$\Sigma F_x = ma_x \Rightarrow H_{2h} - H_{1h} = 0$$
$$H_{1h} = H_{2h}$$

The horizontal components of the hinge forces are equal in magnitude and opposite in direction.

Sum torques about point A. H_{1V}, H_{2V}, and H_{2h} all have zero moment arm and hence zero torque about an axis at this point.

Thus $\Sigma \tau_A = 0 \Rightarrow H_{1h}(1.50m) - w(0.50m) = 0$

$$H_{1h} = w\left(\frac{0.50m}{1.50m}\right) = (330N)\left(\frac{1}{3}\right) = 110N$$

The horizontal component of each hinge force is $\underline{110N}$.

11-19

a)

$$\tau_1 = F_1 \ell_1 = +(6.00N)(3.00m) = +18.0 N\cdot m$$
$$\tau_2 = -F_2\ell_2 = -(6.00N)(\ell + 3.00m) = -18.0 N\cdot m - (6.00N)\ell$$

$$\Sigma \tau = \tau_1 + \tau_2 = +18.0 N\cdot m - 18.0 N\cdot m - (6.00N)\ell = -(6.00N)\ell$$

Want ℓ that makes $\Sigma\tau = -9.60 N\cdot m$ (net torque must be clockwise)

$$-(6.00N)\ell = -9.60 N\cdot m$$
$$\ell = \frac{9.60 N\cdot m}{6.00 N} = \underline{1.6m}$$

b) $|\tau_2| > |\tau_1|$ since F_2 has a larger moment arm; the net torque is clockwise.

11-19 (cont)

c) [diagram: F_1 up on left, F_2 down at distance ℓ, axis at F_2]

$\tau_1 = -F_1 \ell_1 = -(6.00\text{ N})\ell$
$\tau_2 = 0$, since $\vec{F}_2$ is at the axis
$\Sigma \tau = -9.60\text{ N·m}$ gives $-(6.00\text{ N})\ell = -9.60\text{ N·m}$
$\ell = \underline{1.6\text{ m}}$; same as for axis in part (a)

net torque is clockwise, same as in part (b)

11-23

$Y = \dfrac{\ell_0 F_\perp}{A \Delta \ell} \Rightarrow A = \dfrac{\ell_0 F_\perp}{Y \Delta \ell}$ (A is the cross-section area of the wire)

For steel, $Y = 2.0 \times 10^{11}$ Pa (Table 11-1)

Thus $A = \dfrac{(3.00\text{ m})(400\text{ N})}{(2.0 \times 10^{11}\text{ Pa})(0.20 \times 10^{-2}\text{ m})} = 3.0 \times 10^{-6}\text{ m}^2$.

$A = \pi r^2$, so $r = \sqrt{A/\pi} = \sqrt{\dfrac{3.0 \times 10^{-6}\text{ m}^2}{\pi}} = 9.8 \times 10^{-4}\text{ m}$

$d = 2r = 1.96 \times 10^{-3}\text{ m} = \underline{1.96\text{ mm}}$

11-27

[diagram: 0.40 m above $m_1 = 5.0$ kg with tension T_1; 0.40 m between m_1 and $m_2 = 10.0$ kg with tension T_2]

Calculate T_1 and T_2.

Free-body diagram for m_2:
$\Sigma F_y = ma_y$
$T_2 - m_2 g = 0$
$T_2 = 98.0\text{ N}$

Free-body diagram for m_1:
$\Sigma F_y = ma_y$
$T_1 - T_2 - m_1 g = 0$
$T_1 = T_2 + m_1 g$
$T_1 = 98.0\text{ N} + 49.0\text{ N}$
$T_1 = 147\text{ N}$

a) $Y = \dfrac{\text{stress}}{\text{strain}} \Rightarrow \text{strain} = \dfrac{\text{stress}}{Y} = \dfrac{F_\perp}{AY}$

upper wire: strain $= \dfrac{T_1}{AY} = \dfrac{147\text{ N}}{(3.0 \times 10^{-7}\text{ m}^2)(2.0 \times 10^{11}\text{ Pa})} = \underline{2.5 \times 10^{-3}}$

lower wire: strain $= \dfrac{T_2}{AY} = \dfrac{98\text{ N}}{(3.0 \times 10^{-7}\text{ m}^2)(2.0 \times 10^{11}\text{ Pa})} = \underline{1.6 \times 10^{-3}}$

b) strain $= \dfrac{\Delta \ell}{\ell_0} \Rightarrow \Delta \ell = \ell_0 (\text{strain})$

upper wire: $\Delta \ell = (0.40\text{ m})(2.45 \times 10^{-3}) = 9.8 \times 10^{-4}\text{ m} = \underline{0.98\text{ mm}}$
lower wire: $\Delta \ell = (0.40\text{ m})(1.63 \times 10^{-3}) = 6.5 \times 10^{-4}\text{ m} = \underline{0.65\text{ mm}}$

11-29

$B = -\dfrac{\Delta P}{\Delta V / V_0} = -\dfrac{1.8 \times 10^6\text{ Pa}}{(-0.30\text{ cm}^3)/(800\text{ cm}^3)} = \underline{+4.8 \times 10^9\text{ Pa}}$

$k = \dfrac{1}{B} = \dfrac{1}{4.8 \times 10^9\text{ Pa}} = \underline{2.1 \times 10^{-10}\text{ Pa}^{-1}}$

11-33

$$\text{shear stress} = \frac{F_\parallel}{A}$$

If the shear stress on one rivet is 5.00×10^8 Pa then the force $F_\parallel$ on each rivet is $F_\parallel = A(5.00 \times 10^8 \text{ Pa}) = \pi r^2 (5.00 \times 10^8 \text{ Pa}) = \pi (0.150 \times 10^{-2} \text{ m})^2 (5.00 \times 10^8 \text{ Pa})$

$$F_\parallel = 3.53 \times 10^3 \text{ N}$$

Each rivet carries one quarter of the load $\Rightarrow$ total load of

$$4 F_\parallel = \underline{1.41 \times 10^4 \text{ N}}$$

11-35

Same material $\Rightarrow$ same S

$$S = \frac{\text{stress}}{\text{strain}} \Rightarrow \text{strain} = \frac{\text{stress}}{S} = \frac{F_\parallel / A}{S}$$

Same forces $\Rightarrow$ same $F_\parallel$

For the smaller object, $(\text{strain})_1 = \frac{F_\parallel}{A_1 S}$

For the larger object, $(\text{strain})_2 = \frac{F_\parallel}{A_2 S}$

$$\frac{(\text{strain})_2}{(\text{strain})_1} = \left(\frac{F_\parallel}{A_2 S}\right)\left(\frac{A_1 S}{F_\parallel}\right) = \frac{A_1}{A_2}$$

Larger solid has double each edge length, so $A_2 = 4 A_1$, and

$$\frac{(\text{strain})_2}{(\text{strain})_1} = \underline{\frac{1}{4}}$$

11-37

$$\text{tensile stress} = \frac{F_\perp}{A} = \frac{F_\perp}{\pi r^2} = \frac{48.6 \text{ N}}{\pi (0.425 \times 10^{-3} \text{ m})^2} = \underline{8.6 \times 10^7 \text{ Pa}}$$

Problems

11-41

a) Free-body diagram for the car:

$n_r = (1-f)w$ $\quad n_f = fw$

Let the cg be a distance x from the rear axle. Sum torques about the rear axle (point A).

$$\Sigma \tau_A = 0 \Rightarrow n_f d - wx = 0$$
$$fwd - wx = 0$$
$$x = fd, \text{ as was to be shown}$$

b) In Example 11-2, $n_f = 0.53 w \Rightarrow f = 0.53$ and $d = 2.46$ m. The general result derived in part (a) gives $x = fd = (0.53)(2.46 \text{ m}) = 1.30 \text{ m}$, the same as calculated in the example.

11-45

a) Seesaw set so Susie exerts maximum torque about the pivot ⇒ pivot moved as far as possible from Susie ⇒ pivot is 0.30 m to left of center of seesaw.

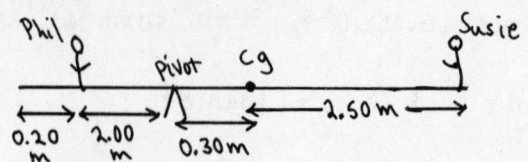

w_p is the weight of Phil, w_s is the weight of Susie, and w_{ss} is the weight of the seesaw. n is the normal force exerted by the pivot.

Free-body diagram for the seesaw:

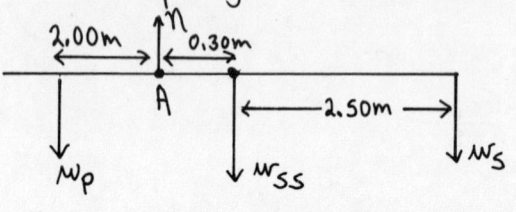

Apply $\Sigma \tau = 0$ with the axis at the pivot (point A), ↻+

$$w_p(2.00\text{m}) - w_{ss}(0.30\text{m}) - w_s(2.80\text{m}) = 0$$

$$w_p = \frac{w_s(2.80\text{m}) + w_{ss}(0.30\text{m})}{2.00\text{m}}$$

$$w_p = \frac{(450\text{N})(2.80\text{m}) + (300\text{N})(0.30\text{m})}{2.00\text{m}} = \underline{675\text{N}}$$

b) The center of gravity of each piece of the system (Phil, Susie, and the seesaw) is located some distance above the pivot. When the seesaw is horizontal the force diagram that takes this into account gives forces with lines of action and moment arms the same as in part (a).

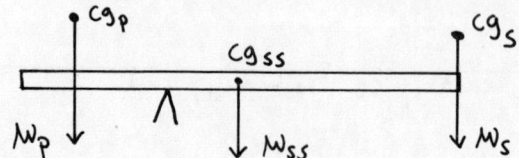

The center of gravity cg_{tot} of the total system (seesaw + Phil + Susie) is located a distance y_{cm} directly above the pivot P. When the seesaw is horizontal the line of action of the total weight Mg and the normal n coincide and there is zero net torque.

Now consider the force diagram when the system is rotated through an angle θ:

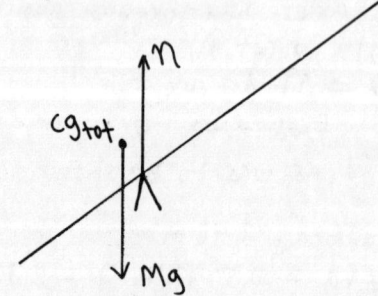

The line of action of the total weight force no longer passes through the pivot so now produces a counterclockwise torque about the pivot. The resultant torque in this direction rotates the system farther from equilibrium.

(The same conclusion is reached if the seesaw is rotated clockwise from the horizontal.)

11-49

a)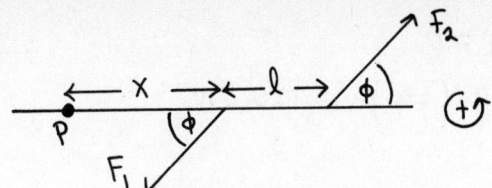

$F_1 = F_2 = F$
For an axis at P,
$\tau_1 = -F_1 x \sin\phi = -Fx\sin\phi$
$\tau_2 = +F_2(x+\ell)\sin\phi = +F(x+\ell)\sin\phi$

$\sum \tau_P = \tau_1 + \tau_2 = F\sin\phi(x+\ell-x) = F\ell\sin\phi$, which is independent of x. Therefore, the resultant torque is the same for an axis at any point along the rod.

b)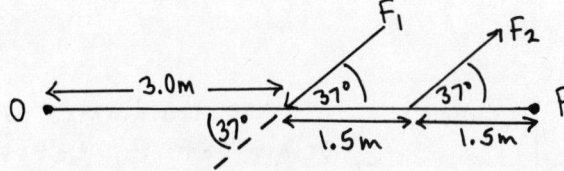

For an axis at O,
$\tau_1 = -F_1(3.0\,m)\sin 37° = -(8.00\,N)(3.0\,m)\sin 37° = -14.44\,N\cdot m$
$\tau_2 = +F_2(4.5\,m)\sin 37° = +(8.00\,N)(4.5\,m)\sin 37° = +21.66\,N\cdot m$

$\sum \tau_O = \tau_1 + \tau_2 = -14.44\,N\cdot m + 21.66\,N\cdot m = \underline{+7.2\,N\cdot m}$

For an axis at P,
$\tau_1 = +F_1(3.0\,m)\sin 37° = +(8.00\,N)(3.0\,m)\sin 37° = +14.44\,N\cdot m$
$\tau_2 = -F_2(1.5\,m)\sin 37° = -(8.00\,N)(1.5\,m)\sin 37° = -7.22\,N\cdot m$

$\sum \tau_P = \tau_1 + \tau_2 = +14.44\,N\cdot m - 7.22\,N\cdot m = \underline{+7.2\,N\cdot m}$

The general result derived in part (a) gives $\sum \tau = F\ell\sin\phi$
$\Rightarrow \sum \tau = (8.00\,N)(1.5\,m)\sin 37° = +7.2\,N\cdot m$

The torque about P is the same as about O, and agrees with the general result derived in part (a).

11-51

Free-body diagram for the bar:

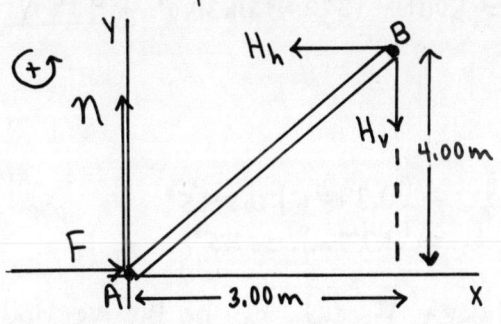

n is the normal force exerted on the bar by the surface. There is no friction force at this surface.

H_h and H_v are the components of the force exerted on the bar by the hinge. The components of the force of the bar on the hinge will be equal in magnitude and opposite in direction.

$\sum F_x = ma_x \Rightarrow F = H_h = 90.0\,N$

$\sum F_y = ma_y \Rightarrow n - H_v = 0$
$H_v = n$, but we don't know either of these forces

11-51 (cont)

$$\Sigma \tau_B = 0 \Rightarrow F(4.00\,m) - n(3.00\,m) = 0$$

$$n = \left(\frac{4.00\,m}{3.00\,m}\right) F = \tfrac{4}{3}(90.0\,N) = 120.0\,N \Rightarrow H_V = 120.0\,N$$

Force of bar on hinge:
- horizontal component 90.0 N, to right
- vertical component 120.0 N, upward

11-61

Free-body diagram for the gate:

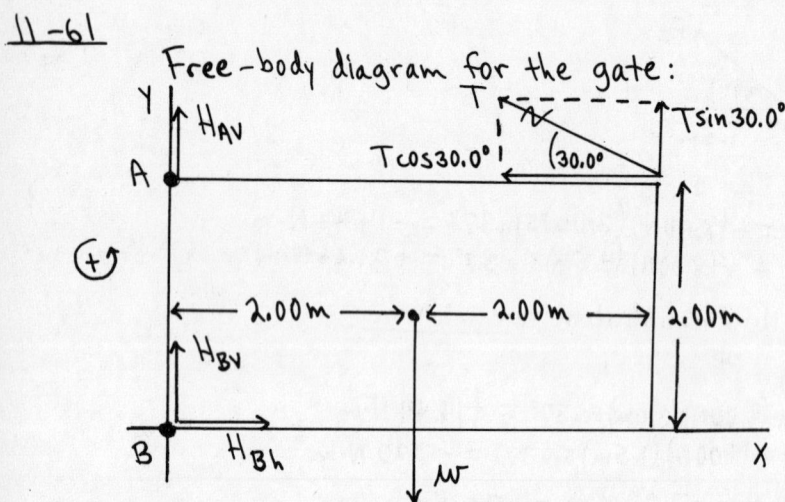

Use coordinates with the origin at B. Let $\vec{H}_A$ and $\vec{H}_B$ be the forces exerted by the hinges A and B. The problem states that $\vec{H}_A$ has no horizontal component.

Replace the tension $\vec{T}$ by its horizontal and vertical components.

a) $\Sigma \tau_B = 0 \Rightarrow +(T\sin 30.0°)(4.00\,m) + (T\cos 30.0°)(2.00\,m) - w(2.00\,m) = 0$

$$T(2\sin 30.0° + \cos 30.0°) = w$$

$$T = \frac{w}{2\sin 30.0° + \cos 30.0°} = \frac{600\,N}{2\sin 30.0° + \cos 30.0°} = \underline{322\,N}$$

b) $\Sigma F_x = ma_x \Rightarrow H_{Bh} - T\cos 30.0° = 0$

$$H_{Bh} = T\cos 30.0° = (322\,N)\cos 30.0° = \underline{279\,N}$$

c) $\Sigma F_y = ma_y \Rightarrow H_{Av} + H_{Bv} + T\sin 30.0° - w = 0$

$$H_{Av} + H_{Bv} = w - T\sin 30.0° = 600\,N - (322\,N)\sin 30.0° = \underline{439\,N}$$

11-63

Free-body diagram for the crate:

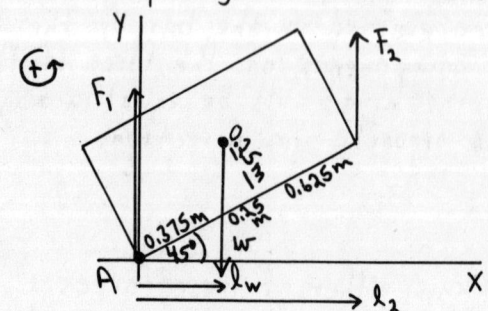

$\ell_w = (0.375\,m)\cos 45°$
$\ell_2 = (1.25\,m)\cos 45°$

Let $\vec{F}_1$ and $\vec{F}_2$ be the vertical forces exerted by you and your friend. Take the origin at the lower left-hand corner of the crate (point A).

162

11-63 (cont)

$\Sigma F_y = ma_y \Rightarrow F_1 + F_2 - w = 0$
$F_1 + F_2 = w = (200 kg)(9.80 m/s^2) = 1960 N$

$\Sigma \tau_A = 0 \Rightarrow F_2 l_2 - w l_w = 0$
$F_2 = w \left(\frac{l_w}{l_2}\right) = 1960 N \left(\frac{0.375m \cos 45°}{1.25m \cos 45°}\right) = 588 N$
Then $F_1 = w - F_2 = 1960 N - 588 N = 1372 N$.

The person below (you) applies a force of **1372 N**. The person above (your friend) applies a force of **588 N**. It is better to be the person above.

11-65

a) Free-body diagram for the pole:

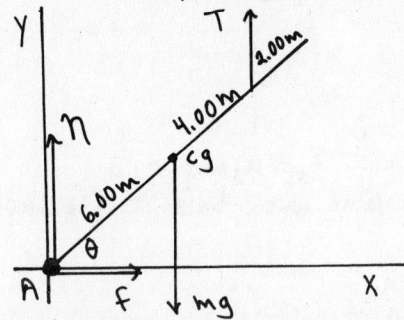

n and f are the vertical and horizontal components of the force the ground exerts on the pole.
$\Sigma F_x = ma_x$
$f = 0$
The force exerted by the ground has no horizontal component.

$\Sigma \tau_A = 0 \quad (+\curvearrowleft)$
$+ T(10.0m) \cos\theta - mg(6.0m) \cos\theta = 0$
$T = mg\left(\frac{6.0m}{10.0m}\right) = 0.6(7500 N) = \underline{4500 N}$

$\Sigma F_y = 0$
$n + T - mg = 0$
$n = mg - T = 7500 N - 4500 N = 3000 N$

The force exerted by the ground is **vertical** (upward) and has magnitude **3000 N**.

b) In the $\Sigma \tau_A = 0$ equation the angle θ divided out. All forces on the pole are vertical and their moment arms are all proportional to $\cos\theta$.

11-67

a) Find the angle where the bale starts to tip:
Starts to tip $\Rightarrow$ only lower left-hand corner of the bale makes contact with the conveyor belt $\Rightarrow$ the line of action of the normal force n passes through the left-hand edge of the bale. Consider $\Sigma \tau_A = 0$ with point A at the lower left-hand corner. Then $\tau_n = 0$ and $\tau_f = 0$, so it must be that $\tau_{mg} = 0$ also. This means that the line of action of the gravity force must pass through point A. Thus the free-body diagram must be as follows:

11-67 (cont)

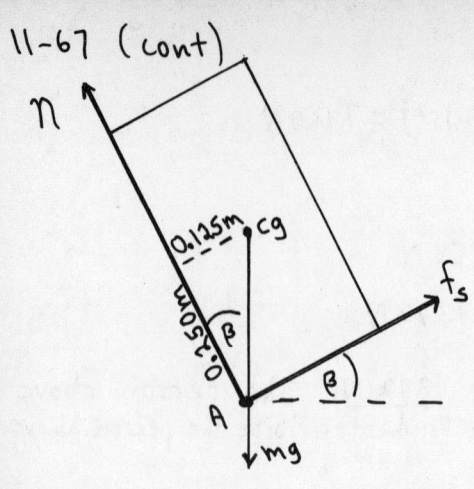

$$\tan\beta = \frac{0.125 m}{0.250 m} = 0.500$$

$$\beta = \underline{26.6°}, \text{ angle where tips}$$

At the angle where the bale is ready to slip down the incline f_s has its maximum possible value, $f_s = \mu_s n$. Free-body diagram for the bale, with the origin of coordinates at the cg:

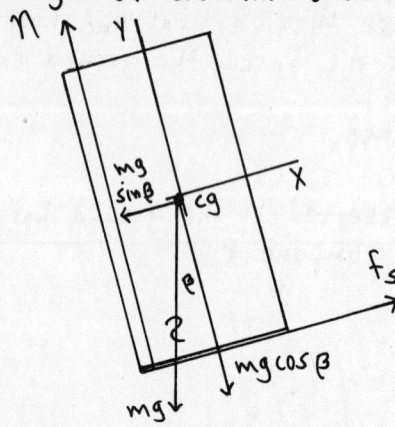

$$\Sigma F_y = ma_y$$
$$n - mg\cos\beta = 0$$
$$n = mg\cos\beta \Rightarrow f_s = \mu_s mg \cos\beta$$
(f_s has maximum value when bale ready to slip)

$$\Sigma F_x = ma_x$$
$$f_s - mg\sin\beta = 0$$
$$\mu_s mg \cos\beta - mg\sin\beta = 0$$
$$\tan\beta = \mu_s$$
$$\mu_s = 0.40 \Rightarrow \beta = 21.8°$$

$\beta = 26.6°$ to tip ; $\beta = 21.8°$ to slip $\Rightarrow$ slips first

b) The magnitude of the friction force didn't enter into the calculation of the tipping angle $\Rightarrow$ still tips at $\beta = 26.6°$
$\mu_s = 0.75 \Rightarrow$ slips at $\beta = \arctan(0.75) = 36.9°$
Now bale will tip over before it slides down the incline.

11-69

a) Free-body diagram for the door:

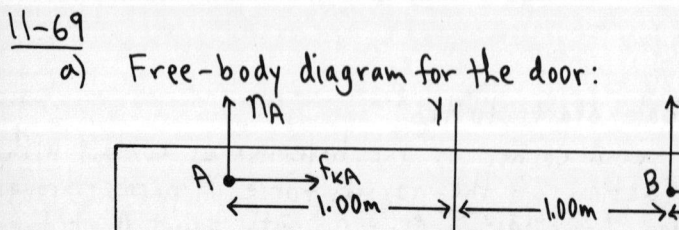

Take the origin of coordinates at the center of the door (at the cg). Let n_A, f_{kA}, n_B, and f_{kB} be the normal and friction forces exerted on the door at each wheel.

11-69 (cont)

$\sum F_y = ma_y$ $\qquad \sum F_x = ma_x$
$n_A + n_B - w = 0$ $\qquad f_{kA} + f_{kB} - F = 0$
$n_A + n_B = w = 800 \text{ N}$ $\qquad F = f_{kA} + f_{kB}$

$f_{kA} = \mu_k n_A$, $f_{kB} = \mu_k n_B$, so $F = \mu_k(n_A + n_B) = \mu_k w = (0.45)(800\text{N}) = 360\text{N}$

$\sum \tau_B = 0$

n_B, f_{kB}, and f_{kA} all have zero moment arm and hence zero torque about this point.

Thus $+ w(1.00\text{m}) - n_A(2.00\text{m}) - F(h) = 0$

$n_A = \dfrac{w(1.00\text{m}) - F(h)}{2.00\text{m}} = \dfrac{(800\text{N})(1.00\text{m}) - (360\text{N})(1.50\text{m})}{2.00\text{m}} = \underline{130\text{N}}$

And then $n_B = 800\text{N} - n_A = 800\text{N} - 130\text{N} = \underline{670\text{N}}$

b) If h is too large the torque of F will cause wheel A to leave the track. When wheel A just starts to lift off the track n_A and f_{kA} both $\to 0$.

The equations in part (a) still apply:
$\cancel{n_A}^0 + n_B - w = 0 \Rightarrow n_B = w = 800\text{N}$

$f_{kB} = \mu_k n_B = (0.45)(800\text{N}) = 360\text{N}$
$F = \cancel{f_{kA}}^0 + f_{kB} = 360\text{N}$

$w(1.00\text{m}) - \cancel{n_A}^0(2.00\text{m}) - Fh = 0$

$h = \dfrac{w(1.00\text{m})}{F} = \dfrac{(800\text{N})(1.00\text{m})}{360\text{N}} = \underline{2.22\text{m}}$

11-71

$Y = \dfrac{F_1 l_0}{A \Delta l}$ (Eq. 11-10 holds since the problem states that stress is proportional to the strain.)

$\Delta l = \dfrac{F_1 l_0}{AY}$

a) Change l_0 but F_1 (same floodlamp), A (same diameter wire), and Y (same material) all stay the same.

$\dfrac{\Delta l}{l_0} = \dfrac{F_1}{AY} = \text{constant}$, so $\dfrac{\Delta l_1}{l_{01}} = \dfrac{\Delta l_2}{l_{02}}$

$\Delta l_2 = \Delta l_1 \left(\dfrac{l_{02}}{l_{01}}\right) = 3\Delta l_1 = 3(0.18\text{mm}) = \underline{0.54\text{mm}}$

b) $A = \pi\left(\dfrac{d}{2}\right)^2 = \tfrac{1}{4}\pi d^2$, so $\Delta l = \dfrac{F_1 l_0}{\tfrac{1}{4}\pi d^2}$

F_1, l_0, Y stay the same $\Rightarrow \Delta l(d^2) = \dfrac{F_1 l_0}{\tfrac{1}{4}\pi} = \text{constant}$

11-71 (cont)

$$\Delta l_1 d_1^2 = \Delta l_2 d_2^2$$

$$\Delta l_2 = \Delta l_1 \left(\frac{d_1}{d_2}\right)^2 = (0.18\,mm)\left(\frac{1}{3}\right)^2 = \underline{0.02\,mm}$$

c) $F_\perp, l_0, A$ stay the same $\Rightarrow \Delta l\, Y = \frac{F_\perp l_0}{A}$ = constant

$$\Delta l_1 Y_1 = \Delta l_2 Y_2$$

$$\Delta l_2 = \Delta l_1 \left(\frac{Y_1}{Y_2}\right) = (0.18\,mm)\left(\frac{7.0\times10^{10}\,Pa}{21\times10^{10}\,Pa}\right) = \underline{0.06\,mm}$$

11-73

Calculate the tension in the wire as the mass passes through the lowest point. Free-body diagram for the mass:

The mass moves in an arc of a circle with radius $R = 0.50\,m$. It has acceleration $\vec{a}_{rad}$ directed in toward the center of the circle, so at this point $\vec{a}_{rad}$ is upward.

$$\Sigma F_y = m a_y$$
$$T - mg = m R \omega^2 \Rightarrow T = m(g + R\omega^2)$$

But ω must be in rad/s: $\omega = (6.00\,rev/s)\left(\frac{2\pi\,rad}{1\,rev}\right) = 37.7\,rad/s$

Then $T = (15.0\,kg)(9.80\,m/s^2 + (0.50\,m)(37.7\,rad/s)^2) = 10{,}807\,N$.

Now calculate the elongation Δl of the wire that this tensile force produces:

$$Y = \frac{F_\perp l_0}{A \Delta l} \Rightarrow \Delta l = \frac{F_\perp l_0}{YA} = \frac{(10{,}807\,N)(0.50\,m)}{(2.0\times10^{11}\,Pa)(0.014\times10^{-4}\,m^2)} = 1.9\times10^{-2}\,m = \underline{1.9\,cm}$$

11-75

a) stress = $\frac{F_\perp}{A}$, so equal stress implies $\frac{T}{A}$ same for each wire.

$$\frac{T_A}{1.00\,mm^2} = \frac{T_B}{4.00\,mm^2} \Rightarrow T_B = 4.00\,T_A$$

The question is where along the rod to hang the weight in order to produce this relation between the tensions in the two wires. Let the weight be suspended at point C, a distance x to the right of wire A. The free-body diagram for the rod is then

$$\Sigma \tau_C = 0$$
$$+T_B(1.05\,m - x) - T_A x = 0$$

But $T_B = 4.00\,T_A \Rightarrow 4.00\,T_A(1.05\,m - x) - T_A x = 0$

$$4.20\,m - 4.00\,x = x$$

$$x = \frac{4.20\,m}{5.00} = \underline{0.84\,m} \text{ (measured from A)}$$

11-75 (cont)

b) $Y = \frac{\text{stress}}{\text{strain}} \Rightarrow \text{strain} = \frac{\text{stress}}{Y} = \frac{F_\perp}{AY}$

Equal strains thus implies $\frac{T_A}{(1.00\text{mm}^2)(2.40\times 10^{11}\text{Pa})} = \frac{T_B}{(4.00\text{mm}^2)(1.20\times 10^{11}\text{Pa})}$

$T_B = \left(\frac{4.00}{1.00}\right)\left(\frac{1.20}{2.40}\right)T_A = 2.00 T_A$

The $\sum \tau_c = 0$ equation still gives $T_B(1.05\text{m}-x) - T_A x = 0$.
But now $T_B = 2.00 T_A$ so $(2.00 T_A)(1.05\text{m}-x) - T_A x = 0$

$2.10\text{m} = 3.00 x \Rightarrow x = \frac{2.10\text{m}}{3.00} = \underline{0.70\text{m}}$ (measured from A)

11-77

Each piece of the composite rod is subjected to a tensile force of $4.00\times 10^4 \text{N}$.

a) $Y = \frac{F_\perp l_0}{A \Delta l} \Rightarrow \Delta l = \frac{F_\perp l_0}{YA}$

$\Delta l_c = \Delta l_s \Rightarrow \frac{F_\perp l_{0,c}}{Y_c A_c} = \frac{F_\perp l_{0,s}}{Y_s A_s}$ (c $\Rightarrow$ copper, s $\Rightarrow$ steel) ; $l_{0,s} = L$

But the $F_\perp$ is the same for both, so
$l_{0,s} = \frac{Y_s}{Y_c} \frac{A_s}{A_c} l_{0,c}$

$L = \left(\frac{2.0\times 10^{11}\text{Pa}}{1.1\times 10^{11}\text{Pa}}\right)\left(\frac{1.00\text{ cm}^2}{2.00\text{ cm}^2}\right)(1.40\text{m}) = \underline{1.27\text{m}}$

b) stress $= \frac{F_\perp}{A} = \frac{T}{A}$

<u>copper</u>: stress $= \frac{T}{A} = \frac{6.00\times 10^4 \text{N}}{2.00\times 10^{-4}\text{m}^2} = \underline{3.00\times 10^8 \text{Pa}}$

<u>steel</u>: stress $= \frac{T}{A} = \frac{6.00\times 10^4 \text{N}}{1.00\times 10^{-4}\text{m}^2} = \underline{6.00\times 10^8 \text{Pa}}$

c) $Y = \frac{\text{stress}}{\text{strain}} \Rightarrow \text{strain} = \frac{\text{stress}}{Y}$

<u>copper</u>: strain $= \frac{3.00\times 10^8 \text{Pa}}{1.1\times 10^{11}\text{Pa}} = \underline{2.7\times 10^{-3}}$

<u>steel</u>: strain $= \frac{6.00\times 10^8 \text{Pa}}{2.0\times 10^{11}\text{Pa}} = \underline{3.0\times 10^{-3}}$

11-79

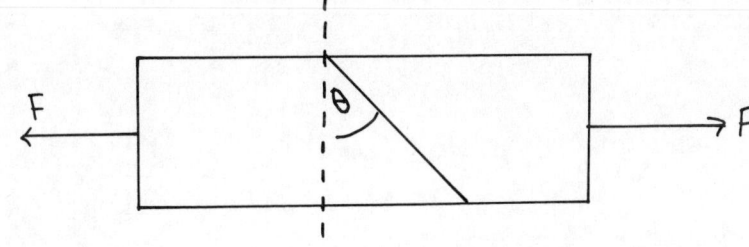

11-79 (cont)

a)

The area of the diagonal face is $\frac{A}{\cos\theta}$.

$F_\perp = F\cos\theta$
$F_\parallel = F\sin\theta$

tensile stress $= \dfrac{F_\perp}{(A/\cos\theta)} = \dfrac{F\cos\theta}{A/\cos\theta} = \dfrac{F\cos^2\theta}{A}$

b) shear stress $= \dfrac{F_\parallel}{(A/\cos\theta)} = \dfrac{F\sin\theta}{(A/\cos\theta)} = \dfrac{F\sin\theta\cos\theta}{A} = \dfrac{F\sin 2\theta}{2A}$ (using a trig identity)

c) From the result of part (a) the tensile stress is a maximum for $\cos\theta = 1 \Rightarrow \underline{\theta = 0°}$

d) From the result of part (b) the shear stress is a maximum for $\sin 2\theta = 1 \Rightarrow 2\theta = 90°$ and $\underline{\theta = 45°}$.

CHAPTER 12

Exercises 1, 3, 7, 9, 13, 17, 21, 23, 25, 29, 33, 35, 37

Problems 41, 43, 45, 51, 53, 57, 59, 63, 65, 69, 71

Exercises

12-1

$$F_{12} = G\frac{m_1 m_2}{r_{12}^2}$$

For the second set of spheres
$$F'_{12} = G\frac{(nm_1)(nm_2)}{(nr_{12})^2} = \frac{n^2}{n^2} G\frac{m_1 m_2}{r_{12}^2} = G\frac{m_1 m_2}{r_{12}^2} ;$$
the force is the same as for the first set of spheres.

12-3

$$F_{S\,on\,m} = G\frac{m_S m_m}{r_{Sm}^2} \quad (S=sun, m=moon) ; \quad F_{E\,on\,m} = G\frac{m_E m_m}{r_{Em}^2} \quad (E=earth)$$

$$\frac{F_{S\,on\,m}}{F_{E\,on\,m}} = \left(G\frac{m_S m_m}{r_{Sm}^2}\right)\left(\frac{r_{Em}^2}{G m_E m_m}\right) = \frac{m_S}{m_E}\left(\frac{r_{Em}}{r_{Sm}}\right)^2$$

r_{Em}, the radius of the moon's orbit around the earth is given in Appendix F as 3.84×10^8 m. The moon is much closer to the earth than it is to the sun, so take the distance r_{Sm} of the moon from the sun to be r_{SE}, the radius of the earth's orbit around the sun.

$$\frac{F_{S\,on\,m}}{F_{E\,on\,m}} = \left(\frac{1.99 \times 10^{30} kg}{5.97 \times 10^{24} kg}\right)\left(\frac{3.84 \times 10^8 m}{1.50 \times 10^{11} m}\right)^2 = \underline{2.18}$$

The force exerted by the sun is larger than the force exerted by the earth. The moon's motion is a combination of orbiting the sun and orbiting the earth.

12-7

Let $\vec{F}_E$ and $\vec{F}_S$ be the gravitational forces exerted on the spaceship by the earth and by the sun. The distance from the earth to the sun is $r = 1.50 \times 10^{11}$ m.

Let the ship be a distance x from the earth; it is then a distance r-x from the sun.

169

12-7 (cont)

$$F_E = F_S \Rightarrow G\frac{m\,m_E}{x^2} = G\frac{m\,m_S}{(r-x)^2}$$

$$\frac{m_E}{x^2} = \frac{m_S}{(r-x)^2} \Rightarrow (r-x)^2 = x^2\frac{m_S}{m_E}$$

$$r-x = x\sqrt{\frac{m_S}{m_E}} \Rightarrow r = x\left(1+\sqrt{\frac{m_S}{m_E}}\right)$$

$$x = \frac{r}{1+\sqrt{\frac{m_S}{m_E}}} = \frac{1.50\times 10^{11}\,m}{1+\sqrt{\frac{1.99\times 10^{30}\,kg}{5.97\times 10^{24}\,kg}}} = \underline{2.59\times 10^8\,m} \quad \text{(from center of earth)}$$

12-9

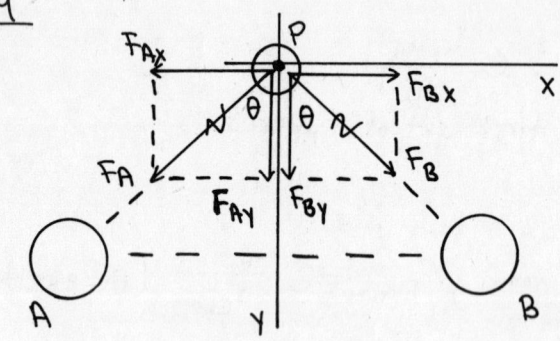

$\sin\theta = 0.80$
$\cos\theta = 0.60$

Take the origin of coordinates at point P.

$$F_A = G\frac{m_A m}{r^2} = (6.673\times 10^{-11}\,N\cdot m^2/kg^2)\frac{(0.26\,kg)(0.010\,kg)}{(0.100\,m)^2} = 1.735\times 10^{-11}\,N$$

$$F_B = G\frac{m_B m}{r^2} = 1.735\times 10^{-11}\,N$$

$$F_{Ax} = -F_A \sin\theta = -(1.735\times 10^{-11}\,N)(0.80) = -1.39\times 10^{-11}\,N$$
$$F_{Ay} = +F_A \cos\theta = +(1.735\times 10^{-11}\,N)(0.60) = +1.04\times 10^{-11}\,N$$

$$F_{Bx} = +F_B \sin\theta = +1.39\times 10^{-11}\,N$$
$$F_{By} = +F_B \cos\theta = +1.04\times 10^{-11}\,N$$

$$\Sigma F_x = ma_x \Rightarrow F_{Ax} + F_{Bx} = ma_x$$
$$0 = ma_x \Rightarrow a_x = 0$$

$$\Sigma F_y = ma_y \Rightarrow F_{Ay} + F_{By} = ma_y$$
$$2(1.04\times 10^{-11}\,N) = (0.010\,kg)\,a_y$$
$$a_y = \underline{2.1\times 10^{-9}\,m/s^2}, \text{ directed downward midway between A and B}$$

12-13

The acceleration due to gravity at the surface of a planet P is given by $g_P = \frac{Gm_P}{R_P^2}$, where m_P is the mass of the planet and R_P is the radius.

For the earth, $g_E = \frac{Gm_E}{R_E^2}$.

For the invented planet $m_P = 8m_E$ and $R_P = 2R_E$, so

170

12-13 (cont)

$$g_p = \frac{Gm_p}{R_p^2} = \frac{G(8m_E)}{(2R_E)^2} = 2\frac{Gm_E}{R_E^2} = 2g_E$$

Since $g_E = 9.80 \text{ m/s}^2$, $g_p = 2(9.80 \text{ m/s}^2) = 19.6 \text{ m/s}^2$.

12-17

Use the measured gravitational force to calculate the gravitational constant G:

$$F_g = G\frac{m_1 m_2}{r^2} \Rightarrow G = \frac{F_g r^2}{m_1 m_2} = \frac{(1.30 \times 10^{-10} \text{ N})(0.0400 \text{ m})^2}{(0.800 \text{ kg})(4.00 \times 10^{-3} \text{ kg})} = 6.50 \times 10^{-11} \text{ N} \cdot \text{m}^2/\text{kg}^2$$

Then use $g = \frac{Gm_E}{R_E^2}$ (Eq. 12-4) to calculate the mass of the earth:

$$g = \frac{Gm_E}{R_E^2} \Rightarrow m_E = \frac{R_E^2 g}{G} = \frac{(6.38 \times 10^6 \text{ m})^2 (9.80 \text{ m/s}^2)}{6.50 \times 10^{-11} \text{ N} \cdot \text{m}^2/\text{kg}^2} = \underline{6.14 \times 10^{24} \text{ kg}}$$

12-21

Example 12-5 gives the escape speed as $v_1 = \sqrt{\frac{2GM}{R}}$, where M and R are the mass and radius of the astronomical object.

$$v_1 = \sqrt{2(6.673 \times 10^{-11} \text{ N} \cdot \text{m}^2/\text{kg}^2)(2.0 \times 10^{15} \text{ kg})/5.0 \times 10^3 \text{ m}} = \underline{7.3 \text{ m/s}}$$

At this speed a person can run 100 m in 14 s; barely possible for the average person.

12-23

The radius of the orbit is $r = h + R_E$.

$$r = 1.50 \times 10^6 \text{ m} + 6.38 \times 10^6 \text{ m} = 7.88 \times 10^6 \text{ m}$$

Free-body diagram for the satellite:

$$\Sigma F_y = ma_y$$
$$F_g = ma_{rad}$$
$$G\frac{mm_E}{r^2} = m\frac{v^2}{r}$$

$$v = \sqrt{\frac{Gm_E}{r}} = \sqrt{\frac{(6.673 \times 10^{-11} \text{ N} \cdot \text{m}^2/\text{kg}^2)(5.97 \times 10^{24} \text{ kg})}{7.88 \times 10^6 \text{ m}}} = \underline{7.11 \times 10^3 \text{ m/s}}$$

12-25

a) Find the orbit radius r:

$$\frac{Gmm_E}{r^2} = m\frac{v^2}{r}$$

12-25 (cont)

$$\Rightarrow r = \frac{GM_E}{v^2} = \frac{(6.673\times 10^{-11}\,N\cdot m^2/kg^2)(5.97\times 10^{24}\,kg)}{(5200\,m/s)^2} = 1.473\times 10^7\,m$$

The period (time for one revolution) is then given by

$$T = \frac{2\pi r}{v} = \frac{2\pi(1.473\times 10^7\,m)}{5200\,m/s} = 1.78\times 10^4\,s = \underline{297\,min}$$

b) $a_{rad} = \dfrac{v^2}{r} = \dfrac{(5200\,m/s)^2}{1.473\times 10^7\,m} = \underline{1.84\,m/s^2}$

12-29

a) The gravitational force exerted on the spacecraft by the sun is $F_g = G\dfrac{m_s m_H}{r^2}$, where m_s is the mass of the sun and m_H is the mass of the Helios B spacecraft.

For a circular orbit $a_{rad} = \dfrac{v^2}{r}$ and $\Sigma F = m_H \dfrac{v^2}{r}$.

If neglect all forces on the spacecraft except the force exerted by the sun,
$F_g = \Sigma F = m_H \dfrac{v^2}{r}$, so $G\dfrac{m_s m_H}{r^2} = m_H \dfrac{v^2}{r}$

$$v = \sqrt{\frac{Gm_s}{r}} = \sqrt{\frac{(6.673\times 10^{-11}\,N\cdot m^2/kg^2)(1.99\times 10^{30}\,kg)}{43\times 10^9\,m}} = 5.6\times 10^4\,m/s = 56\,km/s$$

The actual speed is 71 km/s, so the orbit cannot be circular.

b) The orbit is a circle or an ellipse if it is closed, a parabola or hyperbola if open. The orbit is closed if the total energy (kinetic + potential) is negative, so that the object cannot reach $r\to\infty$.

For Helios B, $K = \tfrac{1}{2}m_H v^2 = \tfrac{1}{2}m_H(71\times 10^3\,m/s)^2 = (2.52\times 10^9\,m^2/s^2)m_H$

$U = -G\dfrac{m_s m_H}{r} = m_H\left(-\dfrac{(6.673\times 10^{-11}\,N\cdot m^2/kg^2)(1.99\times 10^{30}\,kg)}{43\times 10^9\,m}\right) = -(3.09\times 10^9\,m^2/s^2)m_H$

$E = K + U = (2.52\times 10^9\,m^2/s^2)m_H - (3.09\times 10^9\,m^2/s^2)m_H = -(5.7\times 10^8\,m^2/s^2)m_H$

The total energy E is negative, so the orbit is closed. We know from part (a) that it is not circular, so it must be elliptical.

12-33

a) Divide the ring up into small segments dM. The gravitational potential energy of dM and m is $dU = -G\dfrac{m\,dM}{r}$.

The total gravitational potential energy of the ring and particle is $U = \int dU = -Gm\int \dfrac{dM}{r}$

But $r = \sqrt{x^2 + a^2}$ is the same for all segments of the ring, so

12-33 (cont)

$$U = -\frac{Gm}{r}\int dM = -\frac{GmM}{r} = -\frac{GmM}{\sqrt{x^2+a^2}}$$

b) When $x \gg a$, $\sqrt{x^2+a^2} \to \sqrt{x^2} = x$ and $U = -\frac{GmM}{x}$. This is the gravitational potential energy of two point masses separated by a distance x. This is the expected result.

c) $F_x = -\frac{dU}{dx} = -\frac{d}{dx}\left(-\frac{GmM}{\sqrt{x^2+a^2}}\right) = +GmM\frac{d}{dx}(x^2+a^2)^{-1/2} = GmM\left(-\frac{1}{2}(2x)(x^2+a^2)^{-3/2}\right)$

$F_x = -GmM\frac{x}{(x^2+a^2)^{3/2}}$; the minus sign means that the force is attractive.

d) For $x \gg a$, $(x^2+a^2)^{3/2} \to (x^2)^{3/2} = x^3$
Then $F_x = -GmM\frac{x}{x^3} = -\frac{GmM}{x^2}$. This is the force between two point masses separated by a distance x and is the expected result.

e) $x = 0 \Rightarrow F_x = 0$. When the particle is at the center of the ring, symmetrically placed segments of the ring exert equal and opposite forces and the total force exerted by the ring is zero.

12-35

a) $F = W_0 = mg_0 = (4.0\,kg)(25\,m/s^2) = 100\,N$; this is the true weight of the object.

b) From Eq. (12-29), $W = W_0 - \frac{mv^2}{R}$

$T = \frac{2\pi r}{v} \Rightarrow v = \frac{2\pi r}{T} = \frac{2\pi(7.1\times10^7\,m)}{(9.8K)(3600s/1K)} = 1.264\times10^4\,m/s$

$\frac{v^2}{R} = \frac{(1.264\times10^4\,m/s)^2}{7.1\times10^7\,m} = 2.25\,m/s^2$

Then $W = 100\,N - (4.0\,kg)(2.25\,m/s^2) = \underline{91\,N}$

12-37

A black hole with the sun's mass M has the Schwarzschild radius given by Eq. (12-32):

$R_S = \frac{2GM}{c^2} = \frac{2(6.673\times10^{-11}\,N\cdot m^2/kg^2)(1.99\times10^{30}\,kg)}{(2.998\times10^8\,m/s)^2} = 2.95\times10^3\,m$

The ratio of R_S to the current radius R is $\frac{R_S}{R} = \frac{2.95\times10^3\,m}{6.96\times10^8\,m} = \underline{4.24\times10^{-6}}$.

Problems

12-41

a) [diagram: $m = 0.01\,kg$ at origin, $m_1 = 4.00\,kg$ at 1.00 m along x-axis, $m_2 = 3.00\,kg$ at 0.500 m below origin]

Free-body diagram for mass m: [diagram showing F_1 along +x and F_2 along −y]

12-41 (cont)

$$F_1 = G\frac{mm_1}{r_1^2} = (6.673\times 10^{-11} \text{ N·m}^2/\text{kg}^2)\frac{(0.0100\text{kg})(4.00\text{kg})}{(1.00\text{m})^2} = 2.669\times 10^{-12}\text{ N}$$

$$F_2 = G\frac{mm_2}{r_2^2} = (6.673\times 10^{-11} \text{ N·m}^2/\text{kg}^2)\frac{(0.0100\text{kg})(3.00\text{kg})}{(0.500\text{m})^2} = 8.008\times 10^{-12}\text{ N}$$

$F_{1x} = +F_1, \; F_{1y} = 0 \qquad F_{2x} = 0, \; F_{2y} = -F_2$

$F_x = F_{1x} + F_{2x} = F_1 = 2.669\times 10^{-12}\text{ N}$
$F_y = F_{1y} + F_{2y} = -F_2 = -8.008\times 10^{-12}\text{ N}$

$$F = \sqrt{F_x^2 + F_y^2} = \sqrt{(2.669\times 10^{-12}\text{N})^2 + (-8.008\times 10^{-12}\text{N})^2} = \underline{8.44\times 10^{-12}\text{ N}}$$

$$\tan\theta = \frac{F_y}{F_x} = \frac{-8.008\times 10^{-12}\text{N}}{2.669\times 10^{-12}\text{N}} = -3.000 \Rightarrow \underline{\theta = -71.6°}$$

b) Apply $K_1 + U_1 + W_{other} = K_2 + U_2$ to the test mass, with 1 for when it is at the origin and 2 for when it is at infinity.

$W_{other} = W_{you}$, the work you do
$K_1 = 0, \; K_2 = 0$ (for minimum work you do)
$U = -G\frac{mm'}{r}$ for two point masses m and m' separated by distance r, so when the test mass is far from the other masses, $U_2 = 0$.

$$U_1 = -G\frac{m_1 m}{r_1} - G\frac{m_2 m}{r_2} = -Gm\left(\frac{m_1}{r_1} + \frac{m_2}{r_2}\right)$$

$$U_1 = -(6.673\times 10^{-11}\text{ N·m}^2/\text{kg}^2)(0.0100\text{kg})\left(\frac{4.00\text{kg}}{1.00\text{m}} + \frac{3.00\text{kg}}{0.500\text{m}}\right) = -6.67\times 10^{-12}\text{ J}$$

Thus $-6.67\times 10^{-12}\text{ J} + W_{you} = 0$
$\underline{W_{you} = 6.67\times 10^{-12}\text{ J}}$

12-43

$$g = \frac{GM}{r^2} = \frac{(6.673\times 10^{-11}\text{ N·m}^2/\text{kg}^2)(5.00\text{kg})}{(3.00\text{m})^2} = \underline{3.71\times 10^{-11}\text{ N/kg}}$$

$\vec{F}_g$ is directed toward the center of the object, so the gravitational field is directed toward the center of the object.

12-45

To stay above the same point on the surface of the earth the orbital period of the satellite must equal the orbital period of the earth:

$$T = 1\text{d}\left(\frac{24\text{h}}{\text{d}}\right)\left(\frac{3600\text{s}}{1\text{h}}\right) = 8.64\times 10^4\text{ s}$$

Eq. (12-14) gives the relation between the orbit radius and the period

12-45 (cont)

$$T = \frac{2\pi r^{3/2}}{\sqrt{Gm_E}} \Rightarrow T^2 = \frac{4\pi^2 r^3}{Gm_E}$$

$$r = \left(\frac{T^2 Gm_E}{4\pi^2}\right)^{1/3} = \left(\frac{(8.64\times10^4 s)^2 (6.673\times10^{-11} N\cdot m^2/kg^2)(5.97\times10^{24} kg)}{4\pi^2}\right)^{1/3} = 4.23\times10^7 m$$

This is the radius of the orbit; it is related to the height h above the earth's surface and the radius R_E of the earth by $r = h + R_E$.
Thus $h = r - R_E = 4.23\times10^7 m - 6.38\times10^6 m = \underline{3.59\times10^7 m}$.

12-51

Section 12-7 proves that any two spherically symmetric point masses interact as though they were point masses with all the mass concentrated at their centers.

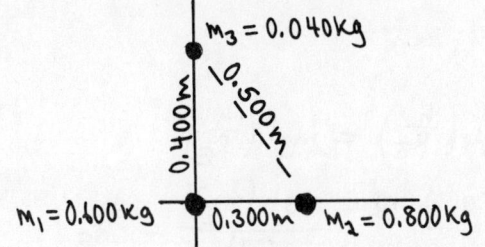
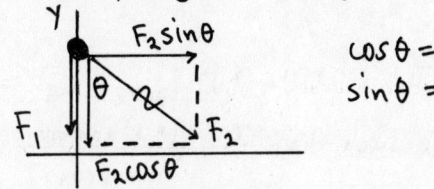

$\cos\theta = 0.800$
$\sin\theta = 0.600$

$$F_1 = G\frac{m_1 m_3}{r_{13}^2} = \frac{(6.673\times10^{-11} N\cdot m^2/kg^2)(0.600 kg)(0.040 kg)}{(0.400 m)^2} = 1.001\times10^{-11} N$$

$$F_2 = G\frac{m_2 m_3}{r_{23}^2} = \frac{(6.673\times10^{-11} N\cdot m^2/kg^2)(0.800 kg)(0.040 kg)}{(0.500 m)^2} = 0.854\times10^{-11} N$$

$F_{1x} = 0, \; F_{1y} = -1.001\times10^{-11} N$
$F_{2x} = +F_2\sin\theta = +(0.854\times10^{-11} N)(0.600) = 0.512\times10^{-11} N$
$F_{2y} = -F_2\cos\theta = -(0.854\times10^{-11} N)(0.800) = -0.683\times10^{-11} N$

$F_x = F_{1x} + F_{2x} = 0 + 0.512\times10^{-11} N = 0.512\times10^{-11} N$
$F_y = F_{1y} + F_{2y} = -1.001\times10^{-11} N - 0.683\times10^{-11} N = -1.684\times10^{-11} N$

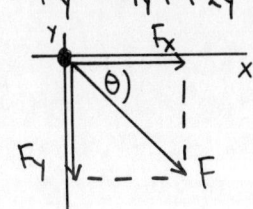

$$F = \sqrt{F_x^2 + F_y^2} = \sqrt{(0.512\times10^{-11} N)^2 + (-1.684\times10^{-11} N)^2} = \underline{1.76\times10^{-11} N}$$

$$\tan\theta = \frac{F_y}{F_x} = \frac{-1.684\times10^{-11} N}{0.512\times10^{-11} N} = -3.29 \Rightarrow \theta = \underline{-73.1°}$$

12-53

$U = mgy$ (Eq. 7-2)
Let $y=0$ at the earth's surface.
Then $U_1 = mgy_1 = 0$ and $U_2 = mgy_2 = mgh$.
$(\Delta U)_{approx} = U_2 - U_1 = mgh$

12-53 (cont)

$$U = -G\frac{m_E m}{r} \quad (\text{Eq. 12-9})$$

This equation has built into it that $U \to 0$ as $r \to \infty$.

$$U_1 = -G\frac{m_E m}{R_E}, \quad U_2 = -G\frac{m_E m}{R_E + h}$$

$$\Delta U = -Gm_E m\left(\frac{1}{R_E + h} - \frac{1}{R_E}\right) = -\frac{Gm_E m}{R_E}\left(\frac{1}{1 + h/R_E} - 1\right)$$

For only a 1% error in $\Delta U = mgh$ the value of h must be small compared to R_E, so use the binomial theorem to expand $\frac{1}{1+h/R_E}$ in powers of $\frac{h}{R_E}$:

$$\frac{1}{1+h/R_E} = \left(1 + \frac{h}{R_E}\right)^{-1} = 1 - \frac{h}{R_E} + \frac{h^2}{R_E^2} - \dots$$

$$\Delta U = -\frac{Gm_E m}{R_E}\left(1 - \frac{h}{R_E} + \frac{h^2}{R_E^2} + \dots - 1\right) = \frac{Gm_E m}{R_E^2} h\left(1 - \frac{h}{R_E}\right)$$

$$g = \frac{Gm_E}{R_E^2}, \quad \text{so } \Delta U = mgh\left(1 - \frac{h}{R_E}\right)$$

Then $\Delta U - (\Delta U)_{approx} = mgh - mgh + mgh\left(\frac{h}{R_E}\right) = mgh\left(\frac{h}{R_E}\right)$

Want h such that $\frac{\Delta U - (\Delta U)_{approx}}{(\Delta U)_{approx}} = 0.010$, so $0.010 = \frac{mgh\left(\frac{h}{R_E}\right)}{mgh} = \frac{h}{R_E}$

$$h = 0.01 R_E = 6.4 \times 10^4 \text{ m} = \underline{64 \text{ km}}$$

12-57

Take point #1 to be where the hammer is released and point #2 to be just above the surface of the earth, so $r_1 = R_E + h$ and $r_2 = R_E$.

$$K_1 + U_1 + W_{other} = K_2 + U_2$$

Only gravity does work, so $W_{other} = 0$.

$$K_1 = 0$$
$$K_2 = \tfrac{1}{2} m v_2^2$$
$$U_1 = -G\frac{m m_E}{r_1} = -\frac{Gm m_E}{h + R_E}$$
$$U_2 = -G\frac{m m_E}{r_2} = -\frac{Gm m_E}{R_E}$$

$$\Rightarrow G\frac{m m_E}{h + R_E} = \tfrac{1}{2} m v_2^2 - G\frac{m m_E}{R_E}$$

$$v_2^2 = 2Gm_E\left(\frac{1}{R_E} - \frac{1}{R_E + h}\right) = \frac{2Gm_E}{R_E(R_E + h)}(R_E + h - R_E) = \frac{2Gm_E h}{R_E(R_E + h)}$$

$$v_2 = \sqrt{\frac{2Gm_E h}{R_E(R_E + h)}}$$

12-59

a)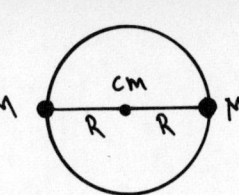
The two stars are separated by a distance $2R$, so
$$F_g = G\frac{M^2}{(2R)^2} = \frac{GM^2}{4R^2}$$

b) $F_g = ma_{rad}$
$$\frac{GM^2}{4R^2} = M\frac{v^2}{R} \Rightarrow v = \sqrt{\frac{GM}{4R}}$$
$$T = \frac{2\pi R}{v} = 2\pi R\sqrt{\frac{4R}{GM}} = 4\pi\sqrt{\frac{R^3}{GM}}$$

c) Apply $K_1 + U_1 + W_{other} = K_2 + U_2$ to the system of the two stars

Separate to infinity $\Rightarrow K_2 = 0, U_2 = 0$
$$K_1 = \tfrac{1}{2}Mv^2 + \tfrac{1}{2}Mv^2 = 2(\tfrac{1}{2}M)\left(\frac{GM}{4R}\right) = \tfrac{1}{4}\frac{GM^2}{R}$$
$$U_1 = -G\frac{M^2}{2R}$$

Energy required is $W_{other} = -(K_1 + U_1) = -\left(\frac{GM^2}{4R} - \frac{GM^2}{2R}\right) = \tfrac{1}{4}\frac{GM^2}{R}$

12-63

$$T = 26\times 10^6\, y \left(\frac{3.156\times 10^7 s}{1 y}\right) = 8.206\times 10^{14}\, s$$

Eq. (12-19): $T = \frac{2\pi a^{3/2}}{\sqrt{Gm_s}}$, $T^2 = \frac{4\pi^2 a^3}{Gm_s}$

$$a = \left(\frac{Gm_s T^2}{4\pi^2}\right)^{1/3} = \left[\frac{(6.673\times 10^{-11}\, N\cdot m^2/kg^2)(1.99\times 10^{30}\, kg)(8.206\times 10^{14}\, s)^2}{4\pi^2}\right]^{1/3} = 1.3\times 10^{16}\, m$$

The orbit radius of Pluto is 5.9×10^{12} m (Appendix F); the semi-major axis for this planet is larger by a factor of 2200.

$$4.3\text{ light years} = 4.3\text{ light years}\left(\frac{9.461\times 10^{15} m}{1\text{ light year}}\right) = 4.1\times 10^{16}\, m$$
The distance to Alpha Centauri is larger by a factor of 3.1.

12-65

Use conservation of energy: $K_1 + U_1 + W_{other} = K_2 + U_2$

The gravity force exerted by the sun is the only force that does work on the comet, so $W_{other} = 0$.
$K_1 = \tfrac{1}{2}mv_1^2$, $v_1 = 2.0\times 10^4$ m/s
$U_1 = -G\frac{m_s m}{r_1}$, $r_1 = 3.0\times 10^{11}$ m
$K_2 = \tfrac{1}{2}m v_2^2$

177

12-65 (cont)

$$U_2 = -G\frac{m_s m}{r_2}, \quad r_2 = 4.0 \times 10^{10} \text{ m}$$

$$\Rightarrow \tfrac{1}{2}mv_1^2 - G\frac{m_s m}{r_1} = \tfrac{1}{2}mv_2^2 - G\frac{m_s m}{r_2}$$

$$v_2^2 = v_1^2 + 2Gm_s\left(\tfrac{1}{r_2} - \tfrac{1}{r_1}\right) = v_1^2 + 2Gm_s\left(\frac{r_1 - r_2}{r_1 r_2}\right)$$

$$v_2 = \sqrt{(2.0 \times 10^4 \text{ m/s})^2 + 2(6.673 \times 10^{-11} \text{ N·m}^2/\text{kg}^2)(1.99 \times 10^{30} \text{ kg})\left(\frac{3.00 \times 10^{11} \text{ m} - 4.0 \times 10^{10} \text{ m}}{(3.0 \times 10^{11} \text{ m})(4.0 \times 10^{10} \text{ m})}\right)}$$

$v_2 = \underline{7.8 \times 10^4 \text{ m/s}}$ (The comet has greater speed when it is closer to the earth.)

12-69

$$K_1 + U_1 + W_{other} = K_2 + U_2$$

$U_1 = -G\frac{m_m m}{r_1}$, where m_m is the mass of the moon and $r_1 = R_m + h$, where R_m is the radius of the moon and $h = 2000 \times 10^3$ m

$$U_1 = -(6.673 \times 10^{-11} \text{ N·m}^2/\text{kg}^2)\frac{(7.35 \times 10^{22} \text{ kg})(3000 \text{ kg})}{1.76 \times 10^6 \text{ m} + 2000 \times 10^3 \text{ m}} = -3.934 \times 10^9 \text{ J}$$

$U_2 = -G\frac{m_m m}{r_2}$, where r_2 is the new orbit radius of 4000×10^3 m

$$U_2 = -(6.673 \times 10^{-11} \text{ N·m}^2/\text{kg}^2)\frac{(7.35 \times 10^{22} \text{ kg})(3000 \text{ kg})}{4000 \times 10^3 \text{ m}} = -3.678 \times 10^9 \text{ J}$$

For a circular orbit $v = \sqrt{\frac{Gm_m}{r}}$

(Eq. 12-12, with the mass of the moon rather than the mass of the earth)

$$K = \tfrac{1}{2}mv^2 = \tfrac{1}{2}m\left(\frac{Gm_m}{r}\right) = \tfrac{1}{2}\frac{Gm_m m}{r}, \text{ so } K = -\tfrac{1}{2}U$$

$K_1 = -\tfrac{1}{2}U_1 = +1.967 \times 10^9 \text{ J}$
$K_2 = -\tfrac{1}{2}U_2 = +1.839 \times 10^9 \text{ J}$

$K_1 + U_1 + W_{other} = K_2 + U_2$ gives
$W_{other} = (K_2 - K_1) + (U_2 - U_1) = (1.839 \times 10^9 \text{ J} - 1.967 \times 10^9 \text{ J}) + (-3.678 \times 10^9 \text{ J} + 3.934 \times 10^9 \text{ J})$
$W_{other} = -1.28 \times 10^8 \text{ J} + 2.56 \times 10^8 \text{ J} = \underline{1.3 \times 10^8 \text{ J}}$

(Note: When the orbit radius increases the kinetic energy decreases, the gravitational potential energy increases, and the total energy increases.)

12-71

Use a coordinate system with the origin at the left-hand end of the rod and the x'-axis along the rod. Divide the rod into small segments of length dx'. (Use x' for the coordinate so not to confuse with the distance x from the end of the rod to the particle.)

12-71 (cont)

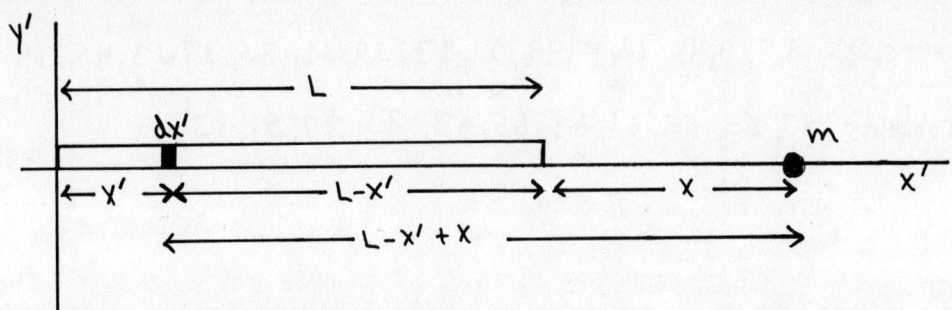

The mass of each segment is $dM = dx' \frac{M}{L}$. Each segment is a distance $L-x'+x$ from mass m, so the force on the particle due to a segment is

$$dF = \frac{G\,dM}{(L-x'+x)^2} = \frac{GMm}{L} \frac{dx'}{(L-x'+x)^2}.$$

$$F = \int_0^L dF = \frac{GMm}{L} \int_0^L \frac{dx'}{(L-x'+x)^2} = \frac{GMm}{L}\left(\frac{1}{L-x'+x}\bigg|_0^L\right)$$

$$F = \frac{GMm}{L}\left(\frac{1}{x} - \frac{1}{L+x}\right)$$

$$F = \frac{GMm}{L} \frac{L+x-x}{x(L+x)} = \frac{GMm}{x(L+x)}$$

$x \gg L \Rightarrow F \to \frac{GMm}{x^2}$, the same as for a pair of point masses.

CHAPTER 13

Exercises 3, 7, 9, 11, 13, 17, 19, 21, 23, 29, 31, 35, 37, 43, 45, 49

Problems 53, 55, 59, 61, 63, 65, 67, 75, 77, 81, 83

Exercises

13-3

a) $f = 262$ Hz
$T = \frac{1}{f} = \frac{1}{262 \text{ Hz}} = \underline{3.82 \times 10^{-3} \text{ s}}$
$\omega = 2\pi f = 2\pi (262 \text{ Hz}) = \underline{1650 \text{ rad/s}}$

b) $f = 4(262 \text{ Hz}) = 1048$ Hz
$T = \frac{1}{f} = \frac{1}{1048 \text{ Hz}} = \underline{9.54 \times 10^{-4} \text{ s}}$ (smaller by factor of 4)
$\omega = 2\pi f = 2\pi (1048 \text{ Hz}) = \underline{6580 \text{ rad/s}}$ (factor of 4 larger)

13-7

a) $T = \frac{1}{f} = \frac{1}{8.00 \text{ Hz}} = \underline{0.125 \text{ s}}$

b) $\omega = 2\pi f = 2\pi (8.00 \text{ Hz}) = \underline{50.3 \text{ rad/s}}$

c) $\omega = \sqrt{\frac{k}{m}} \Rightarrow m = \frac{k}{\omega^2} = \frac{200 \text{ N/m}}{(50.3 \text{ rad/s})^2} = \underline{0.0792 \text{ kg}}$

13-9

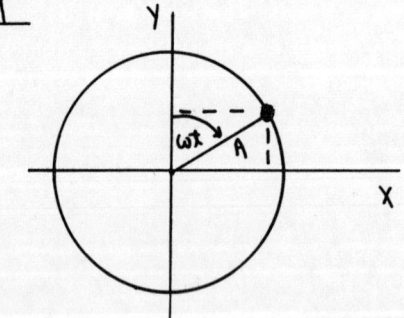

From the diagram, $X = A \sin \omega t$

$T = \frac{2\pi}{\omega} \Rightarrow \omega = \frac{2\pi}{T} = \frac{2\pi}{(\frac{\pi}{2} \text{s})} = 4.00 \text{ rad/s}$

Then $t = \frac{\pi}{10} \text{ s} \Rightarrow \omega t = (4.00 \text{ rad/s})(\frac{\pi}{10} \text{ s}) = 0.400\pi \text{ rad}$

$X = A \sin \omega t = (0.400 \text{ m}) \sin(0.400 \pi \text{ rad}) = \underline{0.380 \text{ m}}$

13-11

$f = 440$ Hz, $A = 3.0$ mm, $\phi = 0$

a) $x = A \cos(\omega t + \phi)$
$\omega = 2\pi f = 2\pi (440 \text{ Hz}) = 2.76 \times 10^3 \text{ rad/s}$
$x = (3.0 \times 10^{-3} \text{ m}) \cos((2.76 \times 10^3 \text{ rad/s}) t)$

b) $v = -\omega A \sin(\omega t + \phi)$
$v_{max} = \omega A = (2.76 \times 10^3 \text{ rad/s})(3.0 \times 10^{-3} \text{ m}) = \underline{8.28 \text{ m/s}}$ (maximum magnitude of velocity)

180

13-11 (cont)

$$a = -\omega^2 A \cos(\omega t + \phi)$$
$$a_{max} = \omega^2 A = (2.76 \times 10^3 \text{ rad/s})^2 (3.0 \times 10^{-3} \text{ m}) = \underline{2.29 \times 10^4 \text{ m/s}^2}$$
(maximum magnitude of acceleration)

c) $v_{max} = \omega A = 2\pi f A$

$v_{max,1} = 2\pi f_1 A_1$
$f_2 = \frac{1}{3} f_1, \quad A_2 = 3 A_1$
$v_{max,2} = 2\pi f_2 A_2 = 2\pi (\frac{1}{3} f_1)(3 A_1) = 2\pi f_1 A_1 = v_{max,1} = \underline{8.28 \text{ m/s}}$ (unchanged)

$a_{max} = \omega^2 A = 4\pi^2 f^2 A$

$a_{max,1} = 4\pi^2 f_1^2 A_1$
$a_{max,2} = 4\pi^2 f_2^2 A_2 = 4\pi^2 (\frac{1}{3} f_1)^2 (3 A_1) = \frac{1}{3} 4\pi^2 f_1^2 A_1 = \frac{1}{3} a_{max,1}$
$a_{max,2} = \frac{1}{3}(2.29 \times 10^4 \text{ m/s}^2) = 7.63 \times 10^3 \text{ m/s}^2$ (smaller by a factor of 3)

13-13

a) Eq. (13-19): $A = \sqrt{x_0^2 + \frac{v_0^2}{\omega^2}} = \sqrt{x_0^2 + \frac{m v_0^2}{k}} = \sqrt{(0.200 \text{ m})^2 + \frac{(3.00 \text{ kg})(-6.00 \text{ m/s})^2}{200 \text{ N/m}}}$

$A = \underline{0.762 \text{ m}}$

b) Eq. (13-18): $\phi = \arctan\left(-\frac{v_0}{\omega x_0}\right)$

$\omega = \sqrt{\frac{k}{m}} = \sqrt{\frac{200 \text{ N/m}}{3.00 \text{ kg}}} = 8.165 \text{ rad/s}$

$\phi = \arctan\left(-\frac{(-6.00 \text{ m/s})}{(8.165 \text{ rad/s})(0.200 \text{ m})}\right) = \arctan(+3.674) = \underline{74.8°}$ (or 1.31 rad)

c) $x = A \cos(\omega t + \phi) \Rightarrow x = (0.762 \text{ m}) \cos[(8.16 \text{ rad/s}) t + 1.31 \text{ rad}]$

13-17

a) $-kx = ma \Rightarrow a = -\frac{k}{m} x$ (Eq. 13-4)
But the maximum $|x|$ is A, so $a_{max} = \frac{k}{m} A = \omega^2 A$.

$f = 6.00 \text{ Hz} \Rightarrow \omega = \sqrt{\frac{k}{m}} = 2\pi f = 2\pi (6.00 \text{ Hz}) = 37.7 \text{ rad/s}$
$a_{max} = \omega^2 A = (37.7 \text{ rad/s})^2 (0.180 \text{ m}) = \underline{256 \text{ m/s}^2}$

$\frac{1}{2} m v^2 + \frac{1}{2} k x^2 = \frac{1}{2} k A^2$
$v = v_{max}$ when $x = 0 \Rightarrow \frac{1}{2} m v_{max}^2 = \frac{1}{2} k A^2$
$v_{max} = \sqrt{\frac{k}{m}} A = \omega A = (37.7 \text{ rad/s})(0.180 \text{ m}) = \underline{6.79 \text{ m/s}}$

b) $a = -\frac{k}{m} x = -\omega^2 x = -(37.7 \text{ rad/s})^2 (0.090 \text{ m}) = \underline{-128 \text{ m/s}^2}$

13-17 (cont)

$\frac{1}{2}mv^2 + \frac{1}{2}kx^2 = \frac{1}{2}kA^2 \Rightarrow v = \pm\sqrt{\frac{k}{m}}\sqrt{A^2 - x^2} = \pm\omega\sqrt{A^2 - x^2}$

$v = \pm(37.7 \text{ rad/s})\sqrt{(0.180\text{m})^2 - (0.090\text{m})^2} = \pm 5.88 \text{ m/s}$

The speed is $\underline{5.88 \text{ m/s}}$.

c) $x = A\cos(\omega t + \phi)$
Let $\phi = -\frac{\pi}{2}$ so that $x = 0$ at $t = 0$.
Then $x = A\cos(\omega t - \frac{\pi}{2}) = A\sin(\omega t)$ [using $\cos(a - \frac{\pi}{2}) = \sin a$]

Find the time t that gives $x = 0.120$ m.
$0.120\text{m} = (0.180\text{m})\sin(\omega t)$
$\sin \omega t = 0.6667$

$t = \frac{\arcsin(0.6667)}{\omega} = \frac{0.7297 \text{ rad}}{37.7 \text{ rad/s}} = \underline{0.0194 \text{ s}}$

Note: It takes one-fourth of a period for the object to go from $x = 0$ to $x = A = 0.180$ m. So the time we have calculated should be less than $T/4$.

$T = \frac{1}{f} = \frac{1}{6.00 \text{ Hz}} = 0.167 \text{ s}$; $\frac{T}{4} = 0.0417 \text{ s}$, and the time we calculated $\underline{\text{is}}$ less than this.

13-19

a) $E = \frac{1}{2}mv^2 + \frac{1}{2}kx^2$
$E = \frac{1}{2}(0.500 \text{ kg})(0.300 \text{ m/s})^2 + \frac{1}{2}(400 \text{ N/m})(0.012 \text{ m})^2 = 0.0225\text{J} + 0.0288\text{J} = \underline{0.0513 \text{ J}}$

b) $E = \frac{1}{2}kA^2 \Rightarrow A = \sqrt{\frac{2E}{k}} = \sqrt{\frac{2(0.0513\text{J})}{400 \text{ N/m}}} = \underline{0.0160 \text{ m}}$

c) $E = \frac{1}{2}mv_{max}^2 \Rightarrow v_{max} = \sqrt{\frac{2E}{m}} = \sqrt{\frac{2(0.0513\text{J})}{0.500 \text{ kg}}} = \underline{0.453 \text{ m/s}}$

13-21

a) $U + K = E$
$U = K \Rightarrow 2U = E$
$2(\frac{1}{2}kx^2) = \frac{1}{2}kA^2 \Rightarrow x = \pm\frac{A}{\sqrt{2}}$

$U = K \Rightarrow 2K = E$
$2(\frac{1}{2}mv^2) = \frac{1}{2}kA^2 \Rightarrow v = \pm\sqrt{\frac{k}{m}}\frac{A}{\sqrt{2}} = \pm\frac{\omega A}{\sqrt{2}}$

b) In one cycle x goes from A to 0 to $-A$ to 0 to $+A$.
Thus $x = +\frac{A}{\sqrt{2}}$ twice and $x = -\frac{A}{\sqrt{2}}$ twice in each cycle.
Therefore, $U = K$ four times each cycle.

c) The time between $U = K$ occurrences is the time Δt_a for $x_1 = +\frac{A}{\sqrt{2}}$ to $x_2 = -\frac{A}{\sqrt{2}}$, time Δt_b for $x_1 = -\frac{A}{\sqrt{2}}$ to $x_2 = +\frac{A}{\sqrt{2}}$, time Δt_c for $x_1 = +\frac{A}{\sqrt{2}}$ to

13-21 (cont)

$x_2 = +\frac{A}{\sqrt{2}}$, and time Δt_d for $x_1 = -\frac{A}{\sqrt{2}}$ to $x_2 = -\frac{A}{\sqrt{2}}$.

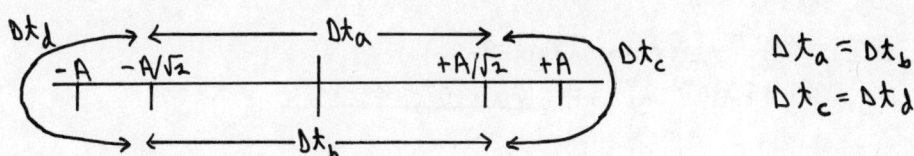

$\Delta t_a = \Delta t_b$
$\Delta t_c = \Delta t_d$

Calculation of Δt_a:

Specify x in $x = A \cos \omega t$ (choose $\phi = 0$ so $x = A$ at $t=0$) and solve for t.

$x_1 = +\frac{A}{\sqrt{2}} \Rightarrow \frac{A}{\sqrt{2}} = A \cos(\omega t_1)$

$\cos \omega t_1 = \frac{1}{\sqrt{2}}$

$\omega t_1 = \arccos(\frac{1}{\sqrt{2}}) = \frac{\pi}{4}$ rad

$t_1 = \frac{\pi}{4\omega}$

$x_2 = -\frac{A}{\sqrt{2}} \Rightarrow -\frac{A}{\sqrt{2}} = A \cos(\omega t_2)$

$\cos \omega t_2 = -\frac{1}{\sqrt{2}}$

$\omega t_2 = \frac{3\pi}{4}$ rad

$t_2 = \frac{3\pi}{4\omega}$

$\Delta t_a = t_2 - t_1 = \frac{3\pi}{4\omega} - \frac{\pi}{4\omega} = \frac{\pi}{2\omega}$ (Note that this is $\frac{T}{4}$, one fourth period.)

Calculation of Δt_d:

$x_1 = -\frac{A}{\sqrt{2}} \Rightarrow t_1 = \frac{3\pi}{4\omega}$

$x_2 = -\frac{A}{\sqrt{2}}$; t_2 is the next time after t_1 that gives $\cos \omega t_2 = -\frac{1}{\sqrt{2}}$

$\Rightarrow \omega t_2 = \omega t_1 + \frac{\pi}{2} = \frac{5\pi}{4}$

$t_2 = \frac{5\pi}{4\omega}$

$\Delta t_d = t_2 - t_1 = \frac{5\pi}{4\omega} - \frac{3\pi}{4\omega} = \frac{\pi}{2\omega}$, so is the same as Δt_a.

Therefore, the occurrences of $K = U$ are equally spaced in time, with a time interval between them of $\frac{\pi}{2\omega}$. (This is one fourth T, as it must be if there are 4 equally spaced occurrences each period.)

13-23

Let d be the distance the spring stretches when the block hangs at rest.

Free-body diagram for the block:
(Take +y to be downward.)

$\sum F_y = ma_y$

$mg - kd = 0 \Rightarrow d = \frac{mg}{k}$

$T = 2\pi \sqrt{\frac{m}{k}} \Rightarrow \frac{m}{k} = \left(\frac{T}{2\pi}\right)^2$

Thus $d = \left(\frac{T}{2\pi}\right)^2 g$

$d = \left(\frac{0.400 s}{2\pi}\right)^2 (9.80 \text{ m/s}^2) = \underline{0.0397 \text{ m}}$

13-29

Ticks four times each second $\Rightarrow 0.25$ s per tick.
Each tick is half a period $\Rightarrow T = 0.50$ s and $f = \frac{1}{T} = \frac{1}{0.50 s} = 2.00$ Hz.

a) thin rim $\Rightarrow I = MR^2$ (from Table 9-2)
$I = (0.800 \times 10^{-3} \text{ kg})(0.45 \times 10^{-2} \text{ m})^2 = \underline{1.6 \times 10^{-8} \text{ kg} \cdot \text{m}^2}$

13-31

Eq. (13-25): $U = U_0 \left[\left(\frac{R_0}{r}\right)^{12} - 2\left(\frac{R_0}{r}\right)^6 \right]$
Let $r = R_0 + x$

$U = U_0 \left[\left(\frac{R_0}{R_0 + x}\right)^{12} - 2\left(\frac{R_0}{R_0 + x}\right)^6 \right] = U_0 \left[\left(\frac{1}{1 + x/R_0}\right)^{12} - 2\left(\frac{1}{1 + x/R_0}\right)^6 \right]$

$\left(\frac{1}{1 + x/R_0}\right)^{12} = \left(1 + \frac{x}{R_0}\right)^{-12}$; $\left|\frac{x}{R_0}\right| \ll 1$

Apply Eq. (13-28) with $n = -12$ and $u = +\frac{x}{R_0}$

$\Rightarrow \left(\frac{1}{1 + x/R_0}\right)^{12} = 1 - \frac{12x}{R_0} + 66 \frac{x^2}{R_0^2} - \ldots$

For $\left(\frac{1}{1 + x/R_0}\right)^6 = \left(1 + \frac{x}{R_0}\right)^{-6}$ apply Eq. (13-28) with $n = -6$ and $u = \frac{x}{R_0}$

$\Rightarrow \left(\frac{1}{1 + x/R_0}\right)^6 = 1 - \frac{6x}{R_0} + 15 \frac{x^2}{R_0^2}$

Thus $U = U_0 \left[1 - \frac{12x}{R_0} + 66 \frac{x^2}{R_0^2} - 2 + \frac{12x}{R_0} - 30 \frac{x^2}{R_0^2} \right] = -U_0 + 36 U_0 \frac{x^2}{R_0^2}$

This is in the form $U = \frac{1}{2} k x^2 - U_0$ with $k = \frac{72 U_0}{R_0^2}$, which is the same as the force constant in Eq. (13-29).

13-35

Let the period on earth be $T_E = 2\pi \sqrt{\frac{L}{g_E}}$, where $g_E = 9.80$ m/s^2, the value on earth.

Let the period on the moon be $T_M = 2\pi \sqrt{\frac{L}{g_M}}$, where $g_M = 1.62$ m/s^2, the value on the moon.

We can eliminate L, which we don't know, by taking a ratio:

$\frac{T_M}{T_E} = 2\pi \sqrt{\frac{L}{g_M}} \cdot \frac{1}{2\pi} \sqrt{\frac{g_E}{L}} = \sqrt{\frac{g_E}{g_M}}$

$T_M = T_E \sqrt{\frac{g_E}{g_M}} = (1.60 \text{ s}) \sqrt{\frac{9.80 \text{ m/s}^2}{1.62 \text{ m/s}^2}} = \underline{3.94 \text{ s}}$

13-37

vertical SHM: $f_b = \frac{1}{2\pi} \sqrt{\frac{k}{m}}$
pendulum motion (small amplitude): $f_p = \frac{1}{2\pi} \sqrt{\frac{g}{L}}$
$f_p = \frac{1}{2} f_b$

13-37 (cont)

$$\frac{1}{2\pi}\sqrt{\frac{g}{L}} = \frac{1}{2}\left(\frac{1}{2\pi}\sqrt{\frac{K}{m}}\right)$$

$$\frac{g}{L} = \frac{1}{4}\frac{K}{m} \Rightarrow L = \frac{4gm}{k} = \frac{4w}{k} = \frac{4(1.00\,N)}{2.00\,N/m} = 2.00\,m$$

But this is the <u>stretched</u> length of the spring, its length when the apple is hanging from it. (Note: Small angle of swing means v is small as apple passes through lowest point, so a_{rad} is small and the component of mg perpendicular to the spring is small. Thus the amount the spring is stretched changes very little as the apple swings back and forth.)

Calculate the distance the spring is stretched from its unstretched length when the apple hangs from it.
Free-body diagram for the apple hanging at rest on the end of the spring:

$a = 0$ $k\Delta L \uparrow$ $\Sigma F_y = ma_y$
            $\bullet \longrightarrow x$ $k\Delta L - mg = 0$
            $\downarrow mg$ $\Delta L = \frac{mg}{k} = \frac{w}{k} = \frac{1.00\,N}{2.00\,N/m} = 0.500\,m$

Thus the unstretched length of the spring is
$2.00\,m - 0.500\,m = \underline{1.50\,m}$

13-43

The ornament is a physical pendulum; $T = 2\pi\sqrt{\frac{I}{mgd}}$ (Eq. 13-39).

$I = \frac{7MR^2}{5}$, the moment of inertia about an axis at the edge of the sphere. d is the distance from the axis to the center of gravity, which is at the center of the sphere, so $d = R$.

Thus $T = 2\pi\sqrt{\frac{7MR^2}{5MgR}} = 2\pi\sqrt{\frac{7}{5}}\sqrt{\frac{R}{g}} = 2\pi\sqrt{\frac{7}{5}}\sqrt{\frac{0.050\,m}{9.80\,m/s^2}} = \underline{0.53\,s}$

13-45

a) Eq. (13-43) says $\omega' = \sqrt{\frac{k}{m} - \frac{b^2}{4m^2}}$

$\omega' = \sqrt{\frac{300\,N/m}{0.400\,kg} - \frac{(9.00\,kg/s)^2}{4(0.400\,kg)^2}} = 25.0\,rad/s$

$f' = \frac{\omega'}{2\pi} = \frac{25.0\,rad/s}{2\pi} = \underline{3.97\,Hz}$

b) critical damping $\Rightarrow b = 2\sqrt{km}$ (Eq. 13-44)
$b = 2\sqrt{(300\,N/m)(0.400\,kg)} = \underline{21.9\,kg/s}$

13-49

Eq. (13-46): $A = \frac{F_{max}}{\sqrt{(k - m\omega_d^2)^2 + b^2\omega_d^2}}$

13-49 (cont)

a) Consider the special case where $k - m\omega_d^2 = 0$, so $A = \dfrac{F_{max}}{b\omega_d}$

$$b = \dfrac{F_{max}}{A\omega_d}$$

Units of $\dfrac{F_{max}}{A\omega_d}$ are $\dfrac{kg \cdot m/s^2}{(m)(s^{-1})} = kg/s$; for units consistency the units of b must be kg/s.

b) Units of $\sqrt{km}$: $((N/m)kg)^{1/2} = \left(\dfrac{Nkg}{m}\right)^{1/2} = \left(\dfrac{(kg \cdot m/s^2)(kg)}{m}\right)^{1/2} = \left(\dfrac{kg^2}{s^2}\right)^{1/2} = \dfrac{kg}{s}$,

the same as the units for b.

c) At resonance $k - m\omega_d^2 = 0$, so $A = \dfrac{F_{max}}{b\omega_d}$.

But $k - m\omega_d^2 = 0$ also says that $\omega_d = \sqrt{\dfrac{k}{m}}$, so $A = \dfrac{F_{max}}{b} \sqrt{\dfrac{m}{k}}$.

(i) $b = 0.2\sqrt{km}$

$A = F_{max}\sqrt{\dfrac{m}{k}} \dfrac{1}{0.2\sqrt{km}} = \dfrac{F_{max}}{0.2k} = 5.0 \dfrac{F_{max}}{k}$

(ii) $b = 0.4\sqrt{km}$

$A = F_{max}\sqrt{\dfrac{m}{k}} \dfrac{1}{0.4\sqrt{km}} = 2.5 \dfrac{F_{max}}{k}$

Both these results agree with what is shown in Fig. 13-24.

Problems

13-53

a) $T = 2\pi\sqrt{\dfrac{m}{k}}$; independent of A so period doesn't change

$f = \dfrac{1}{T}$; doesn't change

$\omega = 2\pi f$; doesn't change

b) $v_{max} = \omega A = 2\pi f A$

$v_{max,1} = 2\pi f A_1$, $v_{max,2} = 2\pi f A_2$ (f doesn't change)

Since $A_2 = \tfrac{1}{2} A_1$, $v_{max,2} = 2\pi f(\tfrac{1}{2} A_1) = \tfrac{1}{2} \cdot 2\pi f A_1 = \tfrac{1}{2} v_{max,1}$; v_{max} is $\tfrac{1}{2}$ as great.

c) $v = \pm\sqrt{\dfrac{k}{m}} \sqrt{A^2 - x^2}$

$x = \pm \dfrac{A_1}{4} \Rightarrow v = \pm\sqrt{\dfrac{k}{m}} \sqrt{A^2 - A_1^2/16}$

With the original amplitude $v_1 = \pm\sqrt{\dfrac{k}{m}}\sqrt{A_1^2 - A_1^2/16} = \pm\sqrt{\dfrac{15}{16}} \sqrt{\dfrac{k}{m}} A_1$

With the reduced amplitude $v_2 = \pm\sqrt{\dfrac{k}{m}}\sqrt{A_2^2 - A_1^2/16} = \pm\sqrt{\dfrac{k}{m}}\sqrt{\left(\dfrac{A_1}{2}\right)^2 - \dfrac{A_1^2}{16}}$

$v_2 = \sqrt{\dfrac{1}{4} - \dfrac{1}{16}}\left(\pm\sqrt{\dfrac{k}{m}} A_1\right) = \pm\sqrt{\dfrac{3}{16}} \sqrt{\dfrac{k}{m}} A_1$

13-53 (cont) $\frac{v_1}{v_2} = \sqrt{\frac{15}{3}} = \sqrt{5}$, so $v_2 = \frac{1}{\sqrt{5}} v_1$; the speed at this x is $\frac{1}{\sqrt{5}}$ times as great

e) $U = \frac{1}{2}kx^2$; same x so same U

$K = \frac{1}{2}mv^2$; $K_1 = \frac{1}{2}mv_1^2$
$K_2 = \frac{1}{2}mv_2^2 = \frac{1}{2}m\left(\frac{v_1}{\sqrt{5}}\right)^2 = \frac{1}{5}\left(\frac{1}{2}mv_1^2\right) = \frac{1}{5}K_1$; $\frac{1}{5}$ times as great

13-55

a) $T = 2\pi\sqrt{\frac{m}{k}}$

We are given information about v at a particular x. The expression relating these two quantities comes from conservation of energy:
$\frac{1}{2}mv^2 + \frac{1}{2}kx^2 = \frac{1}{2}kA^2$
We can solve this equation for $\sqrt{\frac{m}{k}}$, and then use that result to calculate T.
$mv^2 = k(A^2 - x^2)$
$\sqrt{\frac{m}{k}} = \frac{\sqrt{A^2 - x^2}}{v} = \frac{\sqrt{(0.100\text{m})^2 - (0.060\text{m})^2}}{0.360\text{ m/s}} = 0.222\text{ s}$

Then $T = 2\pi\sqrt{\frac{m}{k}} = 2\pi(0.222\text{ s}) = \underline{1.39\text{ s}}$

b) We are asked to relate x and v, so use the conservation of energy equation. $\frac{1}{2}mv^2 + \frac{1}{2}kx^2 = \frac{1}{2}kA^2$

$kx^2 = kA^2 - mv^2$
$x = \sqrt{A^2 - \frac{m}{k}v^2} = \sqrt{(0.100\text{m})^2 - (0.222\text{s})^2(0.120\text{m/s})^2} = \underline{0.0964\text{ m}}$

c) "small object whose mass is much less than the mass of the block"
⇒ doesn't alter the motion of the block.

For the block, $-kx = ma \Rightarrow a = -\frac{k}{m}x$.
The maximum $|x|$ is A, so $a_{max} = \frac{k}{m}A$.

If the small object doesn't slip then the static friction force must be able to give it this much acceleration.
Free-body diagram for the small mass (mass m'):

$\Sigma F_y = ma_y$ $\Sigma F_x = ma_x$
$n - m'g = 0$ $\mu_s n = m'a$
$n = m'g$ $\mu_s m'g = m'a$
                           $a = \mu_s g$

But we require that $a = a_{max} = \frac{k}{m}A$, so $\frac{k}{m}A = \mu_s g$

$\Rightarrow \mu_s = \frac{k}{m}\frac{A}{g} = \left(\frac{1}{0.222\text{s}}\right)^2\left(\frac{0.100\text{m}}{9.80\text{m/s}^2}\right) = \underline{0.207}$

13-59

a) Eq. (13-3): $F = -kx$

Eq. (11-10): $Y = \dfrac{\ell_0 F_\perp}{A \Delta \ell} \Rightarrow F_\perp = \dfrac{YA \Delta \ell}{\ell_0}$

$F_\perp$ is the force applied to the end of the wire; the force F with which the wire pulls back is $F = -F_\perp$. The displacement x of the end of the wire is $x = \Delta \ell$.

Thus $F = -\left(\dfrac{YA}{\ell_0}\right)x$ and $k = \dfrac{YA}{\ell_0}$.

b) From Table 11-1, Y (copper) $= 1.1 \times 10^{11}$ Pa

$k = \dfrac{YA}{\ell_0} = \dfrac{(1.1 \times 10^{11} \text{Pa})(\pi (0.75 \times 10^{-3} \text{m})^2)}{3.00 \text{m}} = \underline{6.5 \times 10^4 \text{ N/m}}$

13-61

[Diagram: dock, raft, with $x = +A$ up, $x = 0$, $x = -A$ down]

Let the raft be at $x = +A$ when $t = 0$. Then $\phi = 0$ and $x(t) = A \cos \omega t$. Calculate the time it takes the raft to move from $x = +A = +0.200$ m to $x = A - 0.100$ m $= 0.100$ m.

Write the equation for $x(t)$ in terms of T rather than ω:

$\omega = \dfrac{2\pi}{T} \Rightarrow x(t) = A \cos\left(\dfrac{2\pi t}{T}\right)$

$x = A$ at $t = 0$

$x = 0.100$ m $\Rightarrow 0.100$ m $= (0.200 \text{ m}) \cos\left(\dfrac{2\pi t}{T}\right)$

$\cos\left(\dfrac{2\pi t}{T}\right) = 0.500 \Rightarrow \dfrac{2\pi t}{T} = \arccos(0.500) = 1.047$ rad

$t = \dfrac{T}{2\pi}(1.047 \text{ rad}) = \dfrac{3.50 \text{ s}}{2\pi}(1.047 \text{ rad}) = 0.583$ s

This is the time for the raft to move down from $x = 0.200$ m to $x = 0.100$ m. But people can also get off while the raft is moving up from $x = 0.100$ m to $x = 0.200$ m, so during each period of the motion the time the people have to get off is $2t = 2(0.583 \text{ s}) = \underline{1.17 \text{ s}}$

13-63

Measure x from the equilibrium position of the object, where the gravity and spring forces balance. Let $+x$ be downward.

a) Conservation of energy $\Rightarrow \tfrac{1}{2}mv^2 + \tfrac{1}{2}kx^2 = \tfrac{1}{2}kA^2$

$x = 0 \Rightarrow \tfrac{1}{2}mv^2 = \tfrac{1}{2}kA^2 \Rightarrow v = A\sqrt{\dfrac{k}{m}}$, just as for horizontal SHM.

We can use the period to calculate $\sqrt{\dfrac{k}{m}}$: $T = 2\pi\sqrt{\dfrac{m}{k}} \Rightarrow \sqrt{\dfrac{k}{m}} = \dfrac{2\pi}{T}$.

13-63 (cont)

Thus $v = \frac{2\pi A}{T} = \frac{2\pi (0.100 m)}{1.80 s} = \underline{0.349 \text{ m/s}}$.

b) $ma = -kx \Rightarrow a = -\frac{k}{m}x$

+x is downward, so here $x = -0.050 m$

$a = -\left(\frac{2\pi}{T}\right)^2 (-0.050 m) = +\left(\frac{2\pi}{1.80 s}\right)^2 (0.050 m) = \underline{0.609 \text{ m/s}^2}$ (positive $\Rightarrow$ downward)

c) This is twice the time it takes to go from $x=0$ to $x=+0.050 m$.

$x(t) = A \cos(\omega t + \phi)$
Let $\phi = -\frac{\pi}{2}$, so $x=0$ at $t=0$.
Then $x = A \cos(\omega t - \frac{\pi}{2}) = A \sin \omega t = A \sin\left(\frac{2\pi t}{T}\right)$

Find the time t that gives $x = +0.050 m$:

$0.050 m = (0.100 m) \sin\left(\frac{2\pi t}{T}\right) \Rightarrow \frac{2\pi t}{T} = \arcsin(0.50)$

$\frac{2\pi t}{T} = \frac{\pi}{6} \Rightarrow t = \frac{T}{12} = \frac{1.80 s}{12} = 0.150 s$

The time asked for in the problem is twice this, $\underline{0.300 s}$.

d) The problem is asking for the distance d that the spring stretches when the object hangs at rest from it.

Free-body diagram for the object:

$a=0$ ↑kd (force of spring)

↓mg

$\Sigma F_x = ma_x$
$mg - kd = 0$
$d = \left(\frac{m}{k}\right) g$

But $\sqrt{\frac{k}{m}} = \frac{2\pi}{T}$ (part(a)) $\Rightarrow \frac{m}{k} = \left(\frac{T}{2\pi}\right)^2$

$d = \left(\frac{T}{2\pi}\right)^2 g = \left(\frac{1.80 s}{2\pi}\right)^2 (9.80 \text{ m/s}^2) = \underline{0.804 m}$

13-65

a) First find the speed of the steak just before it strikes the pan. Use a coordinate system with +y downward.

$v_{0y} = 0$ (released from rest)
$y - y_0 = 0.40 m$
$a_y = +9.80 \text{ m/s}^2$
$v_y = ?$

$v_y^2 = v_{0y}^2 + 2a_y(y-y_0)$
$v_y = +\sqrt{2a_y(y-y_0)} = +\sqrt{2(9.80 \text{ m/s}^2)(0.40 m)}$
$v_y = +2.80 \text{ m/s}$

Apply conservation of momentum to the collision between the steak and the pan. After the collision the steak and the pan are moving together with the common velocity v_2. Let A be the steak and B be the pan.

13-65 (cont)

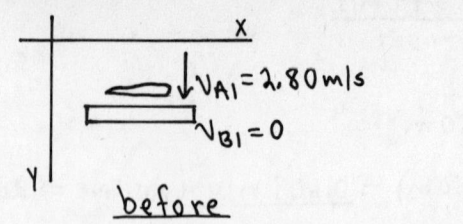
before

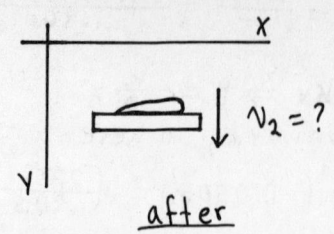
after

P_y conserved $\Rightarrow m_A v_{A1y} + m_B v_{B1y} = (m_A+m_B) v_{2y}$
$$m_A v_{A1} = (m_A+m_B) v_2$$

$v_2 = \left(\dfrac{m_A}{m_A+m_B}\right) v_{A1} = \left(\dfrac{1.8\,kg}{1.8\,kg + 0.20\,kg}\right)(2.80\,m/s) = \underline{2.52\,m/s}$

b) Conservation of energy applied to the SHM gives
$$\tfrac{1}{2} m v_0^2 + \tfrac{1}{2} K x_0^2 = \tfrac{1}{2} K A^2,$$
where v_0 and x_0 are the initial speed and displacement of the object and where the displacement is measured from the equilibrium position of the object.

The weight of the steak will stretch the spring an additional distance d given by
$$Kd = mg \Rightarrow d = \dfrac{mg}{K} = \dfrac{(1.8\,kg)(9.80\,m/s^2)}{500\,N/m} = 0.0353\,m.$$ So just after the steak hits the pan, before the pan has had time to move, the steak+pan is 0.0353 m above the equilibrium position of the combined object
$\Rightarrow x_0 = 0.0353\,m.$

From part (a) $v_0 = 2.52\,m/s$, the speed of the combined object just after the collision.

Then $\tfrac{1}{2} m v_0^2 + \tfrac{1}{2} K x_0^2 = \tfrac{1}{2} K A^2$

$\Rightarrow A = \sqrt{\dfrac{m v_0^2 + K x_0^2}{K}} = \sqrt{\dfrac{2.0\,kg\,(2.52\,m/s)^2 + (500\,N/m)(0.0353\,m)^2}{500\,N/m}} = \underline{0.163\,m}$

c) $T = 2\pi \sqrt{\dfrac{m}{K}} = 2\pi \sqrt{\dfrac{2.0\,kg}{500\,N/m}} = \underline{0.397\,s}$

13-67

a) Measure x from the equilibrium position of the ball bearing and let $+x$ be upward.

Let n be the normal force exerted on the ball bearing by the lens.

Free-body diagram for the ball bearing:

$\Sigma F_x = m a_x$
$n - mg = ma$
$n = m(g+a)$

13-67 (cont)

The lens is moving with simple harmonic motion
$\Rightarrow x(t) = A\cos(\omega t + \phi) = A\cos(2\pi f t + \phi)$
and $a = -\omega^2 x = -(2\pi f)^2 A\cos(2\pi f t + \phi)$
But this must also be a for the ball bearing. Use in the above equation for $n \Rightarrow \boxed{n = m(g - (2\pi f)^2 A\cos(2\pi f t + \phi))}$

b) The ball bounces when it loses contact with the lens during the motion.
Loses contact $\Rightarrow n = 0$, so $g - (2\pi f_b)^2 A\cos(2\pi f_b t + \phi) = 0$.
The smallest frequency f_b where this happens is when $\cos(2\pi f_b t + \phi) = 1$, so
$g - (2\pi f_b)^2 A = 0 \Rightarrow \boxed{g = (2\pi f_b)^2 A}$.

13-75

a) $U = A\left[\frac{1}{r} - \frac{1}{(r-2R_0)}\right]$

$F_r = -\frac{dU}{dr} = -A\left[-\frac{1}{r^2} + \frac{1}{(r-2R_0)^2}\right] = A\left[\frac{1}{r^2} - \frac{1}{(r-2R_0)^2}\right]$, as was to be shown.

b) At equilibrium $F_r = 0 \Rightarrow \frac{1}{r^2} - \frac{1}{(r-2R_0)^2} = 0$
$r^2 = (r-2R_0)^2$
$r = \pm(r-2R_0)$
$r = r - 2R_0$; no solution if $R_0 \neq 0$
$r = -(r-2R_0) \Rightarrow 2r = 2R_0 \Rightarrow r = R_0$, as was to be shown

c) Write $r = R_0 + x$ where $\frac{x}{R_0}$ is small.
Note that $(r-2R_0)^2 = (R_0 + x - 2R_0)^2 = (x - R_0)^2 = (R_0 - x)^2$, so
$F_r = A\left[\frac{1}{(R_0+x)^2} - \frac{1}{(R_0-x)^2}\right] = \frac{A}{R_0^2}\left[\frac{1}{(1+x/R_0)^2} - \frac{1}{(1-x/R_0)^2}\right]$

For $\frac{1}{(1+x/R_0)^2} = (1+\frac{x}{R_0})^{-2}$ apply Eq. (13-28) with $n = -2$ and $u = \frac{x}{R_0}$.
This gives $\frac{1}{(1+x/R_0)^2} \approx 1 - 2\frac{x}{R_0}$.

For $\frac{1}{(1-x/R_0)^2} = (1-\frac{x}{R_0})^{-2}$ apply Eq. (13-28) with $n = -2$ and $u = -\frac{x}{R_0}$.
This gives $\frac{1}{(1-x/R_0)^2} \approx 1 + 2\frac{x}{R_0}$.

Then $F_r \approx \frac{A}{R_0^2}\left[1 - \frac{2x}{R_0} - (1 + \frac{2x}{R_0})\right] = -\frac{4Ax}{R_0^3}$.

$F_r = -kx$ gives $k = \frac{4A}{R_0^3}$.

d) $f = \frac{1}{2\pi}\sqrt{\frac{k}{m}}$.
Using $k = \frac{4A}{R_0^3}$ from part (c) gives $f = \frac{1}{2\pi}\sqrt{\frac{4A}{mR_0^3}} = \frac{1}{\pi}\sqrt{\frac{A}{mR_0^3}}$

13-77

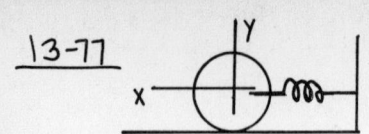

Let the origin of coordinates be at the center of the cylinders when they are at their equilibrium position.

Free-body diagram for the cylinders when they are displaced a distance x to the left:

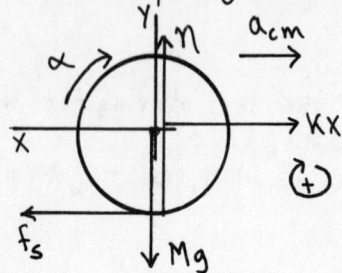

$\sum \tau = I_{cm}\alpha$
$f_s R = (\frac{1}{2}MR^2)\alpha$
$f_s = \frac{1}{2}MR\alpha$

But $R\alpha = a_{cm}$
$\Rightarrow f_s = \frac{1}{2}Ma_{cm}$

$\sum F_x = ma_x$
$f_s - kx = -Ma_{cm}$
$\frac{1}{2}Ma_{cm} - kx = -Ma_{cm}$
$kx = \frac{3}{2}Ma_{cm}$
$\left(\frac{2k}{3M}\right)x = a_{cm}$

Eq. (13-8): $a = -\omega^2 x$ (The minus sign says that x and a have opposite directions, as our diagram shows.)

Our result for a_{cm} is of this form, with $\omega^2 = \frac{2k}{3M} \Rightarrow \omega = \sqrt{\frac{2k}{3M}}$.

$T = \frac{2\pi}{\omega} = 2\pi\sqrt{\frac{3M}{2k}}$.

13-81

a) $T = 2\pi\sqrt{\frac{L}{g}} \Rightarrow L = g\left(\frac{T}{2\pi}\right)^2 = (9.80 \text{ m/s}^2)\left(\frac{4.00 \text{ s}}{2\pi}\right)^2 = \underline{3.97 \text{ m}}$

b) Use a uniform slender rod of mass M and length $L = 0.50$ m. Pivot the rod about an axis that is a distance d above the center of the rod. The rod will oscillate as a physical pendulum with period $T = 2\pi\sqrt{\frac{I}{Mgd}}$. Choose d so that $T = 3.00$ s.

$I = I_{cm} + Md^2 = \frac{1}{12}ML^2 + Md^2 = M\left(\frac{1}{12}L^2 + d^2\right)$

$T = 2\pi\sqrt{\frac{I}{Mgd}} = 2\pi\sqrt{\frac{M(\frac{1}{12}L^2 + d^2)}{Mgd}} = 2\pi\sqrt{\frac{\frac{1}{12}L^2 + d^2}{gd}}$

Note that $T \to \infty$ as $d \to 0$ (pivot at center of rod) and that if pivot is at top of rod then $d = \frac{L}{2}$ and $T = 2\pi\sqrt{\frac{(\frac{1}{12}+\frac{1}{4})L^2}{Lg/2}} = 2\pi\sqrt{\frac{L}{g}\frac{4}{6}} = 2\pi\sqrt{\frac{2L}{3g}}$.

$T = 2\pi\sqrt{\frac{2(0.50 \text{ m})}{3(9.80 \text{ m/s}^2)}} = 1.16$ s, which is less than the desired 4.00 s. Thus it is reasonable to expect that there is a value of d between 0 and $\frac{L}{2}$ for which $T = 4.00$ s.

$T = 2\pi\sqrt{\frac{\frac{1}{12}L^2 + d^2}{gd}}$; solve for d:

$gd\left(\frac{T}{2\pi}\right)^2 = \frac{1}{12}L^2 + d^2$

$d^2 - \left(\frac{T}{2\pi}\right)^2 gd + \frac{L^2}{12} = 0$

$d^2 - \left(\frac{4.00 \text{ s}}{2\pi}\right)^2 (9.80 \text{ m/s}^2)d + \frac{(0.50 \text{ m})^2}{12} = 0$

$d^2 - 3.9718\, d + 0.020833 = 0$

13-81 (cont)

The quadratic formula gives
$$d = \tfrac{1}{2}\left[3.9718 \pm \sqrt{(3.9718)^2 - 4(0.020833)}\right] \text{ m}$$
$d = (1.9859 \pm 1.9806)\text{ m} \Rightarrow d = 3.97\text{ m}$ or $d = 0.0053\text{ m}$

The maximum value d can have is $\tfrac{L}{2} = 0.25\text{ m}$, so the answer we want is
$$d = 0.0053\text{ m} = 0.53\text{ cm}.$$

Therefore, take a slender rod of length 0.50 m and pivot it about an axis that is 0.53 cm above its center.

13-83

a)

$\omega = \sqrt{\dfrac{mgd}{I}}$

$d = X$, the distance from the cg of the object (which is at its geometrical center) from the pivot

I is the moment of inertia about the axis of rotation through O. By the parallel axis theorem $I_0 = md^2 + I_{cm}$. $I_{cm} = \tfrac{1}{12}mL^2$ (Table 9-2), so $I_0 = mx^2 + \tfrac{1}{12}mL^2$.

$$\omega = \sqrt{\frac{mgx}{mx^2 + \tfrac{1}{12}mL^2}} = \sqrt{\frac{gx}{x^2 + L^2/12}}$$

b) Maximum ω as vary X occurs when $\dfrac{d\omega}{dx} = 0$.

$\dfrac{d\omega}{dx} = 0$ gives $\sqrt{g}\,\dfrac{d}{dx}\left(\dfrac{x^{1/2}}{(x^2 + L^2/12)^{1/2}}\right) = 0$

$$\frac{\tfrac{1}{2}x^{-1/2}}{(x^2 + L^2/12)^{1/2}} - \tfrac{1}{2}\frac{2x}{(x^2 + L^2/12)^{3/2}}(x^{1/2}) = 0$$

$$x^{-1/2} - \frac{2x^{3/2}}{x^2 + L^2/12} = 0$$

$x^2 + \dfrac{L^2}{12} = 2x^2 \Rightarrow X = \dfrac{L}{\sqrt{12}}$. Get maximum ω when the pivot is a distance $\dfrac{L}{\sqrt{12}}$ above the center of the rod.

c) To answer this question we need an expression for ω_{max}.

In $\omega = \sqrt{\dfrac{gx}{x^2 + L^2/12}}$ substitute $x = \dfrac{L}{\sqrt{12}}$.

$$\omega_{max} = \sqrt{\frac{g\,\tfrac{L}{\sqrt{12}}}{\tfrac{L^2}{12} + \tfrac{L^2}{12}}} = \sqrt{\frac{g^{1/2}(12)^{-1/4}}{(1/6)^{1/2}}} = \sqrt{\frac{g}{L}}(12)^{-1/4}(6)^{1/2} = \sqrt{\frac{g}{L}}(3)^{1/4}$$

$$\omega_{max}^2 = \frac{g}{L}\sqrt{3}$$

$$L = \frac{g\sqrt{3}}{\omega_{max}^2}$$

$\omega_{max} = 2\pi\text{ rad/s}$ gives $L = \dfrac{(9.80\text{ m/s}^2)\sqrt{3}}{(2\pi\text{ rad/s})^2} = \underline{0.430\text{ m}}$

CHAPTER 14

Exercises 3, 5, 11, 13, 17, 19, 21, 23, 27, 29, 31, 33, 35, 41, 43, 45

Problems 51, 53, 55, 59, 61, 69, 71, 73, 77, 79, 87, 89, 91, 95, 97

Exercises

14-3

$$\rho = \frac{m}{V} \Rightarrow m = \rho V$$

From Table 14-1, $\rho = 2.7 \times 10^3$ kg/m^3.
For a cylinder of length L and radius R, $V = (\pi R^2)L = \pi (1.27 \times 10^{-2} m)^2 (0.762 m)$
$$V = 3.861 \times 10^{-4} m^3$$
Then $m = \rho V = (2.7 \times 10^3 \text{ kg/m}^3)(3.861 \times 10^{-4} m^3) = 1.04$ kg, and
$w = mg = (1.04 \text{ kg})(9.80 \text{ m/s}^2) = \underline{10.2 \text{ N}}$ (about 2.3 lbs). A cart is __not__ needed.

14-5

a) gauge pressure = $p - p_0 = \rho g h$
From Table 14-1 the density of seawater is 1.03×10^3 kg/m^3, so
$p - p_0 = \rho g h = (1.03 \times 10^3 \text{ kg/m}^3)(9.80 \text{ m/s}^2)(600 \text{ m}) = \underline{6.06 \times 10^6 \text{ Pa}}$

b) The force on each side of the window is $F = pA$. Inside the pressure is p_0 and outside in the water the pressure is $p = p_0 + \rho g h$.

inside bell	outside bell
$F_1 = p_0 A \rightarrow$	$\leftarrow F_2 = (p_0 + \rho g h) A$

The net force is
$F_2 - F_1 = (p_0 + \rho g h)A - p_0 A = (\rho g h)A$
$= (6.06 \times 10^6 \text{ Pa}) \pi (7.5 \times 10^{-2} m)^2 = \underline{1.07 \times 10^5 \text{ N}}$.

14-11

The pressure at the bottom of the tank is
$$p = \frac{F_\perp}{A} = \frac{18.0 \times 10^3 \text{ N}}{0.0800 \text{ m}^2} = 2.25 \times 10^5 \text{ Pa}.$$
The density of the kerosene is $\rho = \frac{m}{V} = \frac{205 \text{ kg}}{0.250 \text{ m}^3} = 820$ kg/m^3.

$p = p_0 + \rho g h \Rightarrow h = \frac{p - p_0}{\rho g} = \frac{2.25 \times 10^5 \text{ Pa} - 1.01 \times 10^5 \text{ Pa}}{(820 \text{ kg/m}^3)(9.80 \text{ m/s}^2)} = \underline{2.99 \text{ m}}$

14-13

$p_a = 970$ millibar $= 9.70 \times 10^4$ Pa

a) Apply $p = p_0 + \rho g h$ to the right-hand tube. The top of this tube is open to the air so $p_0 = p_a$. The density of the liquid (mercury) is 13.6×10^3 kg/m^3.

194

14-13 (cont)

Thus $p = 9.70 \times 10^4 \text{ Pa} + (13.6 \times 10^3 \text{ kg/m}^3)(9.80 \text{ m/s}^2)(0.0700 \text{ m}) = \underline{1.06 \times 10^5 \text{ Pa}}$.

b) $p = p_0 + \rho g h = 9.70 \times 10^4 \text{ Pa} + (13.6 \times 10^3 \text{ kg/m}^3)(9.80 \text{ m/s}^2)(0.0300 \text{ m}) = \underline{1.01 \times 10^5 \text{ Pa}}$.

c) Since $y_2 - y_1 = 3.00$ cm the pressure at the mercury surface in the left-hand tube equals that calculated in part (b). Thus the absolute pressure of gas in the tank is 1.01×10^5 Pa.

d) $p - p_a = \rho g h = (13.6 \times 10^3 \text{ kg/m}^3)(9.80 \text{ m/s}^2)(0.300 \text{ m}) = \underline{4.00 \times 10^3 \text{ Pa}}$

14-17

Free-body diagram for the cube:

$\Sigma F_y = 0$
$T + B - mg = 0$
$T = mg - B$
$m = \rho_c V_c$, where ρ_c is the density of the cube and V_c is its volume
$B = \rho_{air} V_c g$

$\Rightarrow T = \rho_c V_c g - \rho_{air} V_c g = (\rho_c - \rho_{air}) V_c g = (2.3 \text{ kg/m}^3 - 1.2 \text{ kg/m}^3)(0.0508 \text{ m})^3 (9.80 \text{ m/s}^2)$
$T = \underline{1.4 \times 10^{-3} \text{ N}}$

14-19

The floating object is the slab of ice plus the woman; the buoyant force must support both. The volume of water displaced equals the volume V_{ice} of the ice.

$a=0$ $B = \rho_{water} V_{ice} g$

$m_{tot} g = (58.0 \text{ kg} + m_{ice}) g$

$\Sigma F_y = ma_y$
$B - m_{tot} g = 0$
$\rho_{water} V_{ice} g = (58.0 \text{ kg} + m_{ice}) g$

But $\rho = \frac{m}{V} \Rightarrow m_{ice} = \rho_{ice} V_{ice}$

$\Rightarrow \rho_{water} V_{ice} = 58.0 \text{ kg} + \rho_{ice} V_{ice}$

$V_{ice} = \dfrac{58.0 \text{ kg}}{\rho_{water} - \rho_{ice}} = \dfrac{58.0 \text{ kg}}{1000 \text{ kg/m}^3 - 920 \text{ kg/m}^3} = \underline{0.725 \text{ m}^3}$

14-21

a)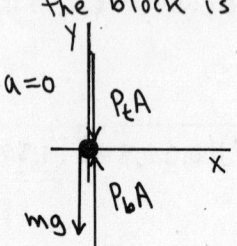

gauge pressure $= p - p_a = \rho g h$

The upper face is 2.00 cm below the top of the oil
$\Rightarrow p - p_a = (750 \text{ kg/m}^3)(9.80 \text{ m/s}^2)(0.0200 \text{ m}) = \underline{147 \text{ Pa}}$

b) The pressure at the interface is $p_{\text{interface}} = p_a + \rho_{\text{oil}} g (0.100 \text{ m})$. The lower face of the block is 2.00 cm below the interface, so the pressure there is $p = p_{\text{interface}} + \rho_{\text{water}} g (0.0200 \text{ m})$. Combining these two equations gives
$p - p_a = \rho_{\text{oil}} g (0.100 \text{ m}) + \rho_{\text{water}} g (0.0200 \text{ m}) = [(750 \text{ kg/m}^3)(0.100 \text{ m}) + (1000 \text{ kg/m}^3)(0.0200 \text{ m})](9.80 \text{ m/s}^2)$
$p - p_a = \underline{931 \text{ Pa}}$

c) Consider the forces on the block. The area of each face of the block is $A = (0.100 \text{ m})^2 = 0.0100 \text{ m}^2$. Let the absolute pressure at the top face be p_t and the pressure at the bottom face be p_b. Then the free-body diagram for the block is

$\Sigma F_y = m a_y$
$p_b A - p_t A - mg = 0$
$(p_b - p_t) A = mg$

Note that $(p_b - p_t) = (p_b - p_a) - (p_t - p_a) = 931 \text{ Pa} - 147 \text{ Pa} = 784 \text{ Pa}$; the difference in absolute pressures equals the difference in gauge pressures.

$m = \dfrac{(p_b - p_t) A}{g} = \dfrac{(784 \text{ Pa})(0.0100 \text{ m}^2)}{9.80 \text{ m/s}^2} = \underline{0.800 \text{ kg}}$

14-23

Use Eq. (14-12): $p - p_a = \dfrac{4 \gamma}{R} = \dfrac{4(25.0 \times 10^{-3} \text{ N/m})}{2.00 \times 10^{-2} \text{ m}} = \underline{5.00 \text{ Pa}}$

14-27

$w + T = 2 \gamma l$
$T = 2 \gamma l - w = 2 \gamma l - mg$, where $m = 1.50 \text{ g} = 1.50 \times 10^{-3} \text{ kg}$
From Table 14-2, $\gamma = 25.0 \times 10^{-3} \text{ N/m}$.
$T = 2(25.0 \times 10^{-3} \text{ N/m})(0.320 \text{ m}) - (1.50 \times 10^{-3} \text{ kg})(9.80 \text{ m/s}^2) = \underline{1.3 \times 10^{-3} \text{ N}}$

14-29

a) $vA = 1.20 \text{ m}^3/\text{s}$

$v = \dfrac{1.20 \text{ m}^3/\text{s}}{A} = \dfrac{1.20 \text{ m}^3/\text{s}}{\pi r^2} = \dfrac{1.20 \text{ m}^3/\text{s}}{\pi (0.200 \text{ m})^2} = \underline{9.55 \text{ m/s}}$

14-29 (cont)

b) $vA = 1.20 \text{ m}^3/\text{s}$
$v\pi r^2 = 1.20 \text{ m}^3/\text{s}$
$r = \sqrt{\dfrac{1.20 \text{ m}^3/\text{s}}{v\pi}} = \sqrt{\dfrac{1.20 \text{ m}^3/\text{s}}{(3.80 \text{ m/s})\pi}} = \underline{0.317 \text{ m}}$

14-31

Apply Bernoulli's equation with points 1 and 2 chosen as shown in the sketch. Let $y=0$ at the bottom of the tank so $y_1 = 12.0 \text{ m}$ and $y_2 = 0$.

$P_2 = P_a$
$P_1 - P_a = 5.00 \text{ atm} \left(\dfrac{1.013 \times 10^5 \text{ Pa}}{1 \text{ atm}}\right) = 5.065 \times 10^5 \text{ Pa}$

$P_1 + \rho g y_1 + \tfrac{1}{2}\rho v_1^2 = P_2 + \rho g y_2 + \tfrac{1}{2}\rho v_2^2$

$A_1 v_1 = A_2 v_2$, so $v_1 = \left(\dfrac{A_2}{A_1}\right) v_2$. But the cross-section area of the tank (A_1) is much larger than the cross-section area of the hole (A_2), so $v_1 \ll v_2$ and the $\tfrac{1}{2}\rho v_1^2$ term can be neglected.

$\Rightarrow \tfrac{1}{2}\rho v_2^2 = (P_1 - P_2) + \rho g y_1$

Use $P_2 = P_a$ and solve for v_2:

$v_2 = \sqrt{\dfrac{2(P_1 - P_a)}{\rho} + 2 g y_1} = \sqrt{\dfrac{2(5.065 \times 10^5 \text{ Pa})}{1030 \text{ kg/m}^3} + 2(9.80 \text{ m/s}^2)(12.0 \text{ m})} = \underline{34.9 \text{ m/s}}$

14-33

$P_1 + \rho g y_1 + \tfrac{1}{2}\rho v_1^2 = P_2 + \rho g y_2 + \tfrac{1}{2}\rho v_2^2$
horizontal $\Rightarrow y_1 = y_2 \Rightarrow P_1 + \tfrac{1}{2}\rho v_1^2 = P_2 + \tfrac{1}{2}\rho v_2^2$

If we solve for v_2 then we can use the discharge rate to calculate A_2.

Also, $v_1 A_1 = 5.00 \times 10^{-3} \text{ m}^3/\text{s} \Rightarrow v_1 = \dfrac{5.00 \times 10^{-3} \text{ m}^3/\text{s}}{1.00 \times 10^{-3} \text{ m}^2} = 5.00 \text{ m/s}$.

Then $\tfrac{1}{2}\rho v_2^2 = \tfrac{1}{2}\rho v_1^2 + (P_1 - P_2)$ gives

$v_2 = \sqrt{v_1^2 + \dfrac{2(P_1 - P_2)}{\rho}} = \sqrt{(5.00 \text{ m/s})^2 + \dfrac{2(1.60 \times 10^5 \text{ Pa} - 1.20 \times 10^5 \text{ Pa})}{1000 \text{ kg/m}^3}} = 10.25 \text{ m/s}$

Then $v_2 A_2 = 5.00 \times 10^{-3} \text{ m}^3/\text{s} \Rightarrow A_2 = \dfrac{5.00 \times 10^{-3} \text{ m}^3/\text{s}}{10.25 \text{ m/s}} = \underline{4.88 \times 10^{-4} \text{ m}^2}$

14-35

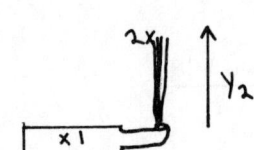

Apply Bernoulli's equation to points 1 and 2 as shown in the sketch. Point 1 is in the mains and point 2 is at the maximum height reached by the stream, so $v_2 = 0$.

14-35 (cont)

$$P_1 + \rho g y_1 + \tfrac{1}{2}\rho v_1^2 = P_2 + \rho g y_2 + \tfrac{1}{2}\rho v_2^2 \rightarrow 0$$

Let $y_1 = 0$, $y_2 = 18.0$ m. The mains have large diameter so $v_1 \approx 0$.

$\Rightarrow P_1 = P_2 + \rho g y_2$

But $P_2 = P_a$, so $P_1 - P_a = \rho g y_2 = (1000 \text{ kg/m}^3)(9.80 \text{ m/s}^2)(18.0 \text{ m}) = \underline{1.76 \times 10^5 \text{ Pa}}$.

14-41

$$\eta = 1.005 \text{ centipoise} = 1.005 \times 10^{-3} \text{ N·s/m}^2$$

a) The volume flow rate $\frac{dV}{dt}$ is given by Eq. (14-26):

$$\frac{dV}{dt} = \frac{\pi}{8}\left(\frac{R^4}{\eta}\right)\left(\frac{P_1-P_2}{L}\right)$$

The absolute pressure at the pump is P_1 and $P_2 = P_a$ is the pressure at the open end of the pipe, so $P_1 - P_2 = P_1 - P_a$, the gauge pressure at the pump.

$$\frac{dV}{dt} = \frac{\pi}{8} \frac{(4.00 \times 10^{-2} \text{ m})^4}{1.005 \times 10^{-3} \text{ N·s/m}^2}\left(\frac{1400 \text{ Pa}}{20.0 \text{ m}}\right) = \underline{0.0700 \text{ m}^3/\text{s}}$$

b) For the same volume flow rate $R^4 \Delta p$ must stay constant, where Δp is the gauge pressure maintained by the pump. Let $R_a = 4.00$ cm and $R_b = 2.00$ cm, so $\Delta P_a = 1400$ Pa and we are asked to calculate ΔP_b.

$$R_a^4 \Delta P_a = R_b^4 \Delta P_b \Rightarrow \Delta P_b = \Delta P_a \left(\frac{R_a}{R_b}\right)^4 = 1400 \text{ Pa}\left(\frac{4.00 \text{ cm}}{2.00 \text{ cm}}\right)^4 = \underline{2.24 \times 10^4 \text{ Pa}}$$

c) Same R, ΔP, and $L \Rightarrow \left(\frac{dV}{dt}\right)\eta = \frac{\pi}{8}R^4\frac{(P_1-P_2)}{L} = $ constant

$\Rightarrow \left(\frac{dV}{dt}\right)_a \eta_a = \left(\frac{dV}{dt}\right)_b \eta_b$ where a refers to 20°C and b to 60°C.

$$\left(\frac{dV}{dt}\right)_b = \left(\frac{dV}{dt}\right)_a \frac{\eta_a}{\eta_b} = 0.0700 \text{ m}^3/\text{s}\left(\frac{1.005 \text{ centipoise}}{0.469 \text{ centipoise}}\right) = \underline{0.150 \text{ m}^3/\text{s}}$$

14-43

The viscous drag force is given by Eq. (14-27): $F = 6\pi\eta r v$.

To compare this to the weight of the sphere, express the weight in terms of the density ρ and radius r of the sphere:

$$m = \rho V = \rho\left(\tfrac{4}{3}\pi r^3\right) \Rightarrow w = \rho g\left(\tfrac{4}{3}\pi r^3\right)$$

$$F = \tfrac{1}{4}w \Rightarrow 6\pi\eta r v = \tfrac{1}{4}\rho g\left(\tfrac{4}{3}\pi r^3\right)$$

$$v = \frac{\rho r^2 g}{18\eta}$$

From Table 14-1, $\rho_{gold} = 19.3 \times 10^3$ kg/m³.

Thus $v = \dfrac{(19.3 \times 10^3 \text{ kg/m}^3)(2.00 \times 10^{-3} \text{ m})^2(9.80 \text{ m/s}^2)}{18(0.986 \text{ N·s/m}^2)} = \underline{0.0426 \text{ m/s}}$

14-45

volume flow rate $\frac{dV}{dt} = \frac{\pi}{8} \left(\frac{R^4}{\eta}\right) \left(\frac{P_1 - P_2}{L}\right)$

a) $\frac{dV}{dt} \sim R^4$ so if triple the diameter (so triple the radius), $\frac{dV}{dt}$ increases by a factor of $(3)^4 = \underline{81}$

b) $\frac{dV}{dt} \sim \eta^{-1}$ so if triple η, $\frac{dV}{dt}$ changes by a factor of $\underline{\frac{1}{3}}$

c) $\frac{dV}{dt} \sim (P_1 - P_2)$ so if triple $(P_1 - P_2)$ then $\frac{dV}{dt}$ increases by a factor of $\underline{3}$.

d) $\frac{dV}{dt} \sim \left(\frac{P_1 - P_2}{L}\right)$ so if triple $\left(\frac{P_1 - P_2}{L}\right)$ (the pressure gradient), then $\frac{dV}{dt}$ increases by a factor of $\underline{3}$

e) $\frac{dV}{dt} \sim L^{-1}$ so if triple L, $\frac{dV}{dt}$ changes by a factor of $\underline{\frac{1}{3}}$

Problems

14-51

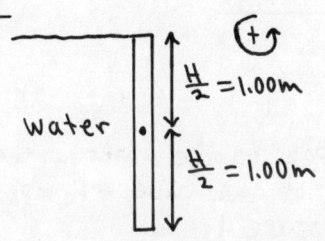

Let T_u be the torque due to the net force of the water on the upper half of the gate and T_ℓ be the torque due to the force on the lower half. With the indicated sign convention T_ℓ is positive and T_u is negative, so the net torque about the hinge is $T = T_\ell - T_u$. Let H be the height of the gate.

Upper-half of gate:
Calculate the torque due to the force on a narrow strip of height dy located a distance y below the top of the gate. Then integrate to get the total torque.

The net force on the strip is $dF = p(y) dA$, where $p(y) = \rho g y$ is the pressure at this depth and $dA = w \, dy$ with $w = 4.00$ m.
$\Rightarrow dF = \rho g y w \, dy$

The moment arm is $\left(\frac{H}{2} - y\right)$, so $d\tau = \rho g w \left(\frac{H}{2} - y\right) y \, dy$.

$T_u = \int_0^{H/2} d\tau = \rho g w \int_0^{H/2} \left(\frac{H}{2} - y\right) y \, dy = \rho g w \left(\frac{H}{4} y^2 - \frac{1}{3} y^3\right)\Big|_0^{H/2}$

$T_u = \rho g w \left(\frac{H^3}{16} - \frac{H^3}{24}\right) = \rho g w \left(\frac{H^3}{48}\right) = (1000 \text{ kg/m}^3)(9.80 \text{ m/s}^2)(4.00 \text{ m}) \frac{(2.00 \text{ m})^3}{48} = 6.533 \times 10^3$ N·m

Lower-half of gate:
The depth of the strip is $\left(\frac{H}{2} + y\right)$ so the force dF is
$dF = p(y) dA = \rho g \left(\frac{H}{2} + y\right) w \, dy$
The moment arm is y, so $d\tau = \rho g w \left(\frac{H}{2} + y\right) y \, dy$.

$T_\ell = \int_0^{H/2} d\tau = \rho g w \int_0^{H/2} \left(\frac{H}{2} + y\right) y \, dy = \rho g w \left(\frac{H}{4} y^2 + \frac{1}{3} y^3\right)\Big|_0^{H/2} = \rho g w \left(\frac{H^3}{16} + \frac{H^3}{24}\right)$

14-51 (cont)

$$\tau_\ell = \rho g w \frac{5H^3}{48} = (1000 \text{ kg/m}^3)(9.80 \text{ m/s}^2)(4.00 \text{ m}) \frac{5(2.00 \text{ m})^3}{48} = 3.267 \times 10^4 \text{ N·m}$$

Then $\tau = \tau_\ell - \tau_u = 3.267 \times 10^4 \text{ N·m} - 6.533 \times 10^3 \text{ N·m} = \underline{2.61 \times 10^4 \text{ N·m}}$.

14-53

$$p = p_0 + \rho g h$$

Use the data to calculate g:

$$g = \frac{p - p_0}{\rho h}$$

$$\rho = \frac{m}{V} = \frac{3.40 \text{ kg}}{0.250 \times 10^{-3} \text{ m}^3} = 1.36 \times 10^4 \text{ kg/m}^3$$

$$g = \frac{112.5 \times 10^3 \text{ Pa} - 102.5 \times 10^3 \text{ Pa}}{(1.36 \times 10^4 \text{ kg/m}^3)(1.50 \text{ m})} = 0.4902 \text{ m/s}^2$$

From Eq. (12-4) applied to the planet rather than to the earth,

$$g = \frac{G m_p}{R_p^2}$$

So $m_p = \frac{g R_p^2}{G} = \frac{(0.4902 \text{ m/s}^2)(3.00 \times 10^6 \text{ m})^2}{6.673 \times 10^{-11} \text{ N·m}^2/\text{kg}^2} = \underline{6.61 \times 10^{22} \text{ kg}}$

14-55

a) Apply $p = p_0 + \rho g h$ to the water in the left-hand arm of the tube. $p_0 = p_a$, so the gauge pressure is

$$p - p_a = \rho g h = (1000 \text{ kg/m}^3)(9.80 \text{ m/s}^2)(0.150 \text{ m})$$
$$p - p_a = \underline{1470 \text{ Pa}}$$

b) The pressure at point 1 equals the pressure at point 2.

$$p_1 = p_a + \rho_w g (0.150 \text{ m})$$
$$p_2 = p_a + \rho_{Hg} g (0.150 \text{ m} - h)$$

$$p_1 = p_2 \Rightarrow \rho_w g (0.150 \text{ m}) = \rho_{Hg} g (0.150 \text{ m} - h)$$

$$0.150 \text{ m} - h = \frac{\rho_w (0.150 \text{ m})}{\rho_{Hg}} = \frac{(1000 \text{ kg/m}^3)(0.150 \text{ m})}{13.6 \times 10^3 \text{ kg/m}^3} = 0.011 \text{ m}$$

$$h = 0.150 \text{ m} - 0.011 \text{ m} = 0.139 \text{ m} = \underline{13.9 \text{ cm}}$$

14-59

Free-body diagram for the barge + coal:

$$\Sigma F_y = m a_y$$
$$B - (m_{barge} + m_{coal}) g = 0$$
$$\rho_w V_{barge} g = (m_{barge} + m_{coal}) g$$
$$m_{coal} = \rho_w V_{barge} - m_{barge}$$

14-59 (cont)

$V_{barge} = (22m)(12m)(40m) = 1.056 \times 10^4 \, m^3$

The mass of the barge is $m_{barge} = \rho_s V_s$, where s refers to steel.

From Table 14-1, $\rho_s = 7800 \, kg/m^3$.

The volume V_s is 0.050m times the total area of the five pieces of steel that make up the barge:

$V_s = (0.050m)[2(22m)(12m) + 2(40m)(12m) + (22m)(40m)] = 118.4 \, m^3$.

Therefore, $m_{barge} = \rho_s V_s = (7800 \, kg/m^3)(118.4 \, m^3) = 9.235 \times 10^5 \, kg$.

Then $m_{coal} = \rho_w V_{barge} - m_{barge} = (1000 \, kg/m^3)(1.056 \times 10^4 \, m^3) - 9.235 \times 10^5 \, kg = \underline{9.64 \times 10^6 \, kg}$

The volume of this mass of coal is $V_{coal} = \frac{m_{coal}}{\rho_{coal}} = \frac{9.64 \times 10^6 \, kg}{1500 \, kg/m^3} = 6420 \, m^3$; this is less than V_{barge} so will fit into the barge.

14-61

a) Free-body diagram for the dirigible:

(The lift corresponds to a mass $m_{lift} = \frac{90,000 \, N}{9.80 \, m/s^2} = 9184 \, kg$. The mass m_{tot} is 9184 kg plus the mass m_{gas} of the gas that fills the dirigible. B is the buoyant force exerted by the air.)

$B = \rho_{air} V g$

$a = 0$

$m_{tot} g = (9184 \, kg + m_{gas}) g$

$\sum F_y = m a_y$

$B - m_{tot} g = 0$

$\rho_{air} V g = (9184 \, kg + m_{gas}) g$

Write m_{gas} in terms of V:

$m_{gas} = \rho_{gas} V$

$\Rightarrow \rho_{air} V = 9184 \, kg + \rho_{gas} V$

$V = \frac{9184 \, kg}{\rho_{air} - \rho_{gas}} = \frac{9184 \, kg}{1.20 \, kg/m^3 - 0.0899 \, kg/m^3} = \underline{8.27 \times 10^3 \, m^3}$

b) Let m_{lift} be the mass that could be lifted.

From part (a), $m_{lift} = (\rho_{air} - \rho_{gas}) V = (1.20 \, kg/m^3 - 0.166 \, kg/m^3)(8.27 \times 10^3 \, m^3) = 8551 \, kg$.

The lift force is $m_{lift} \, g = (8551 \, kg)(9.80 \, m/s^2) = \underline{83,800 \, N}$.

Hydrogen is not used because it is highly explosive in air.

14-69

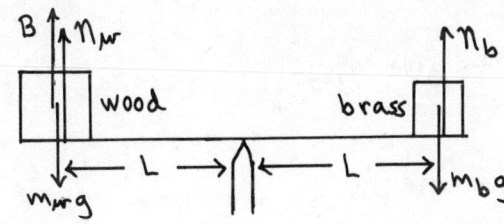

The buoyant force on the brass is neglected, but we include the buoyant force B on the block of wood. n_w and n_b are the normal forces exerted by the balance arm on which the objects sit.

14-69 (cont)

Free-body diagram for the balance arm:

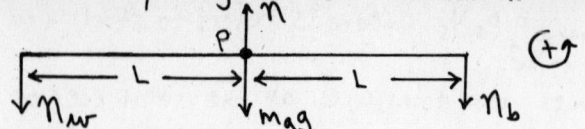

$\Sigma \tau_P = 0$
$n_w L - n_b L = 0$
$n_w = n_b$

Free-body diagram for the brass mass:

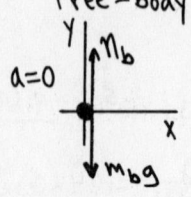

$\Sigma F_y = ma_y$
$n_b - m_b g = 0$
$n_b = m_b g$

Free-body diagram for the block of wood:

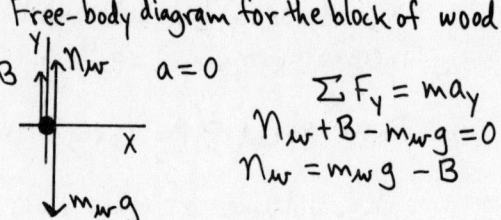

$\Sigma F_y = ma_y$
$n_w + B - m_w g = 0$
$n_w = m_w g - B$

But $n_b = n_w \Rightarrow m_b g = m_w g - B$

And $B = \rho_{air} V_w g = \rho_{air} \left(\frac{m_w}{\rho_w}\right) g \Rightarrow m_b g = m_w g - \rho_{air} \left(\frac{m_w}{\rho_w}\right) g$

$m_w = \frac{m_b}{1 - \rho_{air}/\rho_w} = \frac{0.0800 \text{ kg}}{1 - (1.2 \text{ kg/m}^3 / 150 \text{ kg/m}^3)} = \underline{0.0806 \text{ kg}}$

14-71

Free-body diagram for the piece of alloy:

$\Sigma F_y = ma_y$
$B + T - m_{tot} g = 0$
$B = m_{tot} g - T = 45.0 \text{ N} - 34.0 \text{ N} = 11.0 \text{ N}$

Also, $m_{tot} g = 45.0 \text{ N} \Rightarrow m_{tot} = \frac{45.0 \text{ N}}{9.80 \text{ m/s}^2} = 4.59 \text{ kg}$

We can use the known value of the buoyant force to calculate the volume of the object:

$B = \rho_w V_{obj} g = 11.0 \text{ N}$

$V_{obj} = \frac{11.0 \text{ N}}{\rho_w g} = \frac{11.0 \text{ N}}{(1000 \text{ kg/m}^3)(9.80 \text{ m/s}^2)} = 1.122 \times 10^{-3} \text{ m}^3$

We know two things:

(1) The mass m_g of the gold plus the mass m_a of the aluminum must add to m_{tot}:
$m_g + m_a = m_{tot}$
We write this in terms of the volumes V_g and V_a of the gold and aluminum
$\Rightarrow \boxed{\rho_g V_g + \rho_a V_a = m_{tot}}$

(2) The volumes V_a and V_g must add to give V_{obj}:
$V_a + V_g = V_{obj} \Rightarrow \boxed{V_a = V_{obj} - V_g}$

Use this in the first equation to eliminate V_a
$\Rightarrow \rho_g V_g + \rho_a (V_{obj} - V_g) = m_{tot}$

$V_g = \frac{m_{tot} - \rho_a V_{obj}}{\rho_g - \rho_a} = \frac{4.59 \text{ kg} - (2.7 \times 10^3 \text{ kg/m}^3)(1.122 \times 10^{-3} \text{ m}^3)}{(19.3 \times 10^3 - 2.7 \times 10^3) \text{ kg/m}^3} = 9.40 \times 10^{-5} \text{ m}^3$

Then $w_g = m_g g = \rho_g V_g g = (19.3 \times 10^3 \text{ kg/m}^3)(9.40 \times 10^{-5} \text{ m}^3)(9.80 \text{ m/s}^2) = \underline{17.8 \text{ N}}$

14-73

a) Free-body diagram for the crown:

$\Sigma F_y = ma_y$
$T + B - w = 0$

$T = fw$
$B = \rho_w V_c g$, where ρ_w = density of water, V_c = volume of crown
Then $fw + \rho_w V_c g - w = 0$
$(1-f)w = \rho_w V_c g$

Use $w = \rho_c V_c g$, where ρ_c = density of crown
$(1-f)\rho_c V_c g = \rho_w V_c g$

$\frac{\rho_c}{\rho_w} = \frac{1}{1-f}$, as was to be proven

$f \to 0$ gives $\frac{\rho_c}{\rho_w} = 1$ and $T = 0$. These values are consistent. If the density of the crown equals the density of the water the crown just floats, fully submerged, and the tension should be zero.

When $f \to 1$, $\rho_c \gg \rho_w$ and $T = w$. If $\rho_c \gg \rho_w$ then B is negligible relative to the weight w of the crown and T should equal w.

b) "apparent weight" equals T in rope when the crown is immersed in water
$T = fw$, so need to compute f
$\rho_c = 19.3 \times 10^3$ kg/m^3
$\rho_w = 1.00 \times 10^3$ kg/m^3

$\frac{\rho_c}{\rho_w} = \frac{1}{1-f}$ gives $\frac{19.3 \times 10^3 \text{ kg/m}^3}{1.00 \times 10^3 \text{ kg/m}^3} = \frac{1}{1-f}$

$19.3 = \frac{1}{1-f}$
$1 - f = \frac{1}{19.3} = 0.05181$
$f = 0.9482$

Then $T = fw = (0.9482)(11.6 \text{ N}) = \underline{11.0 \text{ N}}$

c) Now the density of the crown is very nearly the density of lead,
$\rho_c = 11.3 \times 10^3$ kg/m^3.

$\frac{\rho_c}{\rho_w} = \frac{1}{1-f}$ gives $\frac{11.3 \times 10^3 \text{ kg/m}^3}{1.00 \times 10^3 \text{ kg/m}^3} = \frac{1}{1-f}$

$11.3 = \frac{1}{1-f}$
$1 - f = 0.08850$
$f = 0.9115$

Then $T = fw = (0.9115)(11.6 \text{ N}) = \underline{10.6 \text{ N}}$

14-77

In both cases the total buoyant force must equal the weight of the barge plus the weight of the anchor. Thus the total amount of water displaced must be the same when the anchor is in the boat as when it is

14-77 (cont)

over the side. When the anchor is in the water the barge displaces less water, less by the amount the anchor displaces ⇒ the barge **rises** in the water.

The volume of the anchor is $V_{anchor} = \frac{M}{\rho} = \frac{25.0 \text{ kg}}{7860 \text{ kg/m}^3} = 3.181 \times 10^{-3} \text{ m}^3$.

The barge rises in the water a vertical distance h given by $hA = 3.181 \times 10^{-3} \text{ m}^3$, where A is the area of the bottom of the barge.

$h = \frac{3.181 \times 10^{-3} \text{ m}^3}{8.00 \text{ m}^2} = \underline{3.98 \times 10^{-4} \text{ m}}$

14-79

a) Free-body diagram for the brass block:

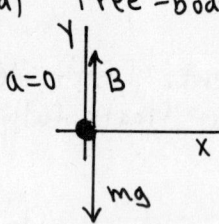

$\Sigma F_y = ma_y$
$B - mg = 0$
$\rho_{Hg} V_{sub} g = \rho_{brass} V_{tot} g$, where V_{sub} is the volume of the block that is below the surface of the mercury and V_{tot} is the total volume of the brass block.

The fraction submerged is $\frac{V_{sub}}{V_{tot}} = \frac{\rho_{brass}}{\rho_{Hg}} = \frac{8.6 \times 10^3 \text{ kg/m}^3}{13.6 \times 10^3 \text{ kg/m}^3} = 0.632$.

The fraction above the mercury surface is $1 - 0.632 = \underline{0.37}$.

b) Let the depth of the water layer be d. Calculate the upward buoyant force B by calculating the gauge pressure $p - p_a$ at the lower face of the block:

$p - p_a = \rho_w g d + \rho_{Hg}(L-d) g$

Then $B = (p - p_a) L^2 = \rho_w g d L^2 + \rho_{Hg}(L-d) g L^2$.

We can write the weight of the block as $mg = \rho_{brass} V g = \rho_{brass} L^3 g$.

Then $B - mg = 0 \Rightarrow \rho_w g d L^2 + \rho_{Hg}(L-d) g L^2 = \rho_{brass} L^3 g$

$\rho_w d + \rho_{Hg}(L-d) = \rho_{brass} L$

$d = \left(\frac{\rho_{Hg} - \rho_{brass}}{\rho_{Hg} - \rho_w}\right) L = \left(\frac{13.6 \times 10^3 \text{ kg/m}^3 - 8.6 \times 10^3 \text{ kg/m}^3}{13.6 \times 10^3 \text{ kg/m}^3 - 1.0 \times 10^3 \text{ kg/m}^3}\right) L = \underline{0.40 L}$

14-87

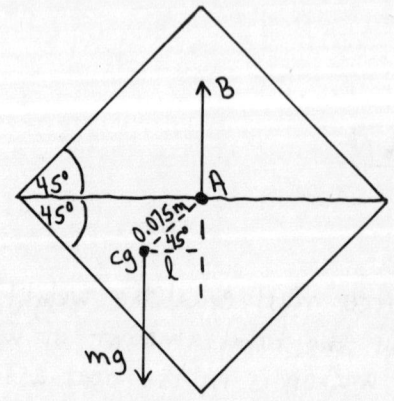

The resultant buoyant force acts at the geometrical center of the submerged portion of the object. The weight of the object acts at the center of gravity of the object. These two points are displaced from each other as in Fig. 14-14-41 b, and this gives to a restoring torque about point A

14-87 (cont)

For an axis at point A
$$\tau_B = 0$$
$$\tau_{mg} = mg\ell = mg(0.075m)\cos 45° = (0.053m)mg$$

The block is floating $\Rightarrow B = mg$. Calculate B in order to find the weight mg of the block. Half of the volume of the block is submerged
$\Rightarrow V_{sub} = \frac{1}{2}(0.30m)^3 = 0.0135 m^3$
Then $B = \rho_w V_{sub} g = (1000 kg/m^3)(0.0135 m^3)(9.80 m/s^2) = 132.3 N$.

Therefore, $\sum \tau_A = \tau_{mg} = (0.053m)(132.3N) = \underline{7.01 N\cdot m}$.

14-89

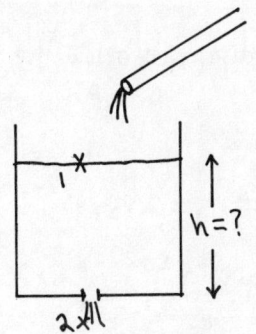

The water level in the vessel will rise until the volume flow rate into the vessel ($1.30 \times 10^{-4} m^3/s$) equals the volume flow rate out the hole in the bottom.

Let points 1 and 2 be chosen as in the sketch.
Bernoulli's equation $\Rightarrow P_1 + \rho g y_1 + \frac{1}{2}\rho v_1^2 = P_2 + \rho g y_2 + \frac{1}{2}\rho v_2^2$

Volume flow rate out hole equals volume flow rate from tube $\Rightarrow v_2 A_2 = 1.30\times 10^{-4} m^3/s \Rightarrow v_2 = \frac{1.30\times 10^{-4} m^3/s}{1.00\times 10^{-4} m^2} = 1.30 m/s$

$A_1 \gg A_2$ and $v_1 A_1 = v_2 A_2 \Rightarrow \frac{1}{2}\rho v_1^2 \ll \frac{1}{2}\rho v_2^2$; neglect the $\frac{1}{2}\rho v_1^2$ term. Measure y from the bottom of the bucket, so $y_2 = 0$ and $y_1 = h$.
$P_1 = P_2 = P_a$ (air pressure)

Then $P_a + \rho g h = P_a + \frac{1}{2}\rho v_2^2$

$h = \frac{v_2^2}{2g} = \frac{(1.30 m/s)^2}{2(9.80 m/s^2)} = \underline{0.086 m}$

14-91

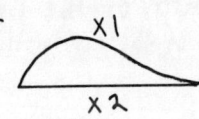

Apply Bernoulli's equation to points 1 and 2, where point 1 is just above the wing and point 2 is just below the wing:
$P_1 + \rho g y_1 + \frac{1}{2}\rho v_1^2 = P_2 + \rho g y_2 + \frac{1}{2}\rho v_2^2$

A "lift" of $2000 N/m^2 \Rightarrow P_2 - P_1 = 2000 Pa$
Solve for $v_1 \Rightarrow v_1 = \sqrt{\frac{2(P_2-P_1)}{\rho} + v_2^2 - 2g(y_1-y_2)}$

Note that $\frac{2(P_2-P_1)}{\rho} = \frac{2(2000 Pa)}{1.20 kg/m^3} = 3333 m^2/s^2$. We aren't given a value for $y_1 - y_2$, but it must be 1m or so. For $y_1 - y_2 = 1m$, $2g(y_1-y_2) = 19.6 m^2/s^2$.
So $2g(y_1-y_2) \ll \frac{2(P_2-P_1)}{\rho}$ and can be neglected.

Thus $v_1 = \sqrt{\frac{2(P_2-P_1)}{\rho} + v_2^2} = \sqrt{3333 m^2/s^2 + (140 m/s)^2} = \underline{151 m/s}$.

14-95

a) Let point 1 be at the end of the pipe and let point 2 be in the stream of liquid at a distance y_2 below the end of the tube.

Consider the free fall of the liquid. Take $+y$ to be downward.
Free fall $\Rightarrow a = g$.
$$v_2^2 = v_1^2 + 2a(y-y_0) \Rightarrow v_2^2 = v_1^2 + 2gy_2 \Rightarrow v_2 = \sqrt{v_1^2 + 2gy_2}$$

Equation of continuity $\Rightarrow v_1 A_1 = v_2 A_2$
$A = \pi r^2 \Rightarrow v_1 \pi r_1^2 = v_2 \pi r_2^2 \Rightarrow v_2 = v_1 \left(\frac{r_1}{r_2}\right)^2$

Use this in the above to eliminate $v_2 \Rightarrow v_1 \frac{r_1^2}{r_2^2} = \sqrt{v_1^2 + 2gy_2}$

$$r_2 = r_1 \frac{\sqrt{v_1}}{(v_1^2 + 2gy_2)^{1/4}}$$

Note that this equation says that r_2 decreases with distance below the end of the pipe.

b) $v_1 = 1.60$ m/s

We want the value of y_2 that gives $r_2 = \frac{1}{2} r_1 \Rightarrow r_1 = 2 r_2$.

The result obtained in part (a) says $r_2^4 (v_1^2 + 2gy_2) = r_1^4 v_1^2$
$$2gy_2 = \left(\frac{r_1}{r_2}\right)^4 v_1^2 - v_1^2$$

$$y_2 = \frac{[(r_1/r_2)^4 - 1] v_1^2}{2g} = \frac{(16-1)(1.60 \text{ m/s})^2}{2(9.80 \text{ m/s}^2)} = \underline{1.96 \text{ m}}$$

14-97

Free-body diagram for the bubble:

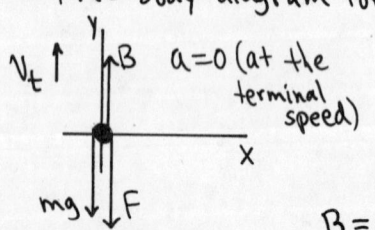

$a = 0$ (at the terminal speed)

Note that the viscous drag force F is downward, since the bubble is traveling upward.

$\Sigma F_y = ma_y$
$B - mg - F = 0$
$B = mg + F$

$B = \rho' V g = \frac{4}{3} \pi r^3 \rho' g$, where ρ' is the density of the liquid
$mg = \rho V g = \frac{4}{3} \pi r^3 \rho g$, where ρ is the density of the air in the bubble
$F = 6\pi \eta r v_t$

Thus $B = mg + F \Rightarrow \frac{4}{3} \pi r^3 \rho' g = \frac{4}{3} \pi r^3 \rho g + 6\pi \eta r v_t$
$6 \eta v_t = \frac{4}{3} r^2 g (\rho' - \rho)$

$v_t = \frac{2 r^2 g}{9 \eta} (\rho' - \rho) = \frac{2(1.50 \times 10^{-3} \text{ m})^2 (9.80 \text{ m/s}^2)}{9(0.150 \text{ N·s/m}^2)} (800 \text{ kg/m}^3 - 1.20 \text{ kg/m}^3) = 2.61 \times 10^{-2}$ m/s

$v_t = \underline{2.61 \text{ cm/s}}$

(Note that the precise value of $\rho = \rho_{air}$ that is used is unimportant since $\rho' \gg \rho$.)

b) $v_t = \frac{2 r^2 g}{9 \eta} (\rho' - \rho) = \frac{2(1.50 \times 10^{-3} \text{ m})^2 (9.80 \text{ m/s}^2)}{9(1.005 \times 10^{-3} \text{ N·s/m}^2)} (1000 \text{ kg/m}^3 - 1.20 \text{ kg/m}^3) = \underline{4.87 \text{ m/s}}$

CHAPTER 15

Exercises 5, 11, 17, 19, 23, 25, 27, 31, 33, 35, 37, 39, 41, 45, 47, 49, 53, 57, 59, 61, 63, 67

Problems 71, 73, 75, 79, 81, 83, 87, 89, 93, 97, 99, 101, 103

Exercises

15-5

$$T_F = \tfrac{9}{5} T_C + 32°$$
$$T_F = T_C \Rightarrow T = \tfrac{9}{5} T + 32°$$
$$\tfrac{4}{5} T = -32° \Rightarrow T = -40°$$
$-40°C = -40°F$; must be winter

15-11

a) pressure varies linearly with temperature $\Rightarrow P_2 = P_1 + \gamma (T_2 - T_1)$

$$\gamma = \frac{P_2 - P_1}{T_2 - T_1} = \frac{7.70 \times 10^4 \, Pa - 5.60 \times 10^4 \, Pa}{100°C - 0.01°C} = 210.0 \, Pa/C°$$

Apply $p = P_1 + \gamma (T - T_1)$ with $T_1 = 0.01°C$ and $p = 0$ to solve for T.
$$0 = P_1 + \gamma (T - T_1)$$
$$T = T_1 - \frac{P_1}{\gamma} = 0.01°C - \frac{5.60 \times 10^4 \, Pa}{210.0 \, Pa/C°} = \underline{-267°C}$$

b) Eq. (15-4) says $\frac{T_2}{T_1} = \frac{P_2}{P_1}$, where T is in kelvins

$$P_2 = P_1 \left(\frac{T_2}{T_1}\right) = 5.60 \times 10^4 \, Pa \left(\frac{100 + 273.15}{0.01 + 273.15}\right) = \underline{7.65 \times 10^4 \, Pa} \; ; \text{ this differs from the}$$
7.70×10^5 Pa that was measured so Eq. (15-4) is not precisely obeyed.

15-17

Let $L_0 = 30.000$ cm ; $T_0 = 20.0°C$
$\Delta T = 45.0°C - 20.0°C = 25.0 \, C°$ gives $\Delta L = 0.019$ cm
Thus $\Delta L = \alpha L_0 \Delta T \Rightarrow \alpha = \frac{\Delta L}{L_0 \Delta T} = \frac{0.019 \, cm}{(30.000 \, cm)(25.0 \, C°)} = \underline{2.5 \times 10^{-5} \, (C°)^{-1}}$

15-19

The diameter of the hole undergoes linear expansion just as does a length of brass. α_{brass} is given in Table 15-1.
$\Delta L = \alpha L_0 \Delta T = (2.0 \times 10^{-5} \, (C°)^{-1})(1.600 \, cm)(175°C - 25°C) = 0.0048$ cm
$L = L_0 + \Delta L = 1.600 \, cm + 0.0048 \, cm = \underline{1.605 \, cm}$

15-23

Consider ΔV for the ethanol. From Table 15-2, β for ethanol is 75×10^{-5} K^{-1}. $\Delta T = 10.0°C - 19.0°C = -9.0 C° = -9.0 K$.
Then $\Delta V = \beta V_0 \Delta T = (75 \times 10^{-5} K^{-1})(2200 L)(-9.0 K) = -15 L$.
The volume of the air space will be $\underline{15 L = 0.015 m^3}$.

15-25

a) $\Delta L = L_0 \alpha \Delta T$
$\Rightarrow \alpha = \dfrac{\Delta L}{L_0 \Delta T} = \dfrac{1.7 \times 10^{-2} m}{(3.00 m)(420°C - 20°C)} = \underline{1.4 \times 10^{-5} (C°)^{-1}}$

b) Eq. (15-12): stress $\dfrac{F}{A} = -Y \alpha \Delta T$
$\Delta T = 20°C - 420°C = -400 C°$ (ΔT always means final temperature minus initial temperature)
$\dfrac{F}{A} = -(2.0 \times 10^{11} Pa)(1.4 \times 10^{-5} (C°)^{-1})(-400 C°) = \underline{+1.1 \times 10^9 Pa}$
$\dfrac{F}{A}$ is positive $\Rightarrow$ the stress is a tensile (stretching) stress.

15-27

The change in mechanical energy equals the decrease in gravitational potential energy, $\Delta U = -mgh$. $|\Delta U| = mgh$
$Q = |\Delta U| \Rightarrow mc \Delta T = mgh$
$\Delta T = \dfrac{gh}{c} = \dfrac{(9.80 m/s^2)(443 m)}{4190 J/kg \cdot K} = 1.04 K = \underline{1.04 C°}$

Note that the answer is independent of the mass of the object. Note also the small change in temperature that corresponds to this large change in height!

15-31

a) $Q = mc \Delta T$
$m = \frac{1}{2}(1.3 \times 10^{-3} kg) = 0.65 \times 10^{-3} kg$
$Q = (0.65 \times 10^{-3} kg)(1020 J/kg \cdot K)(37°C - (-40°C)) = \underline{51 J}$

b) 20 breaths/min $\left(\dfrac{60 min}{1 h}\right) = 1200$ breaths/h
So $Q = (1200)(51 J) = \underline{6.1 \times 10^4 J}$

15-33

kettle
$Q = mc \Delta T$, $c = 910$ J/kg·K (from Table 15-3)
$Q = (1.50 kg)(910 J/kg \cdot K)(90.0°C - 20.0°C) = 9.555 \times 10^4 J$

water
$Q = mc \Delta T$, $c = 4190$ J/kg·K (from Table 15-3)
$Q = (2.50 kg)(4190 J/kg \cdot K)(90.0°C - 20.0°C) = 7.333 \times 10^5 J$

15-33 (cont)

Total $Q = 9.555 \times 10^4 \text{ J} + 7.333 \times 10^5 \text{ J} = \underline{8.29 \times 10^5 \text{ J}}$

15-35

a) $P = \frac{Q}{t}$, so the total heat transferred to the liquid is
$$Q = Pt = (65.0 \text{ W})(120 \text{ s}) = 7800 \text{ J}$$

$Q = mc\Delta T \Rightarrow c = \frac{Q}{m\Delta T} = \frac{7800 \text{ J}}{0.780 \text{ kg}(21.32°C - 18.55°C)} = \underline{3.61 \times 10^3 \text{ J/kg·K}}$

b) Then the actual Q transferred to the liquid is less than 7800 J so the actual c is less than our calculated value; our result in part (a) is an overestimate.

15-37

Heat must be added to do the following:
ice at $-15.0°C \rightarrow$ ice at $0°C$
$$Q_{ice} = mc_{ice}\Delta T = (8.00 \times 10^{-3} \text{ kg})(2000 \text{ J/kg·K})(0°C - (-15.0°C)) = 240 \text{ J}$$
phase transition ice ($0°C$) $\rightarrow$ liquid water ($0°C$) (melting)
$$Q_{melt} = +mL_f = (8.00 \times 10^{-3} \text{ kg})(334 \times 10^3 \text{ J/kg}) = 2.672 \times 10^3 \text{ J}$$
water at $0°C$ (from melted ice) $\rightarrow$ water at $100°C$
$$Q_{water} = mc_{water}\Delta T = (8.00 \times 10^{-3} \text{ kg})(4190 \text{ J/kg·K})(100°C - 0°C) = 3.352 \times 10^3 \text{ J}$$
phase transition water ($100°C$) $\rightarrow$ steam ($100°C$) (boiling)
$$Q_{boil} = +mL_v = (8.00 \times 10^{-3} \text{ kg})(2256 \times 10^3 \text{ J/kg}) = 1.805 \times 10^4 \text{ J}$$

The total Q is $Q = 240 \text{ J} + 2.672 \times 10^3 \text{ J} + 3.352 \times 10^3 \text{ J} + 1.805 \times 10^4 \text{ J} = \underline{2.43 \times 10^4 \text{ J}}$

$2.43 \times 10^4 \text{ J} \left(\frac{1 \text{ cal}}{4.186 \text{ J}}\right) = \underline{5.81 \times 10^3 \text{ cal}}$

$2.43 \times 10^4 \text{ J} \left(\frac{1 \text{ Btu}}{1055 \text{ J}}\right) = \underline{23.0 \text{ Btu}}$

15-39

The heat that must be added to a lead bullet of mass m to melt it is
$$Q = mc\Delta T + mL_F$$
($mc\Delta T$ is the heat required to raise the temperature from $25°C$ to the melting point of $327.3°C$; mL_F is the heat required to make the solid $\rightarrow$ liquid phase change)

The kinetic energy of the bullet if its speed is v is $K = \frac{1}{2}mv^2$.

Then $K = Q \Rightarrow \frac{1}{2}mv^2 = mc\Delta T + mL_F$
$$v = \sqrt{2(c\Delta T + L_F)} = \sqrt{2[(130 \text{ J/kg·K})(327.3°C - 25°C) + 24.5 \times 10^3 \text{ J/kg}]} = \underline{357 \text{ m/s}}$$

15-41

one-ton air conditioner $\Rightarrow$ 1 ton (2000 lb) of ice can be frozen from water at $0°C$ in 24 h.

Find the mass m that corresponds to 2000 lb (weight) of water:

$m = (2000 \text{ lb})\left(\frac{1 \text{ kg}}{2.205 \text{ lb}}\right) = 907$ kg (The kg $\leftrightarrow$ lb equivalence from Appendix E has been used.)

The heat that must be removed from the water to freeze it is
$Q = -mL_F = -(907 \text{ kg})(334 \times 10^3 \text{ J/kg}) = -3.03 \times 10^8$ J.

The power required if this is to be done in 1 hour is

$P = \frac{|Q|}{t} = \frac{3.03 \times 10^8 \text{ J}}{(24 \text{ h})\left(\frac{3600 \text{ s}}{1 \text{ h}}\right)} = \underline{3510 \text{ W}}$

or $P = 3510 \text{ W}\left(\frac{1 \text{ Btu/h}}{0.293 \text{ W}}\right) = \underline{1.20 \times 10^4 \text{ Btu/h}}$

15-45

$Q_{system} = 0$

Calculate Q for each component of the system:

copper pot
$Q_{can} = mc\Delta T = (0.500 \text{ kg})(390 \text{ J/kg·K})(T - 20.0°C) = (195 \text{ J/K})T - 3900 \text{ J}$

water
$Q_{water} = mc\Delta T = (0.170 \text{ kg})(4190 \text{ J/kg·K})(T - 20.0°C) = (712.3 \text{ J/K})T - 1.425 \times 10^4 \text{ J}$

iron
$Q_{iron} = mc\Delta T = (0.200 \text{ kg})(470 \text{ J/kg·K})(T - 75.0°C) = (94.0 \text{ J/K})T - 7050 \text{ J}$

$Q_{system} = 0 \Rightarrow Q_{can} + Q_{water} + Q_{iron} = 0$
$(195 \text{ J/K})T - 3900 \text{ J} + (712.3 \text{ J/K})T - 1.425 \times 10^4 \text{ J} + (94.0 \text{ J/K})T - 7050 \text{ J} = 0$
$(1001 \text{ J/K})T = 2.52 \times 10^4 \text{ J}$
$T = \frac{2.52 \times 10^4 \text{ J}}{1001 \text{ J/K}} = \underline{25.2°C}$

15-47

$Q_{system} = 0$

large block of ice $\Rightarrow$ ice is left $\Rightarrow T_2 = 0°C$

Calculate Q for each component of the system:

ingot
$Q_{ingot} = mc\Delta T = (5.00 \text{ kg})(234 \text{ J/kg·K})(0°C - 850°C) = -9.945 \times 10^5 \text{ J}$

ice
$Q_{ice} = +mL_f$, where m is the mass of ice that changes phase (melts)

$Q_{system} = 0 \Rightarrow Q_{ingot} + Q_{ice} = 0$
$-9.945 \times 10^5 \text{ J} + m(334 \times 10^3 \text{ J/kg}) = 0$
$m = \frac{9.945 \times 10^5 \text{ J}}{334 \times 10^3 \text{ J/kg}} = \underline{2.98 \text{ kg}}$

15-49

$Q_{system} = 0$

Calculate Q for each component of the system:
(Beaker has small mass $\Rightarrow Q = mc\Delta T$ for beaker can be neglected.)

0.250 kg of water (cools from 75.0°C to 40.0°C)
$Q_{water} = mc\Delta T = (0.250\,kg)(4190\,J/kg\cdot K)(40.0°C - 75.0°C) = -3.666\times 10^4\,J$

ice (warms to 0°C; melts; water from melted ice warms to 40.0°C)
$Q_{ice} = mc_{ice}\Delta T + mL_f + mc_{water}\Delta T$
$Q_{ice} = m[(2000\,J/kg\cdot K)(0°C - (-20.0°C)) + 334\times 10^3\,J/kg + (4190\,J/kg\cdot K)(40.0°C - 0°C)]$
$Q_{ice} = (5.416\times 10^5\,J/kg)\,m$

$Q_{system} = 0 \Rightarrow Q_{water} + Q_{ice} = 0$
$-3.666\times 10^4\,J + (5.416\times 10^5\,J/kg)\,m = 0$

$m = \dfrac{3.666\times 10^4\,J}{5.416\times 10^5\,J/kg} = \underline{0.0677\,kg}$

15-53

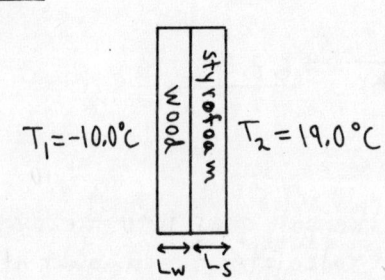

$T_1 = -10.0°C \qquad T_2 = 19.0°C$

Call the temperature at the interface between the wood and the styrofoam T. The heat current in each material is given by $H = KA\left(\dfrac{T_H - T_C}{L}\right)$.

Heat current through the wood:
$H_w = k_w \dfrac{A(T - T_1)}{L_w}$

Heat current through the styrofoam:
$H_s = k_s \dfrac{A(T_2 - T)}{L_s}$

In steady-state heat does not accumulate in either material. The same heat has to pass through both materials in succession, so $H_w = H_s$.

$\Rightarrow k_w \dfrac{A(T - T_1)}{L_w} = k_s \dfrac{A(T_2 - T)}{L_s}$

$k_w L_s (T - T_1) = k_s L_w (T_2 - T)$

$T = \dfrac{k_w L_s T_1 + k_s L_w T_2}{k_w L_s + k_s L_w} = \dfrac{(0.080\,W/m\cdot K)(0.035m)(-10.0°C) + (0.010\,W/m\cdot K)(0.020m)(19.0°C)}{(0.080\,W/m\cdot K)(0.035m) + (0.010\,W/m\cdot K)(0.020m)}$

$T = \dfrac{-0.028\,W\cdot°C/K + 0.0038\,W\cdot°C/K}{0.0030\,W/K} = \underline{-8.1°C}$

b) Heat flow per square meter is $\dfrac{H}{A} = K\left(\dfrac{T_H - T_C}{L}\right)$. We can calculate this either for the wood or for the styrofoam; the results must be the same.

wood
$\dfrac{H_w}{A} = k_w \dfrac{T - T_1}{L_w} = (0.080\,W/m\cdot K)\dfrac{(-8.07°C - (-10.0°C))}{0.020\,m} = 7.7\,W/m^2$

styrofoam
$\dfrac{H_s}{A} = k_s \dfrac{T_2 - T}{L_s} = (0.010\,W/m\cdot K)\dfrac{19.0°C - (-8.1°C)}{0.035\,m} = 7.7\,W/m^2$

15-53 (cont)

Since $K_s < K_w$ it takes a larger temperature gradient $\frac{T_H - T_c}{L}$ across the stryofoam to set up the same heat current as in the wood.

15-57

a) temperature gradient $\frac{T_H - T_c}{L} = \frac{100.0°C - 0.0°C}{0.250 \text{ m}} = 400 \text{ C°/m} = \underline{400 \text{ K/m}}$.

b) $H = KA \frac{T_H - T_c}{L}$

From Table 15-5, $k = 385 \text{ W/m·K}$

$\Rightarrow H = (385 \text{ W/m·K})(1.70 \times 10^{-4} \text{ m}^2)(400 \text{ K/m}) = \underline{26.2 \text{ W}}$

c) $H = 26.2 \text{ W}$ for all sections of the rod.

```
        8.00 cm
   ┌─────────┬──────────┐
T_H│         │          │T_c
   └─────────┴──────────┘
             T
```

Apply $H = KA \frac{\Delta T}{L}$ to the 8.00 cm section:

$H = KA \frac{T_H - T}{L} \Rightarrow T_H - T = \frac{LH}{KA}$

$T = T_H - \frac{LH}{AK} = 100.0°C - \frac{(0.0800 \text{ m})(26.2 \text{ W})}{(1.70 \times 10^{-4} \text{ m}^2)(385 \text{ W/m·K})} = \underline{68.0°C}$

15-59

The heat conducted through the bottom of the pot goes into the water at 100°C to convert it to steam at 100°C. We can calculate the amount of heat from the mass of material that changes phase.

$Q = mL_v = (0.440 \text{ kg})(2256 \times 10^3 \text{ J/kg}) = 9.93 \times 10^5 \text{ J}$

$H = \frac{Q}{t} = \frac{9.93 \times 10^5 \text{ J}}{300 \text{ s}} = 3.31 \times 10^3 \text{ J/s}$

$H = KA \frac{T_H - T_c}{L}$

$T_H - T_c = \frac{HL}{KA} = \frac{(3.31 \times 10^3 \text{ J/s})(1.20 \times 10^{-2} \text{ m})}{(50.2 \text{ W/m·K})(0.150 \text{ m}^2)} = 5.27 \text{ C°}$

$T_H = T_c + 5.27 \text{ C°} = 100°C + 5.27 \text{ C°} = \underline{105°C}$

15-61

$H_{net} = Ae\sigma(T^4 - T_s^4)$ (Eq. 15-26; T must be in kelvins)

Example 15-17 gives $A = 1.2 \text{ m}^2$, $e = 1.0$, and $T = 30°C = 303 \text{ K}$ (body surface temperature)

$T_s = 10.0°C = 283 \text{ K}$

$\Rightarrow H_{net} = (1.2 \text{ m}^2)(1.0)(5.67 \times 10^{-8} \text{ W/m}^2\text{·K}^4)((303 \text{ K})^4 - (283 \text{ K})^4)$

$H_{net} = 573.5 \text{ W} - 436.4 \text{ W} = \underline{137 \text{ W}}$

(Note that this is larger than H_{net} calculated in Example 15-17. The lower temperature of the surroundings increases the rate of heat loss by radiation.)

15-63

$H = Ae\sigma T^4 \Rightarrow A = \dfrac{H}{e\sigma T^4}$

120-W lamp and all electrical energy consumed is radiated $\Rightarrow H = 120$ W

$A = \dfrac{120 \text{ W}}{(0.35)(5.67\times10^{-8} \text{ W/m}^2\cdot\text{K}^4)(2450\text{K})^4} = 1.68\times10^{-4} \text{ m}^2 \left(\dfrac{1\times10^4 \text{ cm}^2}{1 \text{ m}^2}\right) = \underline{1.68 \text{ cm}^2}$

15-67

Eq. (15-27): $H = \dfrac{T_{ic} - T_{amb}}{r_{th}} \Rightarrow T_{amb} = T_{ic} - H r_{th}$

$T_{amb} = 120°C - (32\text{W})(3.0 \text{K/W}) = 120°C - 96C° = \underline{24°C}$

Problems

15-71

a) $V_0 = (0.150 \text{ m})^3 = 3.375\times10^{-3} \text{ m}^3$
From Table 15-2, $\beta = 7.2\times10^{-5} \text{ K}^{-1}$
$\Delta V = V_0 \beta \Delta T = (3.375\times10^{-3}\text{ m}^3)(7.2\times10^{-5}\text{ K}^{-1})(80°C - 20°C) = 1.5\times10^{-5}\text{ m}^3$
(the volume increases)

b) $\rho = \dfrac{m}{V} \Rightarrow \rho_0 = \dfrac{m}{V_0}$ and $\rho = \dfrac{m}{V_0 + \Delta V} = \dfrac{m}{V_0}\left(\dfrac{1}{1+\Delta V/V_0}\right) = \rho_0\left(1+\dfrac{\Delta V}{V_0}\right)^{-1}$

$\dfrac{\Delta V}{V_0} = \beta \Delta T = (7.2\times10^{-5}\text{ K}^{-1})(80°C - 20°C) = 4.32\times10^{-3}$
From Table 14-1, $\rho_0 = 2.7\times10^3 \text{ kg/m}^3$.

$\Delta\rho = \rho - \rho_0 = \rho_0\left(1+\dfrac{\Delta V}{V_0}\right)^{-1} - \rho_0$

To avoid subtracting two numbers that are nearly equal, use the binomial theorem and the fact that $\Delta V/V_0$ is small to write
$\left(1+\dfrac{\Delta V}{V_0}\right)^{-1} = 1 - \dfrac{\Delta V}{V_0} + $ terms that can neglect

$\Delta\rho = \rho_0\left(1-\dfrac{\Delta V}{V_0}\right) - \rho_0 = -\dfrac{\Delta V}{V_0}\rho_0 = -(4.32\times10^{-3})(2.7\times10^3 \text{ kg/m}^3) = \underline{-12 \text{ kg/m}^3}$
(the density decreases)

15-73

a) Heat the ring to make its diameter equal to 3.0040 in. The diameter of the ring undergoes linear expansion
$\Delta L = L_0 \alpha \Delta T \Rightarrow \Delta T = \dfrac{\Delta L}{L_0 \alpha} = \dfrac{0.0040 \text{ in.}}{(3.0000\text{ in.})(1.2\times10^{-5}(C°)^{-1})} = 110 C°$

$T = T_0 + \Delta T = 20.0°C + 110 C° = \underline{130°C}$

b) $L = L_0(1+\alpha \Delta T)$
Want L_s (steel) $= L_b$ (brass) for the same ΔT for both materials
$\Rightarrow L_{0s}(1+\alpha_s \Delta T) = L_{0b}(1+\alpha_b \Delta T)$

15-73 (cont)

$$L_{0s} + L_{0s}\alpha_s \Delta T = L_{0b} + L_{0b}\alpha_b \Delta T$$

$$\Delta T = \frac{L_{0b} - L_{0s}}{L_{0s}\alpha_s - L_{0b}\alpha_b} = \frac{3.0040 \text{ in.} - 3.0000 \text{ in.}}{(3.0000 \text{ in.})(1.2\times 10^{-5}(C°)^{-1}) - (3.0040 \text{ in.})(2.0\times 10^{-5}(C°)^{-1})}$$

$$\Delta T = \frac{0.0040}{3.60\times 10^{-5} - 6.01\times 10^{-5}} C° = -170 C°$$

$$T = T_0 + \Delta T = 20.0°C - 170 C° = \underline{-150°C}$$

15-75

Call the metals A and B. Use the data given to calculate α for each metal.

$$\Delta L = L_0 \alpha \Delta T \Rightarrow \alpha = \frac{\Delta L}{L_0 \Delta T}$$

metal A: $\alpha_A = \frac{\Delta L}{L_0 \Delta T} = \frac{0.0750 \text{ cm}}{(30.0 \text{ cm})(100 C°)} = 2.50\times 10^{-5} (C°)^{-1}$

metal B: $\alpha_B = \frac{\Delta L}{L_0 \Delta T} = \frac{0.0400 \text{ cm}}{(30.0 \text{ cm})(100 C°)} = 1.33\times 10^{-5} (C°)^{-1}$

Now consider the composite rod. Let L_A be the length of metal A in this rod.

```
  ← L_A →←— 30.0 cm - L_A —→
  |  A   |         B        |
```
$\Delta T = 100 C° \Rightarrow \Delta L = 0.055 \text{ cm}$

$$\Delta L = \Delta L_A + \Delta L_B = (\alpha_A L_A + \alpha_B L_B) \Delta T$$

$$\frac{\Delta L}{\Delta T} = \alpha_A L_A + \alpha_B (0.300 \text{ m} - L_A)$$

$$L_A = \frac{\frac{\Delta L}{\Delta T} - (0.300 \text{ m})\alpha_B}{\alpha_A - \alpha_B} = \frac{\frac{0.055\times 10^{-2} \text{ m}}{100 C°} - (0.300 \text{ m})(1.33\times 10^{-5}(C°)^{-1})}{2.50\times 10^{-5} (C°)^{-1} - 1.33\times 10^{-5} (C°)^{-1}}$$

$$L_A = \left(\frac{5.50\times 10^{-6} - 3.99\times 10^{-6}}{2.50\times 10^{-5} - 1.33\times 10^{-5}}\right) \text{m} = 0.129 \text{ m} = \underline{12.9 \text{ cm}}$$

$$L_B = 30.0 \text{ cm} - L_A = 30.0 \text{ cm} - 12.9 \text{ cm} = \underline{17.1 \text{ cm}}$$

15-79

Let β_ℓ and β_m be the coefficients of volume expansion for the liquid and for the metal. Let ΔT be the (negative) change in temperature when the system is cooled to the new temperature.

Change in volume of cylinder when cool: $\Delta V_m = V_0 \beta_m \Delta T$ (negative)
Change in volume of liquid when cool: $\Delta V_\ell = V_0 \beta_\ell \Delta T$ (negative)
The difference $\Delta V_m - \Delta V_\ell$ must be equal to the positive volume change due to the decrease in pressure, which is $-\frac{\Delta p V_0}{B} = -k \Delta p V_0$.

$$\Rightarrow \Delta V_m - \Delta V_\ell = -k \Delta p V_0$$

$$V_0 \Delta T (\beta_m - \beta_\ell) = -k \Delta p V_0$$

$$\Delta T = -\frac{k \Delta p}{\beta_m - \beta_\ell} = \frac{k \Delta p}{\beta_\ell - \beta_m} = \frac{(700\times 10^{-10} Pa^{-1})(1.013\times 10^5 Pa - 1.013\times 10^7 Pa)}{5.30\times 10^{-4} K^{-1} - 3.60\times 10^{-5} K^{-1}} = -14.2 C°$$

$$T = T_0 + \Delta T = 60.0°C - 14.2 C° = \underline{45.8°C}$$

15-81

a) The kinetic energy is $K = \frac{1}{2}mv^2$.
The heat energy required to raise its temperature by 600 C° (but not to melt it) is $Q = mc\Delta T$.

The ratio is $\dfrac{K}{Q} = \dfrac{\frac{1}{2}mv^2}{mc\Delta T} = \dfrac{v^2}{2c\Delta T} = \dfrac{(7700\,m/s)^2}{2(910\,J/kg\cdot K)(600\,C°)} = \underline{54.3}$.

b) The heat generated when friction work (due to the friction force exerted by the air) removes the kinetic energy of the satellite during reentry is very large, and could melt the satellite. Manned space vehicles must have heat shields made of very high melting temperature materials, and reentry must be made slowly.

15-83

a) Eq. (15-14) $\Rightarrow dQ = mc\,dT$
But the problem gives the molar heat capacity $C = Mc$; $dQ = nC\,dT$ (Eq. 15-19).

$$Q = n\int_{T_1}^{T_2} C\,dT = n\int_{T_1}^{T_2} k\,\dfrac{T^3}{\Theta^3}\,dT = \dfrac{nk}{\Theta^3}\int_{T_1}^{T_2} T^3\,dT = \dfrac{nk}{\Theta^3}\left(\tfrac{1}{4}T^4\Big|_{T_1}^{T_2}\right)$$

$Q = \dfrac{nk}{4\Theta^3}(T_2^4 - T_1^4) = \dfrac{(2.00\,mol)(1940\,J/mol\cdot K)}{4(281\,K)^3}\left((50.0\,K)^4 - (10.0\,K)^4\right) = \underline{273\,J}$

b) $C_{av} = \dfrac{1}{n}\dfrac{\Delta Q}{\Delta T} = \dfrac{1}{2.00\,mol}\left(\dfrac{273\,J}{50.0\,K - 10.0\,K}\right) = \underline{3.41\,J/mol\cdot K}$

c) $C = k\left(\dfrac{T}{\Theta}\right)^3 = (1940\,J/mol\cdot K)\left(\dfrac{50.0\,K}{281\,K}\right)^3 = \underline{10.9\,J/mol\cdot K}$

(C is increasing with T, so C at the upper end of the temperature interval is larger than its average value over the interval.)

15-87

Heat comes out of the 0.0°C water and into the ice cube to warm it to 0.0°C. The heat that comes out of the water causes a phase change liquid → solid.

Heat that goes into the ice cube:
$Q_{ice} = mc_{ice}\Delta T = (0.060\,kg)(2000\,J/kg\cdot K)(0.0°C - (-10.0°C)) = 1200\,J$

Heat that comes out of the water when mass m freezes:
$Q_{water} = -mL_f$

$Q_{system} = 0 \Rightarrow Q_{ice} + Q_{water} = 0$
$1200\,J - mL_f = 0$

$m = \dfrac{1200\,J}{L_f} = \dfrac{1200\,J}{334\times 10^3\,J/kg} = 3.59\times 10^{-3}\,kg = \underline{3.59\,g}$

15-89

Assume that all the ice melts and that all the steam condenses. If we calculate a final temperature T that is outside the range 0°C to 100°C then we know that this assumption is incorrect. Calculate Q for each piece of the system and then set the total $Q_{system} = 0$.

__copper can__ (changes temperature from 0.0°C to T; no phase change)
$Q_{can} = mc\Delta T = (0.322 \text{ kg})(390 \text{ J/kg·K})(T - 0.0°C) = (125.6 \text{ J/K})T$

__ice__ (melting phase change and then water produced warms to T)
$Q_{ice} = +mL_f + mc\Delta T = (0.0420 \text{ kg})(334 \times 10^3 \text{ J/kg}) + (0.0420 \text{ kg})(4190 \text{ J/kg·K})(T - 0.0°C)$
$Q_{ice} = 1.403 \times 10^4 \text{ J} + (176.0 \text{ J/K})T$

__steam__ (condenses to liquid and then water produced cools to T)
$Q_{steam} = -mL_v + mc\Delta T = -(0.0120 \text{ kg})(2256 \times 10^3 \text{ J/kg}) + (0.0120 \text{ kg})(4190 \text{ J/kg·K})(T - 100°C)$
$Q_{steam} = -2.707 \times 10^4 \text{ J} + 50.28T - 5028 \text{ J} = -3.210 \times 10^4 \text{ J} + 50.28T$

$Q_{system} = 0 \Rightarrow Q_{can} + Q_{ice} + Q_{steam} = 0$
$(125.6 \text{ J/K})T + 1.403 \times 10^4 \text{ J} + (176.0 \text{ J/K})T - 3.210 \times 10^4 \text{ J} + 50.28T = 0$
$(351.9 \text{ J/K})T = 1.807 \times 10^4 \text{ J}$

$T = \dfrac{1.807 \times 10^4 \text{ J}}{351.9 \text{ J/K}} = \underline{51.4°C}$

This is between 0°C and 100°C so our assumptions about the phase changes being complete were correct.

15-93

Assume that all the ice melts and that all the steam condenses; if this is the case the final temperature must be between 0°C and 100°C. Do the calculation. If the T_2 we calculate is not in this range then our assumption is wrong.

$Q_{system} = 0$

__ice__
$Q_{ice} = +mL_f + mc\Delta T = (0.150 \text{ kg})(334 \times 10^3 \text{ J/kg}) + (0.150 \text{ kg})(4190 \text{ J/kg·K})(T_2 - 0°C)$
$Q_{ice} = 50,100 \text{ J} + (628.5 \text{ J/K})T_2$

__water__
$Q_{water} = mc\Delta T = (0.200 \text{ kg})(4190 \text{ J/kg·K})(T_2 - 40.0°C)$
$Q_{water} = (838 \text{ J/K})T_2 - 33,520 \text{ J}$

__steam__
$Q_{steam} = -L_v + mc\Delta T = -(0.0800 \text{ kg})(2256 \times 10^3 \text{ J/kg}) + (0.0800 \text{ kg})(4190 \text{ J/kg·K})(T_2 - 100°C)$
$Q_{steam} = -180,480 \text{ J} + (335.2 \text{ J/K})T_2 - 33,520 \text{ J}$
$Q_{steam} = -214,000 \text{ J} + (335.2 \text{ J/K})T_2$

15-93 (cont)

$Q_{system} = 0$

$Q_{ice} + Q_{water} + Q_{steam} = 0$

$50,100 J + (628.5 J/K)T_2 + (838 J/K)T_2 - 33,520 J - 214,000 J + (335.2 J/K)T_2 = 0$

$(1801.7 J/K)T_2 = 197,420 J$

$T_2 = 109.6 °C$

This is not possible if the final phase is all liquid water; there must be some steam left. Steam left $\Rightarrow T_2 = \underline{100°C}$

15-97

Use H written in terms of the thermal resistance R:
$H = \frac{A \Delta T}{R}$, where $R = \frac{L}{k}$ and $R = R_1 + R_2 + ...$ (additive).

single pane

$R_s = R_{glass} + R_{film}$, where $R_{film} = 0.15 \, m^2 \cdot K/W$ is the combined thermal resistance of the air films on the room and outdoor surfaces of the window.

$R_{glass} = \frac{L}{k} = \frac{3.5 \times 10^{-3} m}{0.80 \, W/m \cdot K} = 0.004375 \, m^2 \cdot K/W$

$R_s = 0.004375 \, m^2 \cdot K/W + 0.15 \, m^2 \cdot K/W = 0.1544 \, m^2 \cdot K/W$

double pane

$R_d = 2 R_{glass} + R_{air} + R_{film}$, where R_{air} is the thermal resistance of the air space between the panes.

$R_{air} = \frac{L}{k} = \frac{5.0 \times 10^{-3} m}{0.024 \, W/m \cdot K} = 0.2083 \, m^2 \cdot K/W$

$R_d = 2(0.004375 \, m^2 \cdot K/W) + 0.2083 \, m^2 \cdot K/W + 0.15 \, m^2 \cdot K/W = 0.3671 \, m^2 \cdot K/W$

$H_s = \frac{A \Delta T}{R_s}$, $H_d = \frac{A \Delta T}{R_d}$, so $\frac{H_s}{H_d} = \frac{R_d}{R_s}$ (since A and ΔT are same for both)

$\frac{H_s}{H_d} = \frac{0.3671 \, m^2 \cdot K/W}{0.1544 \, m^2 \cdot K/W} = \underline{2.4}$

15-99

a) Heat must be conducted from the water to cool it to 0°C and to cause the phase transition. The entire volume of water is not at the air temperature.

b) Consider a section of ice that has area A. At time t let the thickness be h. Consider a short time interval t to $t + dt$. Let the thickness that freezes in this time be dh. The mass of the section that freezes in the time interval dt is $dm = \rho dV = \rho A \, dh$. The heat that must be conducted away from this mass of water to freeze it is

$dQ = dm \, L_f = (\rho A L_f) dh$

$H = \frac{dQ}{dt} = kA \frac{dT}{h}$, so the heat dQ conducted in time dt through the thickness h that is already there is $dQ = kA \left(\frac{T_H - T_C}{h} \right) dt$.

15-99 (cont)

Equate these two expressions for dQ

$$\Rightarrow \rho A L_f \, dh = kA\left(\frac{T_H - T_c}{h}\right) dt$$

$$h \, dh = \left(\frac{k(T_H - T_c)}{\rho L_f}\right) dt$$

Integrate from $t=0$ to time t. At $t=0$ the thickness h is zero.

$$\int_0^h h \, dh = \frac{k(T_H - T_c)}{\rho L_f} \int_0^t dt$$

$$\tfrac{1}{2} h^2 = \frac{k(T_H - T_c)}{\rho L_f} t \Rightarrow h = \sqrt{\frac{2k(T_H - T_c)}{\rho L_f}} \sqrt{t}$$

The thickness after time t is proportional to $\sqrt{t}$.

c) The expression in part (b) gives

$$t = \frac{h^2 \rho L_f}{2k(T_H - T_c)} = \frac{(0.25\,m)^2 (920\,kg/m^3)(334 \times 10^3\,J/kg)}{2(1.6\,W/m\cdot K)(0°C - (-10°C))} = 6.00 \times 10^5\,s = \underline{167\,hours}$$

d) Find t for $h = 40\,m$: $t = \left(\frac{40\,m}{0.25\,m}\right)^2 (167\,hours) = 4.28 \times 10^6\,hours$. This is about 500 years. With current climate this will not happen.

15-101

Work with a $1.00\,m^2$ area.

The heat current into a $1.00\,m^2$ area of ice due to the absorbed solar radiation is $H = (0.70)(600\,W/m^2)(1.00\,m^2) = 420\,W$.

The heat required to melt a $h = 1.20\,cm$ thick layer of ice that is initially at $0°C$ is $Q = mL_f = \rho h A L_f = (920\,kg/m^3)(0.0120\,m)(1.00\,m^2)(334 \times 10^3\,J/kg)$
$= 3.687 \times 10^6\,J$.

$H = \frac{Q}{t}$, so the time t it takes the heat current from solar radiation to input this amount of heat into the ice is $t = \frac{Q}{H} = \frac{3.687 \times 10^6\,J}{420\,W} = 8.78 \times 10^3\,s = \underline{146\,min}$

15-103

Calculate the net rate of radiation of heat from the can: $H_{net} = A e \sigma (T^4 - T_s^4)$.

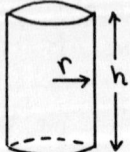

The surface area of the cylindrical can is $A = 2\pi r h + 2\pi r^2$
$A = 2\pi r (h + r) = 2\pi (0.030\,m)(0.150\,m + 0.030\,m) = 0.0339\,m^2$.

$H_{net} = (0.0339\,m^2)(0.200)(5.67 \times 10^{-8}\,W/m^2 \cdot K^4)((4.22\,K)^4 - (77.3\,K)^4)$
$H_{net} = -0.0137\,W$ (minus $\Rightarrow$ heat current into the can)

The heat that is put into the can by radiation in one hour is
$Q = -(H_{net}) t = (0.0137\,W)(3600\,s) = 49.4\,J$

This heat boils a mass m of helium according to the equation $Q = mL_f$, so
$m = \frac{Q}{L_f} = \frac{49.4\,J}{2.09 \times 10^4\,J/kg} = 2.36 \times 10^{-3}\,kg = \underline{2.36\,g}$

CHAPTER 16

Exercises 3, 7, 11, 13, 15, 19, 21, 23, 27, 29, 31, 33, 39

Problems 41, 45, 47, 49, 51, 55, 57, 59, 63, 67, 71

Exercises

16-3

a) $n = \dfrac{m_{tot}}{M} = \dfrac{0.280 \text{ kg}}{4.00 \times 10^{-3} \text{ kg/mol}} = \underline{70.0 \text{ mol}}$

b) $pV = nRT \Rightarrow p = \dfrac{nRT}{V}$
T must be in kelvins; $T = (24 + 273) \text{ K} = 297 \text{ K}$

$p = \dfrac{(70.0 \text{ mol})(8.3145 \text{ J/mol·K})(297 \text{ K})}{25.0 \times 10^{-3} \text{ m}^3} = 6.91 \times 10^6 \text{ Pa}$

$p = (6.91 \times 10^6 \text{ Pa}) \left(\dfrac{1.00 \text{ atm}}{1.013 \times 10^5 \text{ Pa}} \right) = \underline{68.3 \text{ atm}}$

16-7

$pV = nRT$
n, R constant $\Rightarrow \dfrac{pV}{T} = nR = \text{constant} \Rightarrow \dfrac{p_1 V_1}{T_1} = \dfrac{p_2 V_2}{T_2}$

$T_1 = (27 + 273) \text{ K} = 300 \text{ K}$
$p_1 = 1.01 \times 10^5 \text{ Pa}$
$p_2 = 2.44 \times 10^6 \text{ Pa} + 1.01 \times 10^5 \text{ Pa} = 2.54 \times 10^6 \text{ Pa}$ (in the ideal gas equation the pressures must be absolute not gauge pressures)

$T_2 = T_1 \left(\dfrac{p_2}{p_1} \right) \left(\dfrac{V_2}{V_1} \right) = 300 \text{ K} \left(\dfrac{2.54 \times 10^6 \text{ Pa}}{1.01 \times 10^5 \text{ Pa}} \right) \left(\dfrac{54.8 \text{ cm}^3}{548 \text{ cm}^3} \right) = 754 \text{ K}$

$T_2 = (754 - 273)°\text{C} = \underline{481°\text{C}}$

(Note that the units cancel in the $\dfrac{V_2}{V_1}$ volume ratio, so it was not necessary to convert the volumes in cm³ to m³.)

16-11

$pV = nRT$; n, R, p are constant $\Rightarrow \dfrac{V}{T} = \dfrac{nR}{p} = \text{constant} \Rightarrow \dfrac{V_1}{T_1} = \dfrac{V_2}{T_2}$
$T_1 = (24 + 273) \text{ K} = 297 \text{ K}$ (T must be in kelvins)
$V_2 = V_1 \left(\dfrac{T_2}{T_1} \right) = (0.700 \text{ L}) \left(\dfrac{77.3 \text{ K}}{297 \text{ K}} \right) = \underline{0.182 \text{ L}}$

16-13

Eq. (16-5): $\rho = \dfrac{pM}{RT} \Rightarrow p = \dfrac{RT\rho}{M}$
$T = (-56.5 + 273.15) \text{ K} = 216.6 \text{ K}$
for air $M = 28.8 \times 10^{-3}$ kg/mol (Example 16-3)
$p = \dfrac{(8.3145 \text{ J/mol·K})(216.6 \text{ K})(0.364 \text{ kg/m}^3)}{28.8 \times 10^{-3} \text{ kg/mol}} = \underline{2.28 \times 10^4 \text{ Pa}}$

16-15

a) $pV = nRT$

Find the initial pressure p_1:

$$p_1 = \frac{nRT_1}{V} = \frac{(9.0\,\text{mol})(8.3145\,\text{J/mol·K})((23.0+273.15)\,\text{K})}{2.20\times10^{-3}\,\text{m}^3} = 1.007\times10^7\,\text{Pa}$$

$$p_2 = 100\,\text{atm}\left(\frac{1.013\times10^5\,\text{Pa}}{1\,\text{atm}}\right) = 1.013\times10^7\,\text{Pa}$$

$\frac{p}{T} = \frac{nR}{V} = \text{constant}$, so $\frac{p_1}{T_1} = \frac{p_2}{T_2}$

$$T_2 = T_1\left(\frac{p_2}{p_1}\right) = (296.15\,\text{K})\left(\frac{1.013\times10^7\,\text{Pa}}{1.007\times10^7\,\text{Pa}}\right) = 297.9\,\text{K} = \underline{24.8°C}$$

b) The coefficient of volume expansion for a gas is much larger than for a solid, so the expansion of the tank is negligible.

16-19

a) Use the density and the mass of 1 mole to calculate the volume.

$\rho = \frac{m}{V} \Rightarrow V = \frac{m}{\rho}$, where $m = m_{tot}$, the mass of 1.00 mol of water.

$m_{tot} = nM = (1.00\,\text{mol})(18.0\times10^{-3}\,\text{kg/mol}) = 18.0\times10^{-3}\,\text{kg}$

Then $V = \frac{m}{\rho} = \frac{18.0\times10^{-3}\,\text{kg}}{1000\,\text{kg/m}^3} = \underline{1.80\times10^{-5}\,\text{m}^3}$

b) One mole contains $N_A = 6.022\times10^{23}$ molecules, so the volume occupied by one molecule is $\frac{1.80\times10^{-5}\,\text{m}^3/\text{mol}}{6.022\times10^{23}\,\text{molecules/mol}} = 2.989\times10^{-29}\,\text{m}^3/\text{molecule}$

$V = a^3$, where a is the length of each side of the cube occupied by a molecule. $a^3 = 2.989\times10^{-29}\,\text{m}^3 \Rightarrow \underline{a = 3.10\times10^{-10}\,\text{m}}$

c) Atoms and molecules are on the order of 10^{-10} m in diameter, in agreement with the above estimates.

16-21

a) $pV = nRT$, $n = \frac{N}{N_A} \Rightarrow pV = \frac{N}{N_A}RT$

$$p = \left(\frac{N}{V}\right)\left(\frac{R}{N_A}\right)T = \left(\frac{80\,\text{molecules}}{1\times10^{-6}\,\text{m}^3}\right)\left(\frac{8.3145\,\text{J/mol·K}}{6.022\times10^{23}\,\text{molecules/mol}}\right)(7500\,\text{K}) = 8.28\times10^{-12}\,\text{Pa}$$

$p = 8.2\times10^{-17}\,\text{atm}$. This is much lower than the laboratory pressure of $1\times10^{-13}\,\text{atm}$ in Exercise 16-18.

b) The Lagoon Nebula is a very rarefied low pressure gas. The gas would exert <u>very</u> little force on an object passing through it.

16-23

$$v_{rms} = \sqrt{\frac{3RT}{M}} \quad (Eq.\ 16\text{-}19) \Rightarrow \frac{v_{rms}^2}{3R} = \frac{T}{M}, \text{ where } T \text{ must be in kelvins.}$$

Same $v_{rms} \Rightarrow$ same $\frac{T}{M}$ for the two gases $\Rightarrow \frac{T_{O_2}}{M_{O_2}} = \frac{T_{H_2}}{M_{H_2}}$

$$T_{O_2} = T_{H_2}\left(\frac{M_{O_2}}{M_{H_2}}\right) = (27+273)\text{K}\left(\frac{32.0\,g/mol}{2.02\,g/mol}\right) = 4.752 \times 10^3\,\text{K}$$

$T_{O_2} = (4752 - 273)°C = \underline{4480°C}$

16-27

a) $\frac{1}{2}m(v^2)_{av} = \frac{3}{2}kT = \frac{3}{2}(1.38 \times 10^{-23}\,\text{J/molecule·K})(300\,\text{K}) = \underline{6.21 \times 10^{-21}\,\text{J}}$

b) We need the mass m of one atom:

$$m = \frac{M}{N_A} = \frac{28.0 \times 10^{-3}\,\text{kg/mol}}{6.022 \times 10^{23}\,\text{molecules/mol}} = 4.65 \times 10^{-26}\,\text{kg/molecule}$$

Then $\frac{1}{2}m(v^2)_{av} = 6.21 \times 10^{-21}\,\text{J}$ (from part (a)) gives

$$(v^2)_{av} = \frac{2(6.21 \times 10^{-21}\,\text{J})}{m} = \frac{2(6.21 \times 10^{-21}\,\text{J})}{4.65 \times 10^{-26}\,\text{kg}} = \underline{2.67 \times 10^5\,\text{m}^2/\text{s}^2}$$

c) $v_{rms} = \sqrt{(v^2)_{av}} = \sqrt{2.67 \times 10^5\,\text{m}^2/\text{s}^2} = \underline{517\,\text{m/s}}$

d) $p = mv_{rms} = (4.65 \times 10^{-26}\,\text{kg})(517\,\text{m/s}) = \underline{2.40 \times 10^{-23}\,\text{kg·m/s}}$

e) Time between collisions with one wall is

$$t = \frac{0.20\,\text{m}}{v_{rms}} = \frac{0.20\,\text{m}}{517\,\text{m/s}} = 3.87 \times 10^{-4}\,\text{s}$$

In a collision $\vec{v}$ changes direction, so $\Delta p = 2mv_{rms} = 2(2.40 \times 10^{-23}\,\text{kg·m/s})$

$\Delta p = 4.80 \times 10^{-23}\,\text{kg·m/s}$

$$F = \frac{dp}{dt} \Rightarrow F_{av} = \frac{\Delta p}{\Delta t} = \frac{4.80 \times 10^{-23}\,\text{kg·m/s}}{3.87 \times 10^{-4}\,\text{s}} = \underline{1.24 \times 10^{-19}\,\text{N}}$$

f) pressure $= \frac{F}{A} = \frac{1.24 \times 10^{-19}\,\text{N}}{(0.10\,\text{m})^2} = \underline{1.24 \times 10^{-17}\,\text{Pa}}$ (due to one atom)

g) pressure $= 1\,\text{atm} = 1.013 \times 10^5\,\text{Pa}$

Number of atoms needed $= \frac{1.013 \times 10^5\,\text{Pa}}{1.24 \times 10^{-17}\,\text{Pa/atom}} = \underline{8.17 \times 10^{21}\,\text{atoms}}$

h) $pV = NkT$ (Eq. 16-18) $\Rightarrow N = \frac{pV}{kT} = \frac{(1.013 \times 10^5\,\text{Pa})(0.10\,\text{m})^3}{(1.384 \times 10^{-23}\,\text{J/molecule·K})(300\,\text{K})} = \underline{2.45 \times 10^{22}\,\text{atoms}}$

i) From the factor of $\frac{1}{3}$ in $(v_x^2)_{av} = \frac{1}{3}(v^2)_{av}$.

16-29

a) $Q = nC_v \Delta T = n\left(\frac{5}{2}R\right)\Delta T = (6.00\,\text{mol})\left(\frac{5}{2}\right)(8.3145\,\text{J/mol·K})(25.0\,\text{K}) = \underline{3.12 \times 10^3\,\text{J}}$

b) $Q = nC_v \Delta T = n\left(\frac{3}{2}R\right)\Delta T = (6.00\,\text{mol})\left(\frac{3}{2}\right)(8.3145\,\text{J/mol·K})(25.0\,\text{K}) = \underline{1.87 \times 10^3\,\text{J}}$

16-31

a) $\frac{1}{2}R$ contribution to C_V for each degree of freedom $\Rightarrow C_V = 6(\frac{1}{2}R) = 3R$
$C_V = 3(8.3145 \text{ J/mol·K}) = \underline{24.9 \text{ J/mol·K}}$

b) For water vapor the specific heat capacity is $c = 2000$ J/kg·K. The molar heat capacity is $C = Mc = (18.0 \times 10^{-3} \text{ kg/mol})(2000 \text{ J/kg·K}) = \underline{36.0 \text{ J/mol·K}}$.

The difference is 36.0 J/mol·K $- 24.9$ J/mol·K $= 11.1$ J/mol·K, which is about $2.7(\frac{1}{2}R)$; the vibrational degrees of freedom make a significant contribution.

16-33

Eq. (16-33): $f(v) = \frac{8\pi}{m}\left(\frac{m}{2\pi kT}\right)^{3/2} \epsilon e^{-\epsilon/kT}$

At the maximum of $f(\epsilon)$, $\frac{df}{d\epsilon} = 0$.

$\frac{df}{d\epsilon} = \frac{8\pi}{m}\left(\frac{m}{2\pi kT}\right)^{3/2} \frac{d}{d\epsilon}(\epsilon e^{-\epsilon/kT}) = 0$

$\Rightarrow \frac{d}{d\epsilon}(\epsilon e^{-\epsilon/kT}) = 0 \Rightarrow e^{-\epsilon/kT} - \frac{\epsilon}{kT}e^{-\epsilon/kT} = 0$

$(1 - \frac{\epsilon}{kT})e^{-\epsilon/kT} = 0$

$1 - \frac{\epsilon}{kT} = 0 \Rightarrow \epsilon = kT$, as was to be shown

16-39

If the temperature at altitude y is below the freezing point, only cirrus clouds can form. Use $T = T_0 - \alpha y$ to find the y that gives $T = 0.0$ °C.

$y = \frac{T_0 - T}{\alpha} = \frac{15.0°C - 0.0°C}{6.0 C°/km} = \underline{2.5 \text{ km}}$

(Note: The solid-liquid phase transition occurs at 0°C only for $p = 1.01 \times 10^5$ Pa. Use the result of Example 16-4 to estimate the pressure at an altitude of 2.5 km.

$p_2 = p_1 e^{-Mg(y_2 - y_1)/RT}$

$\frac{Mg(y_2 - y_1)}{RT} = 1.10\left(\frac{2500 \text{ m}}{8863 \text{ m}}\right) = 0.310$ (using the calculation in Example 16-4)

Then $p_2 = (1.01 \times 10^5 \text{ Pa}) e^{-0.31} = 0.74 \times 10^5$ Pa.

This pressure is well above the triple point pressure for water. Figure 16-18 shows that the fusion curve has large slope and it takes a large change in pressure to change the phase transition temperature very much. Using 0.0°C introduces little error.)

Problems

16-41

$$pV = nRT \Rightarrow pV = \frac{m}{M}RT$$

T, V, M, R are constant $\Rightarrow \frac{p}{m} = \frac{RT}{MV}$ = constant

So $\frac{p_1}{m_1} = \frac{p_2}{m_2}$, where m is the mass of gas in the tank.

$p_1 = 1.30 \times 10^6$ Pa $+ 1.01 \times 10^5$ Pa $= 1.40 \times 10^6$ Pa
$p_2 = 2.00 \times 10^5$ Pa $+ 1.01 \times 10^5$ Pa $= 3.01 \times 10^5$ Pa

$m_1 = \frac{pVM}{RT}$

$V = hA = h\pi r^2 = (1.00 \text{ m}) \pi (0.060 \text{ m})^2 = 0.01131 \text{ m}^3$

$m_1 = \frac{(1.40 \times 10^6 \text{ Pa})(0.01131 \text{ m}^3)(26.04 \times 10^{-3} \text{ kg/mol})}{(8.3145 \text{ J/mol} \cdot \text{K})((22.0 + 273.15) \text{ K})} = 0.1680$ kg

Then $m_2 = m_1 \left(\frac{p_2}{p_1}\right) = (0.1680 \text{ kg})\left(\frac{3.01 \times 10^5 \text{ Pa}}{1.40 \times 10^6 \text{ Pa}}\right) = 0.0361$ kg

m_2 is the mass that remains in the tank. The mass that has been used is
$m_1 - m_2 = 0.1680 \text{ kg} - 0.0361 \text{ kg} = \underline{0.132 \text{ kg}}$.

16-45

a) Consider the gas in one cylinder. Calculate the volume to which this quantity of gas expands when the pressure is decreased from 2.50×10^6 Pa to 1.01×10^5 Pa.

$pV = nRT$

n, R, T constant $\Rightarrow pV = nRT$ = constant $\Rightarrow p_1 V_1 = p_2 V_2$

$V_2 = V_1 \left(\frac{p_1}{p_2}\right) = (1.10 \text{ m}^3)\left(\frac{2.50 \times 10^6 \text{ Pa}}{1.01 \times 10^5 \text{ Pa}}\right) = 51.98 \text{ m}^3$

The number of cylinders required to fill a 600 m^3 balloon is

$\frac{600 \text{ m}^3}{51.98 \text{ m}^3} = \underline{11.5 \text{ cylinders}}$.

b) The upward force on the balloon is given by Archimedes' principle (Chapter 14):
B = weight of air displaced by balloon = $\rho_{air} V g$

Free-body diagram for the balloon:

$a = 0$, B up, $m_{gas} g$ down, $m_L g$ down

m_{gas} is the mass of the gas that is inside the balloon
m_L is the mass of the load that can be supported by the balloon

$\Sigma F_y = ma_y$
$B - m_L g - m_{gas} g = 0$
$\rho_{air} V g - m_L g - m_{gas} g = 0$
$\boxed{m_L = \rho_{air} V - m_{gas}}$

Calculate m_{gas}, the mass of hydrogen that occupies 600 m^3 at $0°C$ and $p = 1.01 \times 10^5$ Pa.

$pV = nRT = \frac{m_{gas}}{M} RT \Rightarrow m_{gas} = \frac{pVM}{RT} = \frac{(1.01 \times 10^5 \text{ Pa})(600 \text{ m}^3)(2.02 \times 10^{-3} \text{ kg/mol})}{(8.3145 \text{ J/mol} \cdot \text{K})(288 \text{ K})}$

16-45 (cont)

$$m_{gas} = 51.12 \text{ kg}$$

Then $m_L = (1.23 \text{ kg/m}^3)(600 \text{ m}^3) - 51.12 \text{ kg} = 687 \text{ kg}$, and the weight that can be supported is $w_L = m_L g = (687 \text{ kg})(9.80 \text{ m/s}^2) = \underline{6730 \text{ N}}$.

c) $m_L = \rho_{air} V - m_{gas}$

$$m_{gas} = \frac{\rho V m}{RT} = 51.12 \text{ kg}\left(\frac{4.00 \text{ g/mol}}{2.02 \text{ g/mol}}\right) = 101.2 \text{ kg} \text{ (using the results of part (b))}$$

Then $m_L = (1.23 \text{ kg/m}^3)(600 \text{ m}^3) - 101.2 \text{ kg} = 637 \text{ kg}$.

$$w_L = m_L g = (637 \text{ kg})(9.80 \text{ m/s}^2) = \underline{6240 \text{ N}}$$

16-47

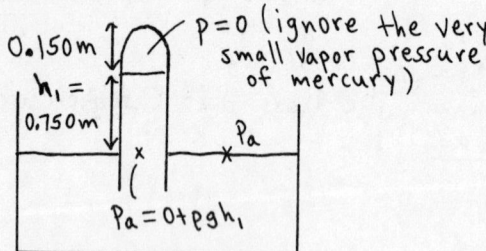 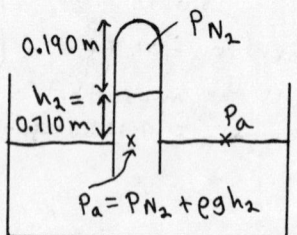

The pressure at points level with the surface of the mercury outside the tube is equal to air pressure, p_a. In the first sketch this gives $p_a = \rho g h_1$ and in the second, $p_a = p_{N_2} + \rho g h_2$. Equating these two expressions for p_a gives $\rho g h_1 = p_{N_2} + \rho g h_2$.

$$p_{N_2} = \rho g (h_1 - h_2) = (13.6 \times 10^3 \text{ kg/m}^3)(9.80 \text{ m/s}^2)(0.750 \text{ m} - 0.710 \text{ m}) = 5.331 \times 10^3 \text{ Pa}$$

We know P, V, and T; calculate the mass m_{tot} of nitrogen:

$pV = nRT$

$$n = \frac{pV}{RT} = \frac{(5.331 \times 10^3 \text{ Pa})(0.190 \text{ m})(0.570 \times 10^{-4} \text{ m}^2)}{(8.3145 \text{ J/mol·K})(300 \text{ K})} = 2.315 \times 10^{-5} \text{ mol}$$

$$m_{tot} = nM = (2.315 \times 10^{-5} \text{ mol})(28.0 \times 10^{-3} \text{ kg/mol}) = \underline{6.48 \times 10^{-7} \text{ kg}}$$

16-49

a)

$P_1 = 5.20 \times 10^5 \text{ Pa}$, $P_2 = P_{air} = 1.00 \times 10^5 \text{ Pa}$

large tank $\Rightarrow v_1 \approx 0$

$$P_1 + \rho g y_1 + \tfrac{1}{2}\rho v_1^2 = P_2 + \rho g y_2 + \tfrac{1}{2}\rho v_2^2$$

$$\tfrac{1}{2}\rho v_2^2 = P_1 - P_2 + \rho g(y_1 - y_2)$$

$h = 3.50 \text{ m}$, 1.00 m, $y = 0$

$$v_2 = \sqrt{\tfrac{2}{\rho}(P_1 - P_2) + 2g(y_1 - y_2)} = \sqrt{\tfrac{2}{1000 \text{ kg/m}^3}(5.20 \times 10^5 \text{ Pa} - 1.00 \times 10^5 \text{ Pa}) + 2(9.80 \text{ m/s}^2)(3.50 \text{ m} - 1.00 \text{ m})}$$

$$\underline{v_2 = 29.8 \text{ m/s}}$$

b) $\underline{h = 3.00 \text{ m}}$

The volume of the air in the tank increases so its pressure decreases.

16-49 (cont)

$pV = nRT =$ constant, so $pV = p_0 V_0$
(p_0 is the pressure for $h = 3.50$ m and p is the pressure for $h = 3.00$ m)
$p(4.00\text{m} - h)A = p_0(4.00\text{m} - h_0)A$

$p = p_0 \frac{h_0}{h} = 5.20 \times 10^5 \text{ Pa} \left(\frac{4.00\text{m} - 3.50\text{m}}{4.00\text{m} - 3.00\text{m}}\right) = 2.60 \times 10^5 \text{ Pa}$

Repeat the calculation of part (a), but now $p_1 = 2.60 \times 10^5$ Pa and $y_1 = 3.00$ m.

$v_2 = \sqrt{\frac{2}{\rho}(p_1 - p_2) + 2g(y_1 - y_2)} = \sqrt{\left(\frac{2}{1000 \text{kg/m}^3}\right)(2.60 \times 10^5 \text{ Pa} - 1.00 \times 10^5 \text{ Pa}) + 2(9.80 \text{ m/s}^2)(3.00\text{m} - 1.00\text{m})}$

$v_2 = \underline{19.0 \text{ m/s}}$

$h = 2.00$ m
$p = p_0 \frac{h_0}{h} = 5.20 \times 10^5 \text{ Pa} \left(\frac{4.00\text{m} - 3.50\text{m}}{4.00\text{m} - 2.00\text{m}}\right) = 1.30 \times 10^5 \text{ Pa}$

$v_2 = \sqrt{\frac{2}{\rho}(p_1 - p_2) + 2g(y_1 - y_2)} = \sqrt{\left(\frac{2}{1000 \text{kg/m}^3}\right)(1.30 \times 10^5 \text{Pa} - 1.00 \times 10^5 \text{Pa}) + 2(9.80 \text{m/s}^2)(2.00\text{m} - 1.00\text{m})}$

$v_2 = \underline{8.92 \text{ m/s}}$

c) $v_2 = 0$ means $\frac{2}{\rho}(p_1 - p_2) + 2g(y_1 - y_2) = 0$
$p_1 - p_2 = -g\rho(y_1 - y_2)$

$y_1 - y_2 = h - 1.00$ m
$p_1 = p_0 \left(\frac{0.50\text{m}}{4.00\text{m} - h}\right) = 5.20 \times 10^5 \text{ Pa} \left(\frac{0.50\text{m}}{4.00\text{m} - h}\right)$

Thus $5.20 \times 10^5 \text{ Pa} \left(\frac{0.50\text{m}}{4.00\text{m} - h}\right) - 1.00 \times 10^5 \text{ Pa} = (9.80 \text{ m/s}^2)(1000 \text{ kg/m}^3)(1.00\text{m} - h)$

$\frac{260}{4.00 - h} - 100 = 9.80 - 9.80 h$

$260 = (4.00 - h)(109.8 - 9.80h)$
$260 = 9.80h^2 - 149h + 439.2$
$9.80h^2 - 149h + 179.2 = 0$
$h^2 - 15.20h + 18.29 = 0$

quadratic formula: $h = \frac{1}{2}(15.20 \pm \sqrt{(15.20)^2 - 4(18.29)}) = (7.60 \pm 6.283)$ m

h must be less than 4.00 m, so the only acceptable value is
$h = 7.60 \text{ m} - 6.28 \text{ m} = \underline{1.32 \text{ m}}$

16-51

a) $U(r) = U_0 \left[\left(\frac{R_0}{r}\right)^{12} - 2\left(\frac{R_0}{r}\right)^6\right]$; Eq. (13-26): $F(r) = 12 \frac{U_0}{R_0}\left[\left(\frac{R_0}{r}\right)^{13} - \left(\frac{R_0}{r}\right)^7\right]$

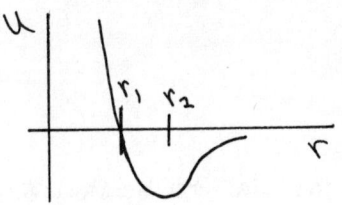
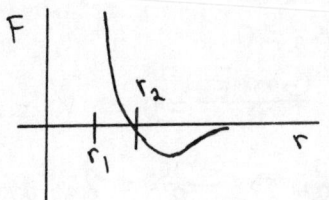

b) equilibrium $\Rightarrow F = 0$; occurs at point r_2. r_2 is where U is a minimum
(stable equilibrium)

16-51 (cont)

c) $U = 0 \Rightarrow \left[\left(\frac{R_0}{r_1}\right)^{12} - 2\left(\frac{R_0}{r_1}\right)^6\right] = 0$

$\left(\frac{r_1}{R_0}\right)^6 = 2 \Rightarrow r_1 = (2)^{1/6} R_0$

$F = 0 \Rightarrow \left(\frac{R_0}{r_2}\right)^{13} - \left(\frac{R_0}{r_2}\right)^7 = 0$

$\left(\frac{r_2}{R_0}\right)^6 = 1 \Rightarrow r_2 = R_0$

Then $\frac{r_1}{r_2} = \frac{2^{1/6} R_0}{R_0} = 2^{1/6}$

d) $W_{other} = \Delta U$

At $r \to \infty$, $U = 0$, so $W = -U(R_0) = -U_0\left[\left(\frac{R_0}{R_0}\right)^{12} - 2\left(\frac{R_0}{R_0}\right)^6\right] = \underline{+U_0}$

16-55

Calculate the number of water molecules N:

$n = \frac{m_{tot}}{M} = \frac{60 \text{ kg}}{18.0 \times 10^{-3} \text{ kg/mol}} = 3.333 \times 10^3 \text{ mol}$

$N = nN_A = (3.333 \times 10^3 \text{ mol})(6.022 \times 10^{23} \text{ molecules/mol}) = 2.0 \times 10^{27}$ molecules

Each water molecule has three atoms, so the number of atoms is
$3(2.0 \times 10^{27}) = \underline{6.0 \times 10^{27} \text{ atoms}}$

16-57

a) $v_{rms} = \sqrt{\frac{3RT}{M}} = \sqrt{\frac{3(8.3145 \text{ J/mol·K})(300 \text{ K})}{28.0 \times 10^{-3} \text{ kg/mol}}} = \underline{517 \text{ m/s}}$

b) $(v_x^2)_{av} = \frac{1}{3}(v^2)_{av} \Rightarrow \sqrt{(v_x^2)_{av}} = \frac{1}{\sqrt{3}}\sqrt{(v^2)_{av}} = \frac{1}{\sqrt{3}} v_{rms} = \frac{1}{\sqrt{3}}(517 \text{ m/s}) = \underline{298 \text{ m/s}}$

16-59

a) Apply conservation of energy: $K_1 + U_1 + W_{other} = K_2 + U_2$, where $U = -\frac{Gmm_E}{r}$.
Let point #1 be at the surface of the earth where the projectile is launched, and let point #2 be far from the earth. Just barely escapes $\Rightarrow v_2 = 0$.

Only gravity does work $\Rightarrow W_{other} = 0$.
$U_1 = -\frac{Gmm_E}{R}$
$r_2 \to \infty \Rightarrow U_2 = 0$
$v_2 = 0 \Rightarrow K_2 = 0$

Thus $K_1 - \frac{Gmm_E}{R} = 0 \Rightarrow K_1 = \frac{Gmm_E}{R}$.

But $g = G\frac{m_E}{R^2} \Rightarrow \frac{Gm_E}{R} = Rg$ and $K_1 = mgR$, as was to be shown.

16-59 (cont)

$\frac{1}{2}m(v^2)_{av} = mgR$ (from part (a))

But also, $\frac{1}{2}(v^2)_{av} = \frac{3}{2}kT \Rightarrow mgR = \frac{3}{2}kT$

$$T = \frac{2mgR}{3k}$$

oxygen
From Example 16-5, $m_{O_2} = 53.1 \times 10^{-27}$ kg/molecule

$T = \frac{2mgR}{3k} = \frac{2(53.1 \times 10^{-27} \text{ kg/molecule})(9.80 \text{ m/s}^2)(6.38 \times 10^6 \text{ m})}{3(1.381 \times 10^{-23} \text{ J/molecule} \cdot \text{K})} = \underline{1.60 \times 10^5 \text{ K}}$

hydrogen
From Example 16-5, $m_{H_2} = 2(1.674 \times 10^{-27} \text{ kg}) = 3.348 \times 10^{-27}$ kg/molecule

$T = \frac{2mgR}{3k} = \frac{2(3.348 \times 10^{-27} \text{ kg/molecule})(9.80 \text{ m/s}^2)(6.38 \times 10^6 \text{ m})}{3(1.381 \times 10^{-23} \text{ J/molecule} \cdot \text{K})} = \underline{1.01 \times 10^4 \text{ K}}$

c) $T = \frac{2mgR}{3k}$

oxygen
$T = \frac{2(53.1 \times 10^{-27} \text{ kg/molecule})(1.63 \text{ m/s}^2)(1.740 \times 10^6 \text{ m})}{3(1.381 \times 10^{-23} \text{ J/molecule} \cdot \text{K})} = \underline{7270 \text{ K}}$

hydrogen
$T = \frac{2(3.348 \times 10^{-27} \text{ kg/molecule})(1.63 \text{ m/s}^2)(1.740 \times 10^6 \text{ m})}{3(1.381 \times 10^{-23} \text{ J/molecule} \cdot \text{K})} = \underline{458 \text{ K}}$

d) The "escape temperatures" are much less for the moon than for the earth. At a given temperature, for the moon a larger fraction of the molecules will have speeds in the Maxwell-Boltzmann distribution larger than the escape speed. After a long time most of the molecules will have escaped from the moon.

16-63

a) The atoms have two degrees of freedom and hence each atom has an average kinetic energy $2(\frac{1}{2}kT) = kT$. Each atom also has potential energy due to the Hooke's law force that binds it at the surface, to its position parallel to the surface. The average potential energy for this two-dimensional harmonic oscillator equals the average kinetic energy, just as for a one-dimensional or three-dimensional oscillator. The average total energy per atom is $kT + kT = 2kT$.

For n moles of atoms the average total energy is $2nRT$ and $C_V = 2R = 2(8.3145 \text{ J/mol} \cdot \text{K}) = \underline{16.6 \text{ J/mol} \cdot \text{K}}$.

b) At low temperatures the average energy per atom is less than $2kT$ because most atoms remain in their lowest energy state. Thus the molar heat capacity will be less than the value found in part (a).

16-67

$$\int_0^\infty v^2 f(v)\, dv = 4\pi \left(\frac{m}{2\pi kT}\right)^{3/2} \int_0^\infty v^4 e^{-mv^2/2kT}\, dv$$

The integral formula with $n=2$ gives $\int_0^\infty v^4 e^{-av^2}\, dv = \frac{3}{8a^2}\sqrt{\frac{\pi}{a}}$.
Apply with $a = \frac{m}{2kT}$,

$$\Rightarrow \int_0^\infty v^2 f(v)\, dv = 4\pi \left(\frac{m}{2\pi kT}\right)^{3/2} \frac{3}{8} \left(\frac{2kT}{m}\right)^2 \sqrt{\frac{2\pi kT}{m}} = \frac{3}{2}\frac{2kT}{m} = \frac{3kT}{m}$$

Equation (16-16) says $\frac{1}{2}m(v^2)_{av} = \frac{3}{2}kT \Rightarrow (v^2)_{av} = \frac{3kT}{m}$, in agreement with our calculation.

16-71

$$\text{relative humidity} = \frac{\text{partial pressure of water vapor at temperature } T}{\text{vapor pressure of water at temperature } T}$$

The experiment shows that the dew point is $10.0\,°C$, so the partial pressure of water vapor at $40.0\,°C$ is equal to the vapor pressure at $10.0\,°C$, which is 1.23×10^3 Pa.

Thus the relative humidity $= \frac{1.23 \times 10^3 \text{ Pa}}{7.34 \times 10^3 \text{ Pa}} = 0.168 = \underline{16.8\%}$.

CHAPTER 17

Exercises 3, 5, 7, 11, 15, 17, 21, 23, 25

Problems 29, 31, 35, 37, 41, 43, 45

Exercises

17-3

$W = \int_{V_1}^{V_2} p\, dV$

Constant pressure $\Rightarrow W = p \int_{V_1}^{V_2} dV = p(V_2 - V_1)$

The problem gives T rather than p and V, so use the ideal gas law to rewrite the expression for W:

$pV = nRT \Rightarrow p_1 V_1 = nRT_1,\ p_2 V_2 = nRT_2;$ subtract $\Rightarrow p(V_2 - V_1) = nR(T_2 - T_1)$

Thus $W = nR(T_2 - T_1)$ is an alternative expression for the work in a constant pressure process for an ideal gas.

Then $W = nR(T_2 - T_1) = (3.00\text{ mol})(8.3145\text{ J/mol·K})(27°C - 147°C) = \underline{-2.99 \times 10^3 \text{ J}}$

17-5

a)

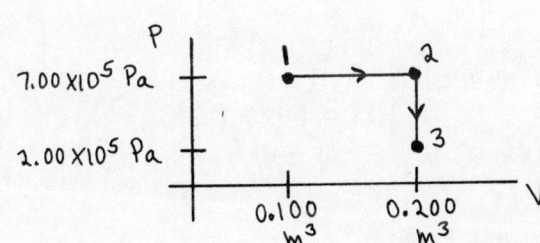

b) $1 \to 2$

p is constant, so $W = p\Delta V = (7.00 \times 10^5 \text{ Pa})(0.200\text{ m}^3 - 0.100\text{ m}^3) = 7.00 \times 10^4 \text{ J}$

$2 \to 3$

$\Delta V = 0$, so $W = 0$

$W_{tot} = W_{1 \to 2} + W_{2 \to 3} = \underline{7.00 \times 10^4 \text{ J}}$

17-7

a) For the water $\Delta T > 0$, so by $Q = mc\Delta T$ heat has been added to the water. This heat energy comes from the burning fuel-oxygen mixture $\Rightarrow$ Q for the system (fuel and oxygen) is negative.

b) Constant volume $\Rightarrow W = 0$.

c) The 1st law (Eq. 17-4) says $\Delta U = Q - W$.

$Q < 0,\ W = 0 \Rightarrow \Delta U < 0$

17-7 (cont)

The internal energy of the fuel-oxygen mixture decreased. In this process internal energy from the fuel-oxygen mixture was transferred to the water, raising its temperature.

17-11

a) $W = \int_{V_1}^{V_2} p\, dV = p(V_2 - V_1)$ for this constant pressure process.
$W = (1.70 \times 10^5 \text{ Pa})(0.80 \text{ m}^3 - 1.20 \text{ m}^3) = \underline{-6.8 \times 10^4 \text{ J}}$
(The volume decreases in the process, so W is negative.)

b) $\Delta U = Q - W$
$Q = \Delta U + W = -1.10 \times 10^5 \text{ J} + (-6.8 \times 10^4 \text{ J}) = \underline{-1.78 \times 10^5 \text{ J}}$
Q negative means heat flows out of the gas.

c) $W = \int_{V_1}^{V_2} p\, dV = p(V_2 - V_1)$ (constant pressure) and $\Delta U = Q - W$ apply to any system, not just to an ideal gas. We did not use the ideal gas equation, either directly or indirectly, in any of the calculations, so the results are the same whether the gas is ideal or not.

17-15

a)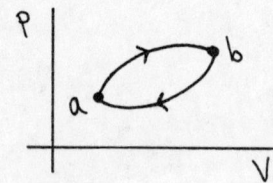

For one cycle $\Delta U = 0$
$|Q| = 6400 \text{ J}$
$\Delta U = Q - W$, so $Q = W$.

The magnitude of positive work done from a to b is larger than the magnitude of the negative work done from b to a, so the net work done in one cycle is positive. Then from $Q = W$, Q must be positive.
$Q = +6400 \text{ J}$; system <u>absorbs</u> heat.

b) $W = Q = \underline{+6400 \text{ J}}$

c)

Now the net work done is negative. Since $Q = W$, Q is negative; system liberates heat.

$|W|$ is the same as in part (b) and $Q = W$, so $Q = \underline{-6400 \text{ J}}$.

17-17

a) $W = \int_{V_1}^{V_2} p\, dV$
$pV = nRT \Rightarrow p = \frac{nRT}{V}$
$W = \int_{V_1}^{V_2} \frac{nRT}{V} dV = nRT \int_{V_1}^{V_2} \frac{dV}{V} = nRT \ln\left(\frac{V_2}{V_1}\right)$

17-17 (cont)

$W = (0.200 \text{ mol})(8.3145 \text{ J/mol·K})(300 \text{ K}) \ln\left(\frac{\frac{1}{5}V_1}{V_1}\right) = 498.8 \text{ J} \ln\left(\frac{1}{5}\right) = \underline{-803 \text{ J}}$

(W for the gas is negative, since the volume decreases.)

b) $\Delta U = nC_v \Delta T$ for any ideal gas process.
$\Delta T = 0$ (isothermal) $\Rightarrow \Delta U = 0$

c) $\Delta U = Q - W$
$\Delta U = 0 \Rightarrow Q = W = \underline{-803 \text{ J}}$
(Q negative $\Rightarrow$ the gas liberates 803 J of heat to the surroundings.)

Note: $Q = nC_v \Delta T$ is only for a constant volume process so doesn't apply here.
$Q = nC_p \Delta T$ is only for a constant pressure process so doesn't apply here.

17-21

a) $Q = nC_p \Delta T$

Need to calculate C_p from $\gamma = \frac{C_p}{C_v}$ and $C_p = C_v + R$.
Combine these equations $\Rightarrow \gamma = \frac{C_v + R}{C_v} = 1 + \frac{R}{C_v}$

$C_v = \frac{R}{\gamma - 1} = \frac{8.3145 \text{ J/mol·K}}{1.127 - 1} = 65.47 \text{ J/mol·K}$

$C_p = C_v + R = 65.47 \text{ J/mol·K} + 8.315 \text{ J/mol·K} = 73.78 \text{ J/mol·K}$

$Q = nC_p \Delta T = 1.50 (73.78 \text{ J/mol·K})(25.0°C - 20.0°C) = \underline{553 \text{ J}}$

b) Constant pressure $\Rightarrow W = p\Delta V = nR\Delta T = (1.50 \text{ mol})(8.3145 \text{ J/mol·K})(25.0°C - 20.0°C)$

$W = 62.4 \text{ J}$

$\Delta U = Q - W = 553 \text{ J} - 62 \text{ J} = \underline{491 \text{ J}}$

17-23

For an adiabatic process for an ideal gas
$T_1 V_1^{\gamma-1} = T_2 V_2^{\gamma-1}, \quad p_1 V_1^\gamma = p_2 V_2^\gamma, \quad \text{and} \quad pV = nRT$

Air is mostly diatomic (O_2, N_2, H_2) $\Rightarrow \gamma = 1.4$ (Example 17-7).

$T_1 V_1^{\gamma-1} = T_2 V_2^{\gamma-1} \Rightarrow T_2 = T_1 \left(\frac{V_1}{V_2}\right)^{\gamma-1} = (293 \text{ K})\left(\frac{V_1}{0.100 V_1}\right)^{1.4-1} = (293 \text{ K})(10.0)^{0.4}$

$T_2 = 736 \text{ K} = \underline{463°C}$

(Note: In the relation $T_1 V_1^{\gamma-1} = T_2 V_2^{\gamma-1}$ the temperature must be in kelvins.)

$p_1 V_1^\gamma = p_2 V_2^\gamma \Rightarrow p_2 = p_1 \left(\frac{V_1}{V_2}\right)^\gamma = (1.00 \text{ atm})\left(\frac{V_1}{0.100 V_1}\right)^{1.4} = (1.00 \text{ atm})(10.0)^{1.4} = \underline{25.1 \text{ atm}}$

Alternatively, can use $pV = nRT$ to calculate p_2:

n, R constant $\Rightarrow \frac{pV}{T} = nR = \text{constant} \Rightarrow \frac{p_1 V_1}{T_1} = \frac{p_2 V_2}{T_2}$

$p_2 = p_1 \left(\frac{V_1}{V_2}\right)\left(\frac{T_2}{T_1}\right) = (1.00 \text{ atm})\left(\frac{V_1}{0.100 V_1}\right)\left(\frac{736 \text{ K}}{293 \text{ K}}\right) = 25.1 \text{ atm}$, which checks.

17-25

a)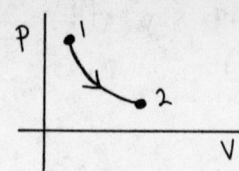

b) Adiabatic $\Rightarrow Q=0$
Then $\Delta U = Q - W \Rightarrow W = -\Delta U = -nC_v \Delta T = nC_v(T_1 - T_2)$ (Eq. 17-25)
$C_v = 20.85$ J/mol·K (Table 17-1)
$W = (0.750 \text{ mol})(20.85 \text{ J/mol·K})(313.1 \text{ K} - 283.1 \text{ K}) = \underline{+469 \text{ J}}$
(W positive for $\Delta V > 0$ (expansion))

c) Adiabatic process $\Rightarrow \underline{Q=0}$.

d) $\Delta U = Q - W$
$Q = 0 \Rightarrow \Delta U = -W = \underline{-469 \text{ J}}$
(In an adiabatic expansion of an ideal gas T decreases and U decreases.)

Problems

17-29

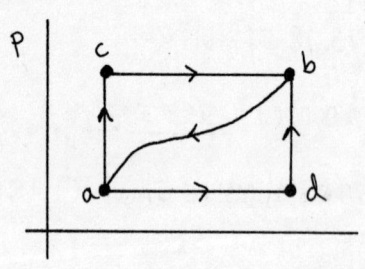

$Q_{acb} = +90.0$ J (positive since heat flows in)

$W_{acb} = +70.0$ J (positive since $\Delta V > 0$)

a) $\Delta U = Q - W$
ΔU is path independent; Q and W depend on the path.

$\Delta U = U_b - U_a$
This can be calculated for any path from a to b, in particular for path acb:
$\Delta U_{a \to b} = Q_{acb} - W_{acb} = 90.0 \text{ J} - 70.0 \text{ J} = \underline{20.0 \text{ J}}$

Now apply $\Delta U = Q - W$ to path adb; $\Delta U = 20.0$ J for this path also.
$W_{adb} = +15.0$ J (positive since $\Delta V > 0$)
$\Delta U_{a \to b} = Q_{adb} - W_{adb} \Rightarrow Q_{adb} = \Delta U_{a \to b} + W_{adb} = +20.0 \text{ J} + 15.0 \text{ J} = \underline{+35.0 \text{ J}}$

b) Apply $\Delta U = Q - W$ to path ba:
$\Delta U_{b \to a} = Q_{ba} - W_{ba}$
$W_{ba} = -45.0$ J (negative since $\Delta V < 0$)
$\Delta U_{b \to a} = U_a - U_b = -(U_b - U_a) = -\Delta U_{a \to b} = -20.0 \text{ J}$
Then $Q_{ba} = \Delta U_{b \to a} + W_{ba} = -20.0 \text{ J} - 45.0 \text{ J} = \underline{-65.0 \text{ J}}$
($Q_{ba} < 0 \Rightarrow$ the system liberates heat)

17-29 (cont)

c) $U_a = 0$, $U_d = 8.0 \text{ J}$
$\Delta U_{a \to b} = U_b - U_a = +20.0 \text{ J} \Rightarrow U_b = 20.0 \text{ J}$

process a → d
$\Delta U_{a \to d} = Q_{ad} - W_{ad}$
$\Delta U_{a \to d} = U_d - U_a = +8.0 \text{ J}$

$W_{adb} = +15.0 \text{ J}$ and $W_{adb} = W_{ad} + W_{db}$. But the work W_{db} for the process d → b is zero since $\Delta V = 0$ for that process. Therefore $W_{ad} = W_{adb} = +15.0 \text{ J}$.

Then $Q_{ad} = \Delta U_{a \to d} + W_{ad} = +8.0 \text{ J} + 15.0 \text{ J} = \underline{+23.0 \text{ J}}$ (positive ⇒ heat absorbed)

process d → b
$\Delta U_{d \to b} = Q_{db} - W_{db}$

$W_{db} = 0$, as already noted
$\Delta U_{d \to b} = U_b - U_d = 20.0 \text{ J} - 8.0 \text{ J} = 12.0 \text{ J}$

Then $Q_{db} = W_{db} + \Delta U_{d \to b} = \underline{+12.0 \text{ J}}$ (positive ⇒ heat absorbed)

17-31

$\Delta U = Q - W$
$Q = +3.65 \times 10^5 \text{ J}$ (positive since heat energy goes into system)
$\Delta U = 0 \Rightarrow W = Q = 3.65 \times 10^5 \text{ J}$

constant pressure ⇒ $W = \int_{V_1}^{V_2} p \, dV = p(V_2 - V_1) = p \Delta V$

Then $\Delta V = \dfrac{W}{p} = \dfrac{3.65 \times 10^5 \text{ J}}{7.20 \times 10^5 \text{ Pa}} = \underline{0.507 \text{ m}^3}$

17-35

The heat produced from the reaction is $Q_{reaction} = m L_{reaction}$, where $L_{reaction}$ is the heat of reaction of the chemicals.
$Q_{reaction} = W + \Delta U_{spray}$

For a mass m of spray,
$W = \tfrac{1}{2} m v^2 = \tfrac{1}{2} m (19 \text{ m/s})^2 = (180.5 \text{ J/kg}) m$
and $\Delta U_{spray} = Q_{spray} = m c \Delta T = m(4190 \text{ J/kg·K})(100°C - 20°C) = (335,200 \text{ J/kg}) m$.
Then $Q_{reaction} = (180 \text{ J/kg} + 335,200 \text{ J/kg}) m = (335,380 \text{ J/kg}) m$
and $Q_{reaction} = m L_{reaction} \Rightarrow \cancel{m} L_{reaction} = (335,380 \text{ J/kg}) \cancel{m}$
$\Rightarrow L_{reaction} = \underline{3.4 \times 10^5 \text{ J/kg}}$

(Note that the amount of energy converted to work is negligible, for the two significant figures to which the answer should be expressed.)

17-37

a)
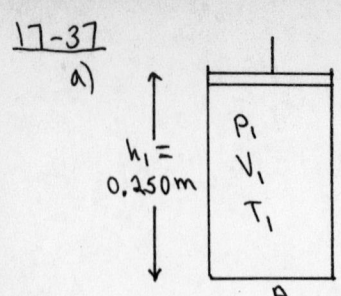 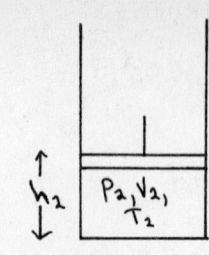

$P_1 = 1.01 \times 10^5$ Pa
$P_2 = 5.10 \times 10^5$ Pa $+ P_{air} = 6.11 \times 10^5$ Pa

$V_1 = h_1 A$
$V_2 = h_2 A$

adiabatic process $\Rightarrow P_1 V_1^\gamma = P_2 V_2^\gamma$; $\gamma = 1.4$ for air

$P_1 h_1^\gamma A^\gamma = P_2 h_2^\gamma A^\gamma$

$h_2 = h_1 \left(\frac{P_1}{P_2}\right)^{1/\gamma} = (0.250 \text{ m})\left(\frac{1.01 \times 10^5 \text{ Pa}}{6.11 \times 10^5 \text{ Pa}}\right)^{1/1.4} = 0.0691 \text{ m}$

The piston has moved a distance $h_1 - h_2 = 0.250 \text{ m} - 0.0691 \text{ m} = \underline{0.181 \text{ m}}$

b) $T_1 V_1^{\gamma-1} = T_2 V_2^{\gamma-1}$
$T_1 h_1^{\gamma-1} A^{\gamma-1} = T_2 h_2^{\gamma-1} A^{\gamma-1}$

$T_2 = T_1 \left(\frac{h_1}{h_2}\right)^{\gamma-1} = 300.1 \text{ K} \left(\frac{0.250 \text{ m}}{0.0691 \text{ m}}\right)^{0.4} = 501.9 \text{ K} = \underline{229°C}$

c) $W = nC_v(T_1 - T_2)$ (Eq. 17-25)
$W = (20.0 \text{ mol})(20.8 \text{ J/mol·K})(300.1 \text{ K} - 501.9 \text{ K}) = \underline{-8.39 \times 10^4 \text{ J}}$

17-41

a) <u>isothermal</u> ($\Delta T = 0$)
$\Delta U = Q - W$
$W = +400 \text{ J}$

For any process of an ideal gas, $\Delta U = nC_v \Delta T$. So for an ideal gas, if $\Delta T = 0$ then $\Delta U = 0$. $\Rightarrow Q = W = \underline{+400 \text{ J}}$

b) <u>adiabatic</u> ($Q = 0$)
$\Delta U = Q - W$; $W = +400 \text{ J}$
$Q = 0 \Rightarrow \Delta U = -W = \underline{-400 \text{ J}}$

17-43

a) In the adiabatic expansion the pressure decreases.

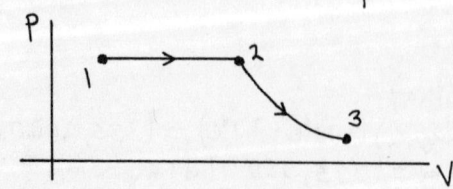

b) <u>process $1 \to 2$</u> ($\Delta p = 0$)
$pV = nRT$ and p constant $\Rightarrow \frac{V_1}{T_1} = \frac{V_2}{T_2}$

$T_2 = T_1 \left(\frac{V_2}{V_1}\right) = (300 \text{ K})\left(\frac{2V_1}{V_1}\right) = \underline{600 \text{ K}}$

234

17-43 (cont)

process 2→3 (Q=0)
$T_2 = 600$ K
Temperature returns to its initial value ⇒ $T_3 = T_1 = 300$ K
$Q_{12} = nC_p \Delta T$, since $\Delta p = 0$ for this process.
$Q_{12} = (0.750 \text{ mol})(20.78 \text{ J/mol·K})(600\text{K} - 300\text{K}) = 4.68 \times 10^3$ J
$Q_{23} = 0$ since process 2→3 is adiabatic
Then $Q = Q_{12} + Q_{23} = 4.68 \times 10^3$ J

c) $\Delta U_{12} = nC_v \Delta T = nC_v(T_2 - T_1)$
$\Delta U_{23} = nC_v \Delta T = nC_v(T_3 - T_2)$
$\Delta U_{tot} = \Delta U_{12} + \Delta U_{23} = nC_v(T_2 - T_1 + T_3 - T_2) = nC_v(T_3 - T_1) = 0$, since $T_1 = T_3$;
$\Delta U_{tot} = 0$ since the ideal gas ends up with the same temperature as at the start

d) $\Delta U = Q - W$
$W = Q - \Delta U$
$\Delta U = 0 \Rightarrow W = Q = \underline{4.68 \times 10^3 \text{ J}}$

e) In process 1→2 the volume doubles, so $V_2 = 0.0400$ m³.

Process 2→3 is adiabatic and we know T_2 and T_3. To find what happens to the volume use $T_2 V_2^{\gamma-1} = T_3 V_3^{\gamma-1}$.

$V_3^{\gamma-1} = V_2^{\gamma-1}\left(\frac{T_2}{T_3}\right) \Rightarrow V_3 = V_2\left(\frac{T_2}{T_3}\right)^{1/(\gamma-1)}$

For helium $\gamma = 1.67$ (from Table 17-1), so $V_3 = (0.0400 \text{ m}^3)\left(\frac{600 \text{ K}}{300 \text{ K}}\right)^{\frac{1}{(1.67-1)}} = \underline{0.113 \text{ m}^3}$
(Note that $V_3 > V_2$ as it should be, since process 2→3 is an expansion.)

17-45

a) initial expansion (state 1 → state 2)
$P_1 = 4.00 \times 10^5$ Pa, $T_1 = 300$ K, $P_2 = 4.00 \times 10^5$ Pa, $V_2 = 2V_1$

$pV = nRT \Rightarrow \frac{T}{V} = \frac{p}{nR} = $ constant, so $\frac{T_1}{V_1} = \frac{T_2}{V_2} \Rightarrow T_2 = T_1\left(\frac{V_2}{V_1}\right) = 300\text{ K}\left(\frac{2V_1}{V_1}\right) = 600$ K

$\Delta p = 0 \Rightarrow W = p\Delta V = nR\Delta T = (0.500 \text{ mol})(8.3145 \text{ J/mol·K})(600\text{K} - 300\text{K}) = \underline{1250 \text{ J}}$
$Q = nC_p \Delta T = (0.500 \text{ mol})(29.17 \text{ J/mol·K})(600\text{K} - 300\text{K}) = \underline{4380 \text{ J}}$
$\Delta U = Q - W = 4380 \text{ J} - 1250 \text{ J} = \underline{3130 \text{ J}}$

b) At the beginning of the final process (cooling at constant volume), T = 600 K. The gas returns to its original volume and pressure, so also to its original temperature of 300 K.
$\Delta V = 0 \Rightarrow \underline{W = 0}$
$Q = nC_v \Delta T = (0.500 \text{ mol})(20.85 \text{ J/mol·K})(300\text{K} - 600\text{K}) = \underline{-3130 \text{ J}}$
$\Delta U = Q - W = \underline{-3130 \text{ J}}$

c) For any ideal gas process $\Delta U = nC_v \Delta T$. For an isothermal process $\Delta T = 0$, so $\Delta U = 0$.

CHAPTER 18

Exercises 5, 7, 9, 11, 15, 17, 21, 23, 25, 29, 35, 37

Problems 39, 41, 43, 45, 53

Exercises

18-5

a) $e = \dfrac{\text{work output}}{\text{heat energy input}} = \dfrac{W}{Q_H} = \dfrac{2000 \text{ J}}{8000 \text{ J}} = 0.250 = \underline{25.0\%}$

b) $W = Q = |Q_H| - |Q_c|$
Heat discarded is $|Q_c| = |Q_H| - W = 8000 \text{ J} - 2000 \text{ J} = \underline{6000 \text{ J}}$

c) Q_H is supplied by the burning fuel; $Q_H = mL_c$ where L_c is the heat of combustion
$m = \dfrac{Q_H}{L_c} = \dfrac{8000 \text{ J}}{4.60 \times 10^4 \text{ J/g}} = \underline{0.174 \text{ g}}$

d) $W = 2000 \text{ J}$ per cycle
In $t = 1.00 \text{ s}$ the engine goes through 80 cycles.
$P = \dfrac{W}{t} = \dfrac{80(2000 \text{ J})}{1.00 \text{ s}} = 1.60 \times 10^5 \text{ W}$
$P = (1.60 \times 10^5 \text{ W})\left(\dfrac{1 \text{ hp}}{746 \text{ W}}\right) = \underline{214 \text{ hp}}$

18-7

a) Process $a \to b$ is adiabatic $\Rightarrow T_a V_a^{\gamma-1} = T_b V_b^{\gamma-1}$
$V_b = V$, $V_a = rV$ (read from Fig. 18-3)
$T_a = 22.0°C = 295 \text{ K}$

$T_b = T_a \left(\dfrac{V_a}{V_b}\right)^{\gamma-1} = (295 \text{ K})\left(\dfrac{rV}{V}\right)^{1.40-1} = (295 \text{ K})(8.00)^{0.4} = 678 \text{ K} = \underline{405°C}$

b) $pV = nRT \Rightarrow \dfrac{pV}{T} = \text{constant} \Rightarrow \dfrac{p_a V_a}{T_a} = \dfrac{p_b V_b}{T_b}$

$p_b = p_a \left(\dfrac{V_a}{V_b}\right)\left(\dfrac{T_b}{T_a}\right) = 8.50 \times 10^4 \text{ Pa} \left(\dfrac{rV}{V}\right)\left(\dfrac{678 \text{ K}}{295 \text{ K}}\right) = \underline{1.56 \times 10^6 \text{ Pa}}$

18-9

a) $Q_H < 0$, refrigerator, $W < 0$, $Q_c > 0$

$Q_c = +9.00 \times 10^4 \text{ J}$
$Q_H = -1.40 \times 10^5 \text{ J}$
$W = Q_c + Q_H = +9.00 \times 10^4 \text{ J} - 1.40 \times 10^5 \text{ J} = -5.00 \times 10^4 \text{ J}$
(W negative $\Rightarrow$ power consumed not produced by device)
$P = \dfrac{W}{t} = \dfrac{-5.00 \times 10^4 \text{ J}}{60.0 \text{ s}} = \underline{-833 \text{ W}}$

b) $K = \left|\dfrac{Q_c}{W}\right| = \dfrac{9.00 \times 10^4 \text{ J}}{5.00 \times 10^4 \text{ J}} = \underline{1.80}$

18-11

a) Performance coefficient $K = \frac{Q_c}{|W|}$ (Eq. 18-9)

$\Rightarrow |W| = \frac{Q_c}{K} = \frac{3.00 \times 10^4 \text{ J}}{2.20} = \underline{1.36 \times 10^4 \text{ J}}$

b)

$Q_H < 0$ ↑
refrigerator ← $W < 0$
$Q_c > 0$ ↑

$W = Q_c + Q_H$
$Q_H = W - Q_c = -1.36 \times 10^4 \text{ J} - 3.00 \times 10^4 \text{ J} = \underline{-4.36 \times 10^4 \text{ J}}$
(negative because heat goes out of the system)

18-15

a) $T_H = 500 \text{ K}$
$Q_H > 0$ ↓
engine → $W > 0$
$Q_c < 0$ ↓
T_c

For a Carnot cycle,

$\frac{|Q_c|}{|Q_H|} = \frac{T_c}{T_H}$ (Eq. 18-13)

$T_c = T_H \frac{|Q_c|}{|Q_H|} = 500 \text{K} \left(\frac{335 \text{ J}}{620 \text{ J}}\right) = \underline{270 \text{ K}}$

b) $e(\text{Carnot}) = 1 - \frac{T_c}{T_H} = 1 - \frac{270 \text{ K}}{500 \text{ K}} = 0.460 = \underline{46.0\%}$

c) $W = Q_c + Q_H = -335 \text{ J} + 620 \text{ J} = \underline{285 \text{ J}}$

or

$e = \frac{W}{Q_H} \Rightarrow W = e Q_H = (0.460)(620 \text{ J}) = 285 \text{ J}$, which checks

18-17

a)

$Q_H < 0$ ↑
refrigerator ← $W < 0$
$Q_c > 0$ ↑

$T_H = 22.0°C = 295 \text{ K}$
$T_c = 0.0°C = 273 \text{ K}$

The amount of heat taken out of the water to make the liquid→solid phase change is $Q = -mL_f = -(40.0 \text{ kg})(334 \times 10^3 \text{ J/kg}) = -1.336 \times 10^7 \text{ J}$.
This amount of heat must go into the working substance of the refrigerator
$\Rightarrow Q_c = +1.336 \times 10^7 \text{ J}$.

Carnot cycle $\Rightarrow \frac{|Q_c|}{|Q_H|} = \frac{T_c}{T_H}$

$|Q_H| = |Q_c|\left(\frac{T_H}{T_c}\right) = 1.336 \times 10^7 \text{ J} \left(\frac{295 \text{ K}}{273 \text{ K}}\right) = \underline{1.44 \times 10^7 \text{ J}}$

b) $W = Q_c + Q_H = +1.336 \times 10^7 \text{ J} - 1.444 \times 10^7 \text{ J} = \underline{-1.08 \times 10^6 \text{ J}}$
(W is negative because this much energy must be supplied to the refrigerator rather than obtained from it.)

18-21

$\Delta S = \dfrac{Q}{T}$ for an isothermal process

$Q = -mL_v = -(0.600 \text{ kg})(2256 \times 10^3 \text{ J/kg}) = -1.354 \times 10^6 \text{ J}$

$T = 100°C = (100 + 273) \text{ K} = 373 \text{ K}$

$\Delta S = \dfrac{Q}{T} = \dfrac{-1.354 \times 10^6 \text{ J}}{373 \text{ K}} = \underline{-3.63 \times 10^3 \text{ J/K}}$

18-23

First use $Q_{system} = 0$ to find the final temperature of the mixed water.
The 2.00 kg of water warms from 20.0°C to T:
$Q = mc\,\Delta T = (2.00 \text{ kg})(4190 \text{ J/kg·K})(T - 20.0°C)$
The 1.00 kg of water cools from 70.0°C to T:
$Q = mc\,\Delta T = (1.00 \text{ kg})(4190 \text{ J/kg·K})(T - 70.0°C)$
$Q_{system} = 0 \Rightarrow (2.00 \text{ kg})(4190 \text{ J/kg·K})(T - 20.0°C) + (1.00 \text{ kg})(4190 \text{ J/kg·K})(T - 70.0°C) = 0$
$2(T - 20.0°C) + (T - 70.0°C) = 0$
$3T = 110°C$
$T = 36.67°C$

Now we can calculate the entropy change.
The 2.00 kg of water:
$\Delta S = \int_1^2 \dfrac{dQ}{T}$

$dQ = mc\,dT \Rightarrow \Delta S = mc \int_1^2 \dfrac{dT}{T} = mc \ln\left(\dfrac{T_2}{T_1}\right)$ (Example 18-6)

$\Delta S = (2.00 \text{ kg})(4190 \text{ J/kg·K}) \ln\left(\dfrac{(36.67 + 273.15) \text{ K}}{(20.0 + 273.15) \text{ K}}\right) = 463.5 \text{ J/K}$

The 1.00 kg of water:
$\Delta S = mc \ln\left(\dfrac{T_2}{T_1}\right) = (1.00 \text{ kg})(4190 \text{ J/kg·K}) \ln\left(\dfrac{(36.67 + 273.15) \text{ K}}{(70.0 + 273.15) \text{ K}}\right) = -428.1 \text{ J/K}$

The total entropy change of the system is
$\Delta S_{tot} = 463.5 \text{ J/K} - 428.1 \text{ J/K} = \underline{35 \text{ J/K}}$

18-25

Reversible $\Rightarrow \Delta S = \int_1^2 \dfrac{dQ}{T}$
Isothermal $\Rightarrow T$ is constant $\Rightarrow \Delta S = \dfrac{Q}{T}$

Calculate Q:
$\Delta U = Q - W$
$\Delta U = nC_v \Delta T = 0$ since $\Delta T = 0$
$W = \int_{V_1}^{V_2} p\,dV\,;\ pV = nRT \Rightarrow p = \dfrac{nRT}{V}$, where T is constant
$W = nRT \int_{V_1}^{V_2} \dfrac{dV}{V} = nRT \ln\left(\dfrac{V_2}{V_1}\right) = (2.00 \text{ mol})(8.3145 \text{ J/mol·K})(300 \text{ K}) \ln\left(\dfrac{0.0450 \text{ m}^3}{0.0200 \text{ m}^3}\right)$
$W = 4.045 \times 10^3 \text{ J}$. Then $Q = \Delta U + W = 0 + 4.045 \times 10^3 \text{ J} = 4.045 \times 10^3 \text{ J}$
$\Delta S = \dfrac{Q}{T} = \dfrac{4.045 \times 10^3 \text{ J}}{300 \text{ K}} = \underline{+13.5 \text{ J/K}}$

18-29

a) The velocity distribution of Eq. (16-32) depends only on T, so in an isothermal process it does not change.

b) Following the reasoning of Example 18-11, the number of possible positions available to each molecule is altered by a factor of $\frac{1}{3}$ (becomes smaller). Hence the number of microscopic states the gas occupies at volume $V/3$ is $w_2 = (\frac{1}{3})^N w_1$, where N is the number of molecules and w_1 is the number of possible microscopic states at the start of the process, where the volume is V. Then, by Eq. (18-23),

$$\Delta S = k \ln\left(\frac{w_2}{w_1}\right) = k \ln\left(\frac{1}{3}\right)^N = N k \ln\left(\frac{1}{3}\right) = n N_A k \ln\left(\frac{1}{3}\right) = n R \ln\left(\frac{1}{3}\right)$$

$$\Delta S = (2.00 \text{ mol})(8.3145 \text{ J/mol·K}) \ln\left(\frac{1}{3}\right) = \underline{-18.3 \text{ J/K}}$$

c) $\Delta S = \frac{Q}{T}$

Need to calculate Q:
$\Delta T = 0 \Rightarrow \Delta U = 0$, since is an ideal gas
Then by $\Delta U = Q - W$, $Q = W$.
For an isothermal process, $W = \int_{V_1}^{V_2} p \, dV = \int_{V_1}^{V_2} \frac{nRT}{V} dV = nRT \ln\left(\frac{V_2}{V_1}\right)$

Thus $Q = nRT \ln\left(\frac{V_2}{V_1}\right)$ and

$\Delta S = \frac{Q}{T} = nR \ln\frac{V_2}{V_1} = (2.00 \text{ mol})(8.3145 \text{ J/mol·K}) \ln\left(\frac{\frac{1}{3}V_1}{V_1}\right) = \underline{-18.3 \text{ J/K}}$

Same as result obtained in part (b).

18-35

a) Total energy collected in one hour:
$(200 \text{ W/m}^2)(3600 \text{ s})(8.0 \text{ m}^2)(0.60) = 3.456 \times 10^6 \text{ J}$

This amount of energy is transferred to the water, to raise its temperature.
$Q = mc\Delta T$

$m = \frac{Q}{c\Delta T} = \frac{3.456 \times 10^6 \text{ J}}{(4190 \text{ J/kg·K})(55.0°C - 15.0°C)} = 20.62 \text{ kg}$

$V = \frac{m}{\rho} = \frac{20.62 \text{ kg}}{1000 \text{ kg/m}^3} = 0.0206 \text{ m}^3 = \underline{20.6 \text{ L}}$

b) The answer in part (a) is for one hour. The volume of 55.0°C water produced in one day (24 hours) is $24(20.6 \text{ L}) = 494 \text{ L}$.
The number of inhabitants is $\frac{494 \text{ L}}{75 \text{ L/inhabitant}} = 6.6$, so $\underline{6 \text{ inhabitants}}$ can be supported, but not 7.

18-37

a) Calculate the mass of water required and then from this the volume of water:

$Q = mc\Delta T \Rightarrow m = \frac{Q}{c\Delta T} = \frac{4.00 \times 10^9 \text{ J}}{(4190 \text{ J/kg·K})(49.0°C - 21.0°C)} = 3.409 \times 10^4 \text{ kg}$

$\rho = \frac{m}{V} \Rightarrow V = \frac{m}{\rho} = \frac{3.409 \times 10^4 \text{ kg}}{1.00 \times 10^3 \text{ kg/m}^3} = \underline{34.1 \text{ m}^3}$

18-37 (cont)

b) Repeat the above calculation, but now with Glauber salt in place of water. The essential difference is that Glauber salt undergoes a phase transition in this temperature range (at 32.0°C).

$Q = mc_{solid}(32.0°C - 21.0°C) + mL_f + mc_{liquid}(49.0°C - 32.0°C)$

$Q = m[(1930 \text{ J/kg·K})(11.0 \text{ K}) + 2.42 \times 10^5 \text{ J/kg} + (2850 \text{ J/kg·K})(17.0 \text{ K})]$

$Q = m[3.117 \times 10^5 \text{ J/kg}]$

Then $m = \dfrac{Q}{3.117 \times 10^5 \text{ J/kg}} = \dfrac{4.00 \times 10^9 \text{ J}}{3.117 \times 10^5 \text{ J/kg}} = \underline{1.283 \times 10^4 \text{ kg}}$

$V = \dfrac{m}{\rho} = \dfrac{1.283 \times 10^4 \text{ kg}}{1600 \text{ kg/m}^3} = \underline{8.0 \text{ m}^3}$

(The space requirement is smaller by about a factor of 4 when Glauber salt is used instead of water.)

Problems

18-39

a) Eq.(18-6): $e = 1 - \dfrac{1}{r^{\gamma-1}} = 1 - \dfrac{1}{9.5^{(1.4-1)}} = 0.5936$

$e = \dfrac{Q_H + Q_C}{Q_H}$ and we are given $Q_H = 200 \text{ J}$; calculate Q_C.

$Q_C = (e-1)Q_H = (0.5936 - 1)(200 \text{ J}) = \underline{-81.3 \text{ J}}$ (negative, since corresponds to heat leaving)

Then $W = Q_C + Q_H = -81.3 \text{ J} + 200 \text{ J} = \underline{119 \text{ J}}$ (Positive, in agreement with Fig. 18-3)

b) For each cylinder of area $A = \pi\left(\dfrac{d}{2}\right)^2$ the piston moves 0.909 m and the volume changes from rV to V.

$\ell_1 A = rV$
$\ell_2 A = V$
and $\ell_1 - \ell_2 = 90.9 \times 10^{-3} \text{ m}$

$\ell_1 A - \ell_2 A = rV - V$
$(\ell_1 - \ell_2) A = (r-1) V$

$V = \dfrac{(\ell_1 - \ell_2)A}{r-1} = \dfrac{(90.9 \times 10^{-3} \text{ m}) \pi \left(\dfrac{87.1 \times 10^{-3} \text{ m}}{2}\right)^2}{9.5 - 1} = 6.372 \times 10^{-5} \text{ m}^3$

At point a the volume is $rV = 9.5(6.372 \times 10^{-5} \text{ m}^3) = \underline{6.05 \times 10^{-4} \text{ m}^3}$

c) **point a:** $T_a = 300 \text{ K}$, $p_a = 8.50 \times 10^4 \text{ Pa}$, and $V_a = 6.05 \times 10^{-4} \text{ m}^3$

point b:
$V_b = \dfrac{V_a}{r} = \underline{6.37 \times 10^{-5} \text{ m}^3}$

18-39 (cont)

process a→b is adiabatic, so $T_a V_a^{\gamma-1} = T_b V_b^{\gamma-1}$
$$T_a (rV)^{\gamma-1} = T_b V^{\gamma-1}$$
$$T_b = T_a r^{\gamma-1} = 300K \,(9.5)^{1.4-1} = \underline{738 K}$$

$pV = nRT \Rightarrow \frac{pV}{T} = nR = $ constant, so $\frac{P_a V_a}{T_a} = \frac{P_b V_b}{T_b}$

$$P_b = P_a \left(\frac{V_a}{V_b}\right)\left(\frac{T_b}{T_a}\right) = (8.50 \times 10^4 \,Pa)\left(\frac{rV}{V}\right)\left(\frac{738K}{300K}\right) = \underline{1.99 \times 10^6 \,Pa}$$

point c

process b→c is at constant volume, so $V_c = V_b = \underline{6.37 \times 10^{-5} \,m^3}$

$Q_H = nC_V \Delta T = nC_V (T_c - T_b)$. The problem specifies $Q_H = 200J$; use to calculate T_c.
First use the p,V,T values at point a to calculate the number of moles n.

$$n = \frac{pV}{RT} = \frac{(8.50 \times 10^4 \,Pa)(6.05 \times 10^{-4} \,m^3)}{(8.3145 \,J/mol \cdot K)(300 \,K)} = 0.02062 \,mol$$

Then $T_c - T_b = \frac{Q_H}{nC_V} = \frac{200 J}{(0.02062 \,mol)(20.5 \,J/mol \cdot K)} = 473.2 \,K$, and
$T_c = T_b + 473.2\,K = 738\,K + 473\,K = \underline{1210 \,K}$

$\frac{P}{T} = \frac{nR}{V} = $ constant $\Rightarrow \frac{P_b}{T_b} = \frac{P_c}{T_c}$

$$P_c = P_b \left(\frac{T_c}{T_b}\right) = (1.99 \times 10^6 \,Pa)\left(\frac{1210K}{738K}\right) = \underline{3.26 \times 10^6 \,Pa}$$

point d:

$$V_d = V_a = \underline{6.05 \times 10^{-4} \,m^3}$$

process c→d is adiabatic, so $T_d V_d^{\gamma-1} = T_c V_c^{\gamma-1}$
$$T_d (rV)^{\gamma-1} = T_c V^{\gamma-1}$$

$$T_d = \frac{T_c}{r^{\gamma-1}} = \frac{1210K}{9.5^{1.4-1}} = \underline{492\,K}$$

$\frac{P_c V_c}{T_c} = \frac{P_d V_d}{T_d}$

$$P_d = P_c \left(\frac{V_c}{V_d}\right)\left(\frac{T_d}{T_c}\right) = (3.26 \times 10^6 \,Pa)\left(\frac{V}{rV}\right)\left(\frac{492K}{1210K}\right) = \underline{1.40 \times 10^5 \,Pa}$$

Can look at process d→a as a check:
$Q_c = nC_V (T_a - T_d) = (0.02062 \,mol)(20.5 \,J/mol \cdot K)(300K - 492K) = -81 J$, which agrees with part (a).

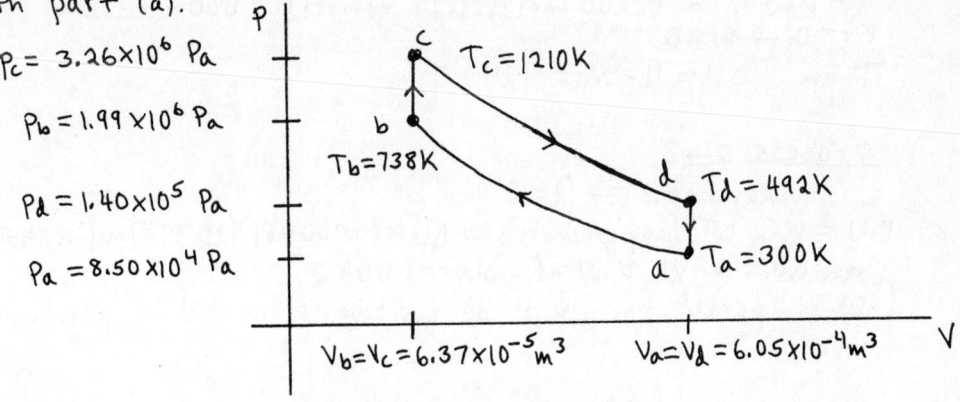

$P_c = 3.26 \times 10^6 \,Pa$
$P_b = 1.99 \times 10^6 \,Pa$
$P_d = 1.40 \times 10^5 \,Pa$
$P_a = 8.50 \times 10^4 \,Pa$

$T_c = 1210 K$
$T_b = 738 K$
$T_d = 492 K$
$T_a = 300 K$

$V_b = V_c = 6.37 \times 10^{-5} m^3 \qquad V_a = V_d = 6.05 \times 10^{-4} m^3$

18-39 (cont)

d) From part (a) the efficiency of this Otto cycle is $e = 0.594 = 59.4\%$.

The efficiency of a Carnot cycle operating between 1210 K and 300 K is
$$e(\text{Carnot}) = 1 - \frac{T_C}{T_H} = 1 - \frac{300 \text{ K}}{1210 \text{ K}} = 0.752 = 75.2\%, \text{ which is larger.}$$

18-41

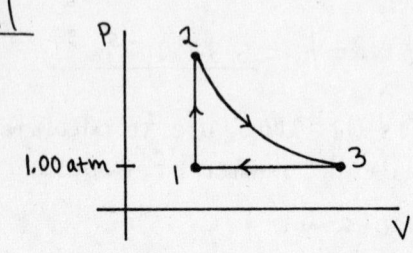

$T_1 = 300 \text{ K}$ $\gamma = 1.40$
$T_2 = 600 \text{ K}$ $C_V = 20.79 \text{ J/mol·K}$
$T_3 = 455 \text{ K}$ $C_p = 29.10 \text{ J/mol·K}$
(since $C_V = \frac{R}{\gamma - 1}$ and $C_p = C_V + R$)

a) **point 1**
$P_1 = 1.00 \text{ atm} = \underline{1.013 \times 10^5 \text{ Pa}}$ (given)

$pV = nRT \Rightarrow V_1 = \frac{nRT_1}{P_1} = \frac{(0.200 \text{ mol})(8.3145 \text{ J/mol·K})(300 \text{ K})}{1.013 \times 10^5 \text{ Pa}} = \underline{4.92 \times 10^{-3} \text{ m}^3}$

point 2
process $1 \to 2$ is at constant volume $\Rightarrow V_2 = V_1 = \underline{4.92 \times 10^{-3} \text{ m}^3}$

$pV = nRT$ and n, R, V constant $\Rightarrow \frac{P_1}{T_1} = \frac{P_2}{T_2}$

$P_2 = P_1 \left(\frac{T_2}{T_1}\right) = 1.00 \text{ atm}\left(\frac{600 \text{ K}}{300 \text{ K}}\right) = 2.00 \text{ atm} = \underline{2.027 \times 10^5 \text{ Pa}}$

point 3
Consider the process $3 \to 1$, since it is simpler than $2 \to 3$.
process $3 \to 1$ is at constant pressure $\Rightarrow P_3 = P_1 = 1.00 \text{ atm} = \underline{1.013 \times 10^5 \text{ Pa}}$

$pV = nRT$ and n, R, p constant $\Rightarrow \frac{V_1}{T_1} = \frac{V_3}{T_3}$

$V_3 = V_1 \left(\frac{T_3}{T_1}\right) = (4.92 \times 10^{-3} \text{ m}^3)\left(\frac{455 \text{ K}}{300 \text{ K}}\right) = \underline{7.47 \times 10^{-3} \text{ m}^3}$

b) **process $1 \to 2$**
constant volume ($\Delta V = 0$)
$Q = nC_V \Delta T = (0.200 \text{ mol})(20.79 \text{ J/mol·K})(600 \text{ K} - 300 \text{ K}) = 1247 \text{ J}$
$\Delta V = 0 \Rightarrow W = 0$
Then $\Delta U = Q - W = 1247 \text{ J}$

process $2 \to 3$
adiabatic $\Rightarrow Q = 0$
$\Delta U = nC_V \Delta T$ (any process) $\Rightarrow \Delta U = (0.200 \text{ mol})(20.79 \text{ J/mol·K})(455 \text{ K} - 600 \text{ K}) = -603 \text{ J}$
Then $\Delta U = Q - W \Rightarrow W = Q - \Delta U = +603 \text{ J}$.
(It is correct for W to be positive since ΔV is positive.)

18-41 (cont)

process 3→1

constant pressure ⇒ $W = p\Delta V = (1.013 \times 10^5 \text{Pa})(4.92 \times 10^{-3} \text{m}^3 - 7.47 \times 10^{-3} \text{m}^3) = -258 \text{ J}$

or $W = nR\Delta T = (0.200 \text{ mol})(8.3145 \text{ J/mol·K})(300 \text{ K} - 455 \text{ K}) = -258 \text{ J}$, which checks.

(It is correct for W to be negative, since ΔV is negative for this process.)

$Q = nC_p \Delta T = (0.200 \text{ mol})(29.10 \text{ J/mol·K})(300 \text{ K} - 455 \text{ K}) = -902 \text{ J}$

$\Delta U = Q - W = -902 \text{ J} - (-258 \text{ J}) = -644 \text{ J}$

or $\Delta U = nC_v \Delta T = (0.200 \text{ mol})(20.79 \text{ J/mol·K})(300 \text{ K} - 455 \text{ K}) = -644 \text{ J}$, which checks

c) $W_{net} = W_{1 \to 2} + W_{2 \to 3} + W_{3 \to 1} = 0 + 603 \text{ J} - 258 \text{ J} = \underline{+345 \text{ J}}$

d) $Q_{net} = Q_{1 \to 2} + Q_{2 \to 3} + Q_{3 \to 1} = 1247 \text{ J} + 0 - 902 \text{ J} = \underline{+345 \text{ J}}$

Note: For a cycle $\Delta U = 0$, so by $\Delta U = Q - W$ it must be that $Q_{net} = W_{net}$ for a cycle. We can also check that $\Delta U_{net} = 0$:

$\Delta U_{net} = \Delta U_{1 \to 2} + \Delta U_{2 \to 3} + \Delta U_{3 \to 1} = 1250 \text{ J} - 603 \text{ J} - 644 \text{ J} = 0$ ✓

e) $e = \frac{\text{work output}}{\text{heat energy input}} = \frac{W}{Q_H} = \frac{345 \text{ J}}{1247 \text{ J}} = 0.277 = \underline{27.7\%}$

$e(\text{Carnot}) = 1 - \frac{T_c}{T_H} = 1 - \frac{300 \text{ K}}{600 \text{ K}} = \underline{0.500}$

18-43

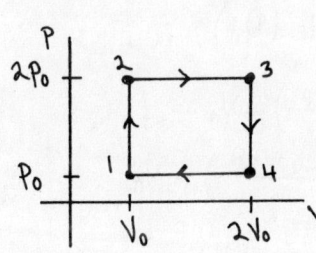

$C_v = 20.76 \text{ J/mol·K}$

For an ideal gas $C_p = C_v + R$.

Calculate Q and W for each process:

process 1→2

$\Delta V = 0 \Rightarrow W = 0$

$\Delta V = 0 \Rightarrow Q = nC_v \Delta T = nC_v(T_2 - T_1)$

But $pV = nRT$ and V constant $\Rightarrow p_1 V = nRT_1$ and $p_2 V = nRT_2$.

Thus $(p_2 - p_1)V = nR(T_2 - T_1) \Rightarrow V\Delta p = nR\Delta T$ (true when V is constant).

Then $Q = nC_v \Delta T = nC_v \frac{V \Delta p}{nR} = \frac{C_v}{R} V \Delta p = \frac{C_v}{R} V_0 (2p_0 - p_0) = \frac{C_v}{R} p_0 V_0$.

($Q > 0 \Rightarrow$ heat is absorbed by the gas.)

process 2→3

$\Delta p = 0 \Rightarrow W = p\Delta V = p(V_3 - V_2) = 2p_0(2V_0 - V_0) = 2p_0 V_0$ (W is positive since V increases.)

$\Delta p = 0 \Rightarrow Q = nC_p \Delta T = nC_p(T_3 - T_2)$

But $pV = nRT$ and p constant $\Rightarrow pV_2 = nRT_2$ and $pV_3 = nRT_3$.

So $p(V_3 - V_2) = nR(T_3 - T_2) \Rightarrow p\Delta V = nR\Delta T$ (true when p is constant)

18-43 (cont)

Thus $Q = nC_p \dfrac{p\Delta V}{nR} = \dfrac{C_p}{R} p\Delta V = \dfrac{C_p}{R}(2p_0)(2V_0 - V_0) = 2\dfrac{C_p}{R} p_0 V_0$.
($Q > 0 \Rightarrow$ heat is absorbed by the gas.)

process 3→4
$\Delta V = 0 \Rightarrow W = 0$.

$\Delta V = 0 \Rightarrow Q = nC_v \Delta T = nC_v \dfrac{V\Delta p}{nR} = \dfrac{C_v}{R}(2V_0)(p - 2p_0) = -2\dfrac{C_v}{R} p_0 V_0$
($Q < 0 \Rightarrow$ heat rejected by the gas.)

process 4→1
$\Delta p = 0 \Rightarrow W = p\Delta V = p(V_1 - V_4) = p_0(V_0 - 2V_0) = -p_0 V_0$
(W is negative since V decreases)

$\Delta p = 0 \Rightarrow Q = nC_p \Delta T = nC_p \dfrac{p\Delta V}{nR} = \dfrac{C_p}{R} p\Delta V = \dfrac{C_p}{R} p_0(V_0 - 2V_0) = -\dfrac{C_p}{R} p_0 V_0$
($Q < 0 \Rightarrow$ heat is rejected by the gas.)

total work performed by the gas during the cycle:
$W_{tot} = W_{1\to 2} + W_{2\to 3} + W_{3\to 4} = 0 + 2p_0 V_0 + 0 - p_0 V_0 = p_0 V_0$
(Note that W_{tot} equals the area enclosed by the cycle in the pV-diagram.)

total heat absorbed by the gas during the cycle (Q_H):
Heat is absorbed in processes 1→2 and 2→3.
$Q_H = Q_{1\to 2} + Q_{2\to 3} = \dfrac{C_v}{R} p_0 V_0 + 2\dfrac{C_p}{R} p_0 V_0 = \left(\dfrac{C_v + 2C_p}{R}\right) p_0 V_0$
But $C_p = C_v + R \Rightarrow Q_H = \dfrac{C_v + 2(C_v + R)}{R} p_0 V_0 = \left(\dfrac{3C_v + 2R}{R}\right) p_0 V_0$

total heat rejected by the gas during the cycle (Q_c):
Heat is rejected in processes 3→4 and 4→1.
$Q_c = Q_{3\to 4} + Q_{4\to 1} = -2\dfrac{C_v}{R} p_0 V_0 - \dfrac{C_p}{R} p_0 V_0 = -\left(\dfrac{2C_v + C_p}{R}\right) p_0 V_0$
$C_p = C_v + R \Rightarrow Q_c = -\left(\dfrac{2C_v + C_v + R}{R}\right) p_0 V_0 = -\left(\dfrac{3C_v + R}{R}\right) p_0 V_0$

As a check on the calculations note that
$Q_c + Q_H = -\left(\dfrac{3C_v + R}{R}\right) p_0 V_0 + \left(\dfrac{3C_v + 2R}{R}\right) p_0 V_0 = p_0 V_0 = W$, as it should.

efficiency

$e = \dfrac{W}{Q_H} = \dfrac{p_0 V_0}{\left(\dfrac{3C_v + 2R}{R}\right)(p_0 V_0)} = \dfrac{R}{3C_v + 2R}$

$e = \dfrac{8.3145 \text{ J/mol·K}}{3(20.76 \text{ J/mol·K}) + 2(8.3145 \text{ J/mol·K})} = 0.105 = \underline{10.5\%}$

18-45

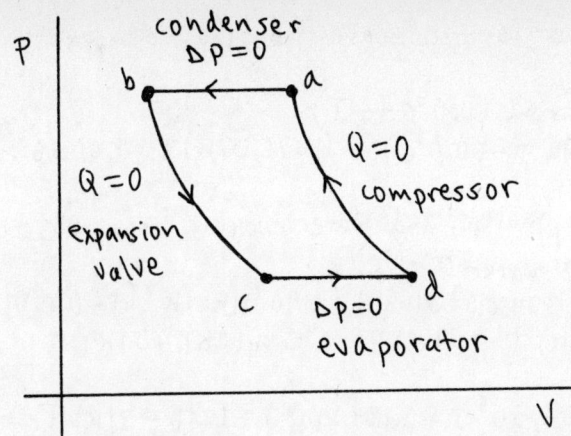

a) process c→d
$$\Delta U = U_d - U_c = 1657 \times 10^3 \text{ J} - 1005 \times 10^3 \text{ J} = 6.52 \times 10^5 \text{ J}$$

$$W = \int_{V_c}^{V_d} p\, dV = p\Delta V \text{ (since is constant pressure process)}$$

$$W = (363 \times 10^3 \text{ Pa})(0.4513 \text{ m}^3 - 0.2202 \text{ m}^3) = +8.389 \times 10^4 \text{ J}$$ (positive since process is an expansion)

$$\Delta U = Q - W \Rightarrow Q = \Delta U + W = 6.52 \times 10^5 \text{ J} + 8.389 \times 10^4 \text{ J} = \underline{7.36 \times 10^5 \text{ J}}$$
(Q positive ⇒ heat goes into the coolant)

b) process a→b
$$\Delta U = U_b - U_a = 1171 \times 10^3 \text{ J} - 1969 \times 10^3 \text{ J} = -7.98 \times 10^5 \text{ J}$$

$$W = p\Delta V = (2305 \times 10^3 \text{ Pa})(0.00946 \text{ m}^3 - 0.0682 \text{ m}^3) = -1.354 \times 10^5 \text{ J}$$
(negative since $\Delta V < 0$ for the process)

$$Q = \Delta U + W = -7.98 \times 10^5 \text{ J} - 1.354 \times 10^5 \text{ J} = \underline{-9.33 \times 10^5 \text{ J}}$$
(negative ⇒ heat comes out of coolant)

c) The coolant cannot be treated as an ideal gas, so we can't calculate W for the adiabatic processes. But $\Delta U = 0$ (for cycle) ⇒ $W_{net} = Q_{net}$

$Q = 0$ for the two adiabatic processes, so
$Q_{net} = Q_{cd} + Q_{ab} = 7.36 \times 10^5 \text{ J} - 9.33 \times 10^5 \text{ J} = -1.97 \times 10^5 \text{ J}$
⇒ $W_{net} = \underline{-1.97 \times 10^5 \text{ J}}$ (negative since work is done on the coolant (working substance))

d) $K = \dfrac{Q_c}{|W|} = \dfrac{+7.36 \times 10^5 \text{ J}}{1.97 \times 10^5 \text{ J}} = \underline{3.74}$

18-53

First use the methods of Chapter 15 to calculate the final temperature T of the system:

0.400 kg of water (cools from 60.0°C to T)
$$Q = mc\Delta T = (0.400 \text{ kg})(4190 \text{ J/kg} \cdot \text{K})(T - 60.0°C) = (1676 \text{ J/K})T - 1.0056 \times 10^5 \text{ J}$$

0.0600 kg of ice (warms to 0°C, melts, and water warms from 0°C to T)
$$Q = mc_{ice}(0°C - (-15.0°C)) + mL_f + mc_{water}(T - 0°C)$$
$$Q = 0.0600 \text{ kg}[(2000 \text{ J/kg} \cdot \text{K})(15.0C°) + 334 \times 10^3 \text{ J/kg} + (4190 \text{ J/kg} \cdot \text{K})(T - 0°C)]$$
$$Q = 1800 \text{ J} + 2.004 \times 10^4 \text{ J} + (251.4 \text{ J/K})T = 2.184 \times 10^4 \text{ J} + (251.4 \text{ J/K})T$$

$$Q_{system} = 0 \Rightarrow (1676 \text{ J/K})T - 1.0056 \times 10^5 \text{ J} + 2.184 \times 10^4 \text{ J} + (251.4 \text{ J/K})T = 0$$
$$(1.927 \times 10^3 \text{ J/K})T = 7.872 \times 10^4 \text{ J}$$

$$T = \frac{7.872 \times 10^4 \text{ J}}{1.927 \times 10^3 \text{ J/K}} = 40.85°C = 314.0 \text{ K}$$

Now we can calculate the entropy changes:

ice

The process takes ice at $-15.0°C$ and produces water at $40.85°C$. Calculate ΔS for a reversible process between these two states, in which heat is added very slowly. ΔS is path independent, so ΔS for a reversible process is the same as ΔS for the actual (irreversible process) as long as the initial and final states are the same.

$$\Delta S = \int_1^2 \frac{dQ}{T}, \text{ where } T \text{ must be in Kelvins}$$

For a temperature change $dQ = mc \, dT \Rightarrow \Delta S = \int_{T_1}^{T_2} \frac{mc \, dT}{T} = mc \ln\left(\frac{T_2}{T_1}\right)$.

For a phase change, since it occurs at constant T,
$$\Delta S = \int_1^2 \frac{dQ}{T} = \frac{Q}{T} = \frac{\pm mL}{T}$$

Therefore $\Delta S_{ice} = mc_{ice} \ln\left(\frac{273 \text{ K}}{258 \text{ K}}\right) + \frac{mL_f}{273 \text{ K}} + mc_{water} \ln\left(\frac{314 \text{ K}}{273 \text{ K}}\right)$

$\Delta S_{ice} = (0.0600 \text{ kg})\left[(2000 \text{ J/kg} \cdot \text{K}) \ln\left(\frac{273 \text{ K}}{258 \text{ K}}\right) + \frac{334 \times 10^3 \text{ J/kg}}{273 \text{ K}} + (4190 \text{ J/kg} \cdot \text{K}) \ln\left(\frac{314 \text{ K}}{273 \text{ K}}\right)\right]$

$\Delta S_{ice} = 6.78 \text{ J/K} + 73.41 \text{ J/K} + 35.18 \text{ J/K} = 115.4 \text{ J/K}$

water

$$\Delta S_{water} = mc \ln\left(\frac{T_2}{T_1}\right) = (0.400 \text{ kg})(4190 \text{ J/kg} \cdot \text{K}) \ln\left(\frac{314 \text{ K}}{333 \text{ K}}\right) = -98.5 \text{ J/K}$$

For the system, $\Delta S = \Delta S_{ice} + \Delta S_{water} = 115.4 \text{ J/K} - 98.5 \text{ J/K} = \underline{+17 \text{ J/K}}$

Note: Our calculation gives $\Delta S > 0$, as it must for an irreversible process of an isolated system.

CHAPTER 19

Exercises 3, 5, 9, 13, 17, 19, 21, 23, 25, 29, 31

Problems 33, 35, 43

Exercises

19-3

a) $v = 344$ m/s
$v = \lambda f \Rightarrow \lambda = \dfrac{v}{f}$

$f = 20.0$ Hz $\Rightarrow \lambda = \dfrac{344 \text{ m/s}}{20.0 \text{ Hz}} = \underline{17.2 \text{ m}}$

$f = 20{,}000$ Hz $\Rightarrow \lambda = \dfrac{344 \text{ m/s}}{20{,}000 \text{ Hz}} = \underline{0.0172 \text{ m}}$

b) $v = 1480$ m/s

$f = 20.0$ Hz $\Rightarrow \lambda = \dfrac{1480 \text{ m/s}}{20.0 \text{ Hz}} = \underline{74.0 \text{ m}}$

$f = 20{,}000$ Hz $\Rightarrow \lambda = \dfrac{1480 \text{ m/s}}{20{,}000 \text{ Hz}} = \underline{0.074 \text{ m}}$

19-5

a)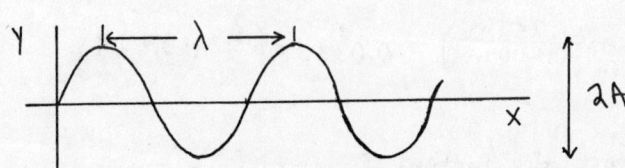

The distance between crests is the wavelength; $\lambda = 7.0$ m.
Time between largest upward displacement to largest downward displacement is one-half the period $\Rightarrow \dfrac{T}{2} = 2.0$ s $\Rightarrow T = 4.0$ s.
$f = \dfrac{1}{T} = \dfrac{1}{4.0 \text{ s}} = 0.250$ Hz

Then $v = f\lambda = (0.250 \text{ Hz})(7.0 \text{ m}) = \underline{1.75 \text{ m/s}}$.

b) The distance between the highest and lowest points is twice the amplitude, so $2A = 0.600$ m $\Rightarrow A = \underline{0.300 \text{ m}}$.

c) v unchanged but A becomes 0.200 m.

d) No. Expect longitudinal as well as transverse surface waves.

19-9

$v = 12.0$ m/s, $A = 0.0500$ m, $\lambda = 0.400$ m

a) $v = f\lambda \Rightarrow f = \dfrac{v}{\lambda} = \dfrac{12.0 \text{ m/s}}{0.400 \text{ m}} = \underline{30.0 \text{ Hz}}$

$T = \dfrac{1}{f} = \dfrac{1}{30.0 \text{ Hz}} = \underline{0.0333 \text{ s}}$

247

19-9 (cont)

$$k = \frac{2\pi}{\lambda} = \frac{2\pi \text{ rad}}{0.400 \text{ m}} = \underline{15.7 \text{ rad/m}}$$

b) For a wave traveling in the +x-direction $y(x,t) = A \sin 2\pi \left(\frac{t}{T} - \frac{x}{\lambda}\right)$ (Eq. 19-4)
At $x=0$, $y(0,t) = A \sin 2\pi \left(\frac{t}{T}\right)$, so $y=0$ at $t=0$ and y is positive for t slightly greater than zero. This equation describes the wave specified in the problem.

Substitute in numerical values $\Rightarrow y(x,t) = (0.0500 \text{ m}) \sin 2\pi \left(\frac{t}{0.0333 \text{ s}} - \frac{x}{0.400 \text{ m}}\right)$.

Or, $y(x,t) = (0.0500 \text{ m}) \sin\left((188 \text{ rad/s}) t - (15.7 \text{ m}^{-1}) x\right)$.

c) From part (b) $y = (0.0500 \text{ m}) \sin 2\pi \left(\frac{t}{0.0333 \text{ s}} - \frac{x}{0.400 \text{ m}}\right)$.

Plug in $x = 0.250 \text{ m}$ and $t = 0.150 \text{ s}$

$\Rightarrow y = (0.0500 \text{ m}) \sin 2\pi \left[\frac{0.150 \text{ s}}{0.0333 \text{ s}} - \frac{0.250 \text{ m}}{0.400 \text{ m}}\right] = (0.0500 \text{ m}) \sin[2\pi(3.875)] = -0.0354 \text{ m}$

$y = \underline{-3.54 \text{ cm}}$

d) In part (c) $t = 0.150 \text{ s}$.

$y = 0$ means $\sin 2\pi \left(\frac{t}{T} - \frac{x}{\lambda}\right) = 0$

$\sin \theta = 0$ for $\theta = 0, \pi, 2\pi, \ldots = n\pi$ for $n = 0, 1, 2, \ldots$

$\Rightarrow 2\pi \left(\frac{t}{T} - \frac{x}{\lambda}\right) = n\pi$

$\frac{t}{T} - \frac{x}{\lambda} = \frac{n}{2}$

$t = T\left(\frac{n}{2} + \frac{x}{\lambda}\right) = (0.0333 \text{ s}) \left(\frac{n}{2} + \frac{0.250 \text{ m}}{0.400 \text{ m}}\right) = 0.0333 \text{ s} \left(\frac{n}{2} + 0.625\right)$

n	t
7	0.1375 s (before the instant in part (c))
8	0.1542 s (the first occurrence of $y=0$ after the instant in part (c))

Thus the elapsed time is $0.1542 \text{ s} - 0.1500 \text{ s} = \underline{0.0042 \text{ s}}$.

9-13

a) The maximum y is $\underline{4 \text{ mm}}$ (read from graph).

b) For either x the time for one full cycle is $\underline{0.04 \text{ s}}$; this is the period.

c) Since $y=0$ for $x=0$ and $t=0$ and since the wave is traveling in the +x-direction then $y(x,t) = A \sin\left[2\pi \left(\frac{t}{T} - \frac{x}{\lambda}\right)\right]$.

From the graph, if the wave is traveling in the +x-direction and if $x=0$ and $x=0.090 \text{ m}$ are within one wavelength the peak at $t = 0.01 \text{ s}$ for $x=0$ moves so that it occurs at $t = 0.036 \text{ s}$ (read from graph, so is approximate) for $x = 0.090 \text{ m}$. This peak for $x=0$ is the first peak past $t=0$ so corresponds to the first maximum in $\sin\left[2\pi\left(\frac{t}{T} - \frac{x}{\lambda}\right)\right]$ and hence occurs at $2\pi\left(\frac{t}{T} - \frac{x}{\lambda}\right) = \frac{\pi}{2}$. If this same peak moves to $t_1 = 0.036 \text{ s}$ at $x_1 = 0.090 \text{ m}$, then

$2\pi \left(\frac{t_1}{T} - \frac{x_1}{\lambda}\right) = \frac{\pi}{2}$

9-13 (cont)

Solve for λ: $\frac{t_1}{T} - \frac{x_1}{\lambda} = \frac{1}{4}$

$\frac{x_1}{\lambda} = \frac{t_1}{T} - \frac{1}{4} = \frac{0.036s}{0.040s} - \frac{1}{4} = 0.65$

$\lambda = \frac{x_1}{0.65} = \frac{0.090 \text{ m}}{0.65} = \underline{0.14 \text{ m}}$

Then $v = f\lambda = \frac{\lambda}{T} = \frac{0.14 \text{ m}}{0.040 s} = \underline{3.5 \text{ m/s}}$.

d) If the wave is traveling in the $-x$-direction, then $y(x,t) = A\sin\left(2\pi\left(\frac{t}{T} + \frac{x}{\lambda}\right)\right)$ and the peak at $t = 0.050$ s at $x=0$ corresponds to the peak at $t_1 = 0.036$ s for $x_1 = 0.090$ m. This peak at $x=0$ is the second peak past the origin so corresponds to $2\pi\left(\frac{t}{T} + \frac{x}{\lambda}\right) = \frac{5\pi}{2}$. If this same peak moves to $t_1 = 0.036$ s for $x_1 = 0.090$ m, then $2\pi\left(\frac{t_1}{T} + \frac{x_1}{\lambda}\right) = \frac{5\pi}{2}$.

$\frac{t_1}{T} + \frac{x_1}{\lambda} = \frac{5}{4}$

$\frac{x_1}{\lambda} = \frac{5}{4} - \frac{t_1}{T} = \frac{5}{4} - \frac{0.036s}{0.040s} = 0.35$

$\lambda = \frac{x_1}{0.35} = \frac{0.090 \text{ m}}{0.35} = \underline{0.26 \text{ m}}$

Then $v = \lambda f = \frac{\lambda}{T} = \frac{0.26 \text{ m}}{0.040 s} = \underline{6.5 \text{ m/s}}$.

e) No. Wouldn't know which point in wave at $x=0$ moved to which point at $x = 0.090$ m.

19-17

a) The tension F in the string is the weight of the hanging mass:
$F = mg = (1.20 \text{ kg})(9.80 \text{ m/s}^2) = 11.8 \text{ N}$
$v = \sqrt{\frac{F}{\mu}} = \sqrt{\frac{11.8 \text{ N}}{0.0180 \text{ kg/m}}} = \underline{25.6 \text{ m/s}}$

b) $v = f\lambda \Rightarrow \lambda = \frac{v}{f} = \frac{25.6 \text{ m/s}}{220 \text{ Hz}} = \underline{0.116 \text{ m}}$

c) $v = \sqrt{\frac{F}{\mu}}$, where $F = mg$. Doubling m increases v by a factor of $\sqrt{2}$.

$\lambda = \frac{v}{f}$. f remains 220 Hz and v increases by a factor of $\sqrt{2}$, so λ increases by a factor of $\sqrt{2}$.

19-19

Calculate the wave speed for transverse waves on the rubber tube. The tension in the tube is equal to the weight of the suspended mass:
$F = mg = (5.0 \text{ kg})(9.80 \text{ m/s}^2) = 49.0 \text{ N}$.
The mass per unit length for the tube is $\mu = \frac{m}{L} = \frac{0.900 \text{ kg}}{12.0 \text{ m}} = 0.0750 \text{ kg/m}$.
$v = \sqrt{\frac{F}{\mu}} = \sqrt{\frac{49.0 \text{ N}}{0.0750 \text{ kg/m}}} = 25.6 \text{ m/s}$

The time is the distance traveled divided by the speed $\Rightarrow t = \frac{12.0 \text{ m}}{25.6 \text{ m/s}} = \underline{0.469 \text{ s}}$

19-21

Calculate the time it takes each sound wave to travel the $L = 80.0\,m$ length of the pipe:

wave in air: $t = \dfrac{80.0\,m}{344\,m/s} = 0.2326\,s$

wave in the metal: $v = \sqrt{\dfrac{Y}{\rho}} = \sqrt{\dfrac{1.10 \times 10^{11}\,Pa}{8900\,kg/m^3}} = 3.52 \times 10^3\,m/s$

$t = \dfrac{80.0\,m}{3.52 \times 10^3\,m/s} = 0.0228\,s$

The time interval between the two sounds is $\Delta t = 0.2326\,s - 0.0228\,s = \underline{0.210\,s}$

19-23

Calculate the wave speed: $v = f\lambda = (250\,Hz)(8.00\,m) = 2000\,m/s$.

$v = \sqrt{\dfrac{B}{\rho}}$ (Eq. 19-21) $\Rightarrow B = \rho v^2 = (900\,kg/m^3)(2000\,m/s)^2 = \underline{3.60 \times 10^9\,Pa}$

19-25

The speed of sound v at the altitude of the plane satisfies the equation
$850\,km/h = 0.80\,v$

$v = \dfrac{850\,km/h}{0.80} = 1062.5\,km/h \left(\dfrac{10^3\,m}{1\,km}\right)\left(\dfrac{1\,h}{3600\,s}\right) = 295.1\,m/s$

If the air is assumed to be an ideal gas, then v is related to T by Eq. (19-27):

$v = \sqrt{\dfrac{\gamma RT}{M}} \Rightarrow T = \dfrac{Mv^2}{\gamma R}$

For air, $\gamma = 1.40$ and $M = 28.8 \times 10^{-3}\,kg/mol$ (Example 19-7).

Then $T = \dfrac{(28.8 \times 10^{-3}\,kg/mol)(295.1\,m/s)^2}{1.40(8.3145\,J/mol\cdot K)} = 215.5\,K = \underline{-58°C}$

19-29

a) Eq. (19-33): $P_{av} = \tfrac{1}{2}\sqrt{\mu F}\,\omega^2 A^2$

$v = \sqrt{\dfrac{F}{\mu}} \Rightarrow \sqrt{\mu} = \dfrac{\sqrt{F}}{v} \Rightarrow P_{av} = \tfrac{1}{2}\dfrac{\sqrt{F}}{v}\sqrt{F}\,\omega^2 A^2 = \tfrac{1}{2}F\dfrac{\omega^2 A^2}{v}$

$\omega = 2\pi f \Rightarrow \dfrac{\omega}{v} = 2\pi \dfrac{f}{v} = \dfrac{2\pi}{\lambda} = k$

$\Rightarrow P_{av} = \tfrac{1}{2}Fk\omega A^2$, as was to be shown.

b) Use Eq. (19-33) since it involves just ω, not k: $P_{av} = \tfrac{1}{2}\sqrt{\mu F}\,\omega^2 A^2$.

P_{av}, μ, A constant $\Rightarrow \sqrt{F}\,\omega^2$ is constant, so $\sqrt{F_1}\,\omega_1^2 = \sqrt{F_2}\,\omega_2^2$

$\omega_2 = \omega_1 \left(\dfrac{F_1}{F_2}\right)^{1/4} = \omega_1 \left(\dfrac{F_1}{2F_1}\right)^{1/4} = \omega_1 (2)^{-1/4}$

ω must be changed by a factor of $(2)^{-1/4}$ (decreased)

19-29 (cont)

In the equation derived in part (a), $P_{av} = \frac{1}{2} F k \omega A^2$, if P_{av} and A are constant then $Fk\omega$ must be constant, so $F_1 k_1 \omega_1 = F_2 k_2 \omega_2$.

$$k_2 = k_1 \left(\frac{F_1}{F_2}\right)\left(\frac{\omega_1}{\omega_2}\right) = k_1 \left(\frac{F_1}{2F_1}\right)\left(\frac{\omega_1}{(2^{-1/4})\omega_1}\right) = k_1 \frac{2^{1/4}}{2} = k_1 \left(\frac{2}{16}\right)^{1/4} = (8^{-1/4}) k_1$$

k must be changed by a factor of $8^{-1/4}$ (decreased).

19-31

a) $v = \sqrt{\frac{Y}{\rho}} = \sqrt{\frac{7.0 \times 10^{10} \text{ Pa}}{2.7 \times 10^3 \text{ kg/m}^3}} = 5092 \text{ m/s}$

$v = f\lambda \Rightarrow \lambda = \frac{v}{f} = \frac{5092 \text{ m/s}}{400 \text{ Hz}} = \underline{12.7 \text{ m}}$

b) $I = \frac{P_{av}}{\pi r^2} = \frac{1}{2}\sqrt{\rho Y} \, \omega^2 A^2$ (Eq. 19-35)

$\omega = 2\pi f = 2513 \text{ rad/s}$

$A = \left(\frac{2 P_{av}}{\pi r^2 \omega^2 \sqrt{\rho Y}}\right)^{1/2} = \left(\frac{2(5.50 \times 10^{-6} \text{ W})}{\pi (0.00900 \text{ m})^2 (2513 \text{ rad/s})^2 \sqrt{(2.7 \times 10^3 \text{ kg/m}^3)(7.0 \times 10^{10} \text{ Pa})}}\right)^{1/2} = \underline{2.23 \times 10^{-8} \text{ m}}$

c) From Eq. (19-9), $v_{max} = \omega A = (2513 \text{ rad/s})(2.23 \times 10^{-8} \text{ m}) = \underline{5.61 \times 10^{-5} \text{ m/s}}$

Problems

19-33

$A = 5.00 \times 10^{-3} \text{ m}, \quad \lambda = 3.60 \text{ m}, \quad v = 24.0 \text{ m/s}$

a) $v = f\lambda \Rightarrow f = \frac{v}{\lambda} = \frac{24.0 \text{ m/s}}{3.60 \text{ m}} = \underline{6.67 \text{ Hz}}$

$\omega = 2\pi f = 2\pi (6.67 \text{ Hz}) = \underline{41.9 \text{ rad/s}}$

$k = \frac{2\pi}{\lambda} = \frac{2\pi \text{ rad}}{3.60 \text{ m}} = \underline{1.75 \text{ rad/m}}$

b) Wave traveling to the right $\Rightarrow y(x,t) = A \sin(\omega t - kx)$.
But to have the $x=0$ end of the string move downward just after $t=0$, this equation must be changed to $y(x,t) = -A\sin(\omega t - kx)$.
Put in the numbers $\Rightarrow y(x,t) = -(5.00 \times 10^{-3} \text{ m}) \sin((41.9 \text{ rad/s})t - (1.75 \text{ rad/m}) x)$.

c) left hand end $\Rightarrow x = 0$
Put this value into the equation of part (b) $\Rightarrow y(0,t) = -(5.00 \times 10^{-3} \text{ m}) \sin((41.9 \text{ rad/s})t)$.

d) Put $x = 0.90$ m into the equation of part (b)
$\Rightarrow y(0.90 \text{ m}, t) = -(5.00 \times 10^{-3} \text{ m}) \sin((41.9 \text{ rad/s})t - (1.75 \text{ rad/m})(0.90 \text{ m}))$
$y(0.90 \text{ m}, t) = -(5.00 \times 10^{-3} \text{ m}) \sin((41.9 \text{ rad/s})t - 1.58 \text{ rad})$

$1.58 \text{ rad} = \frac{\pi}{2} \text{ rad}; \quad \sin(\theta - \frac{\pi}{2}) = -\cos\theta$
Thus $y(0.90 \text{ m}, t) = +(5.00 \times 10^{-3} \text{ m}) \cos((41.9 \text{ rad/s})t)$

e) $y = -A \sin(\omega t - kx)$ (part (b))

19-33 (cont)

The transverse velocity is given by
$$v_y = \frac{\partial y}{\partial t} = -A \frac{\partial}{\partial t} \sin(\omega t - kx) = -A\omega \cos(\omega t - kx)$$

The maximum v_y is $A\omega = (5.00 \times 10^{-3} \text{m})(41.9 \text{ rad/s}) = \underline{0.210 \text{ m/s}}$

f) $y(x,t) = -(5.00 \times 10^{-3} \text{m}) \sin((41.9 \text{ rad/s}) t - (1.75 \text{ rad/m}) x)$

$t = 0.0500 \text{ s}$ and $x = 0.90 \text{ m}$

$\Rightarrow y = -(5.00 \times 10^{-3} \text{m}) \sin((41.9 \text{ rad/s})(0.0500 \text{ s}) - (1.75 \text{ rad/m})(0.90 \text{ m})) = \underline{-2.50 \text{ mm}}$

$v_y = -A\omega \cos(\omega t - kx) = -(0.210 \text{ m/s}) \cos((41.9 \text{ rad/s}) t - (1.75 \text{ rad/m}) x)$

$t = 0.0500 \text{ s}$ and $x = 0.90 \text{ m}$

$\Rightarrow v_y = -(0.210 \text{ m/s}) \cos((41.9 \text{ rad/s})(0.0500 \text{ s}) - (1.75 \text{ rad/m})(0.90 \text{ m})) = \underline{-0.182 \text{ m/s}}$

19-35

a) $v = f\lambda = \frac{\lambda}{T}$

$\lambda = vT = (6400 \text{ m/s})(2.0 \text{ s}) = 1.28 \times 10^4 \text{ m} = \underline{12.8 \text{ km}}$

b) $m = \log\left(\frac{A}{T}\right) + B$

If A increases then m increases. According to Eq. (19-33) the average power carried by the wave is proportional to the square of the amplitude, so larger A means more average power and more destruction.

If T increases then m decreases. Larger T means smaller f and smaller ω. Again by Eq. (19-33), P_{av} is proportional to ω^2 so larger T means smaller P_{av} and less destruction.

c) $m = \log\left(\frac{A}{T}\right) + B$

A is the amplitude in units of μm, so A = 10. T is the period in seconds, so T = 2.0

Then $m = \log\left(\frac{10}{2.0}\right) + 6.8 = 7.5$, a bit larger than for the Loma Prieta earthquake.

19-43

The transverse speed is given by Eq. (19-13): $v_t = \sqrt{\frac{F}{\mu}}$.

The longitudinal wave speed is given by Eq. (19-22): $v_\ell = \sqrt{\frac{Y}{\rho}}$.

$v_\ell = 20 v_t \Rightarrow \sqrt{\frac{Y}{\rho}} = 20 \sqrt{\frac{F}{\mu}} \Rightarrow \boxed{\frac{Y}{\rho} = 400 \frac{F}{\mu}}$

$\mu = \frac{m}{L}, \rho = \frac{m}{V} = \frac{m}{LA}$ where L is the length of the wire and A is its cross section area. Thus $\rho = \frac{\mu}{A} \Rightarrow \mu = A\rho$

Use this in the above equation $\Rightarrow \frac{Y}{\rho} = 400 \frac{F}{A\rho} \Rightarrow \frac{F}{A} = \frac{Y}{400}$.

CHAPTER 20

Exercises 5, 7, 11, 13, 17, 19, 25, 27

Problems 29, 31, 33, 35, 39, 41

Exercises

20-5

Eq. (20-1): $y = (A_{sw} \sin kx) \cos \omega t$

a) node $\Rightarrow y = 0$ for all $t \Rightarrow \sin kx = 0 \Rightarrow kx = n\pi,\ n = 0, 1, 2, \ldots$

$x = \dfrac{n\pi}{k} = \dfrac{n\pi}{1.25\pi\ \text{rad/m}} = (0.80\text{m})n$

x can't be larger than 4.00 m, the length of the wire.

n	x		n	x
0	0 (the left-hand end)		3	2.40 m
1	0.80 m		4	3.20 m
2	1.60 m		5	4.00 m (the right-hand end of the wire)

b) antinode $\Rightarrow \sin kx = \pm 1$, so y will have maximum amplitude

$\sin kx = \pm 1 \Rightarrow kx = \dfrac{n\pi}{2},\ n = 1, 3, 5, \ldots$

$x = \dfrac{n\pi}{2k} = \dfrac{n\pi}{2(1.25\pi\ \text{rad/m})} = (0.40\text{m})n$

n	x		n	x
1	0.40 m		7	2.80 m
3	1.20 m		9	3.60 m
5	2.00 m			

20-7

a) $y(x,t) = (A_{sw} \sin kx) \cos \omega t$ (Eq. 20-1)

$A = 2.50$ cm

$T = 0.500$ s, $f = \dfrac{1}{T} = \dfrac{1}{0.500\text{s}} = 2.00\text{Hz}$, $\omega = 2\pi f = 4.00\pi$ rad/s

antinode to antinode distance is $\dfrac{\lambda}{2} = 12.0$ cm, so $\lambda = 24.0$ cm

$k = \dfrac{2\pi}{\lambda} = \dfrac{2\pi}{24.0\text{cm}} = \dfrac{\pi}{12.0\text{cm}}$

$y(x,t) = (2.50\text{cm}) \sin\left(\dfrac{\pi x}{12.0\text{cm}}\right) \cos\left((4\pi\ \text{s}^{-1})t\right)$

b) $v = f\lambda = (2.00\text{Hz})(0.240\text{m}) = \underline{0.480\ \text{m/s}}$

c) The amplitude of the simple harmonic motion at a particular x is $A_{sw} \sin kx = (2.50\text{cm}) \sin\left(\dfrac{\pi x}{12.0\text{cm}}\right)$, where x is measured from the node at the fixed end of the string.

20-7 (cont)

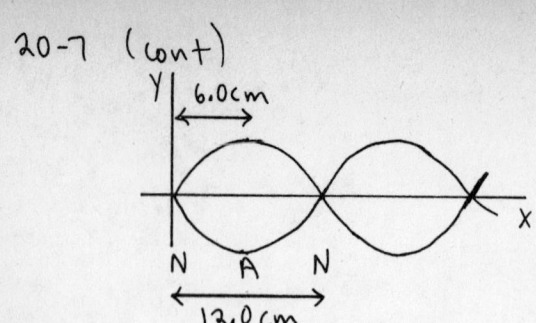

3.0 cm to the right of the first antinode
$\Rightarrow x = 6.0 \text{ cm} + 3.0 \text{ cm} = 9.0 \text{ cm}$

$x = 9.0 \text{ cm} \Rightarrow A_{SW} \sin kx = (2.50 \text{ cm}) \sin\left(\frac{\pi (9.0 \text{ cm})}{12.0 \text{ cm}}\right)$
$= \underline{1.77 \text{ cm}}$

20-11

a) fundamental

$f = 25.0 \text{ Hz}$
From the sketch, $\frac{\lambda}{2} = L \Rightarrow \lambda = 2L = 1.60 \text{ m}$
$v = f\lambda = (25.0 \text{ Hz})(1.60 \text{ m}) = \underline{40.0 \text{ m/s}}$

b) The tension is related to the wave speed by Eq. (19-13)
$v = \sqrt{\frac{F}{\mu}} \Rightarrow F = \mu v^2$

$\mu = \frac{m}{L} = \frac{0.0500 \text{ kg}}{0.800 \text{ m}} = 0.0625 \text{ kg/m}$

$F = \mu v^2 = (0.0625 \text{ kg/m})(40.0 \text{ m/s})^2 = \underline{100 \text{ N}}$

20-13

a) Assume that in each case the string vibrates in its fundamental mode. $f = 440 \text{ Hz}$ when a length $L = 0.600 \text{ m}$ vibrates; use this information to calculate the speed v of waves on the string.
fundamental $\Rightarrow \frac{\lambda}{2} = L \Rightarrow \lambda = 2L = 2(0.600 \text{ m}) = 1.20 \text{ m}$.
Then $v = f\lambda = (440 \text{ Hz})(1.20 \text{ m}) = 528 \text{ m/s}$.

Now find the length $L = x$ of the string that makes $f = 494 \text{ Hz}$.
$\lambda = \frac{v}{f} = \frac{528 \text{ m/s}}{494 \text{ Hz}} = 1.069 \text{ m}$

$L = \frac{\lambda}{2} = 0.534 \text{ m}$, so $x = 0.534 \text{ m} = \underline{53.4 \text{ cm}}$

b) No retuning $\Rightarrow$ same wave speed as in part (a). Find the length of vibrating string needed to produce $f = 294 \text{ Hz}$.
$\lambda = \frac{v}{f} = \frac{528 \text{ m/s}}{294 \text{ Hz}} = 1.80 \text{ m}$

$L = \frac{\lambda}{2} = 0.900 \text{ m}$; string is shorter than this. No, not possible.

20-17

a) The placement of the displacement nodes and antinodes along the pipe is as sketched below. The open end is a displacement antinode, and the closed end is a displacement node.

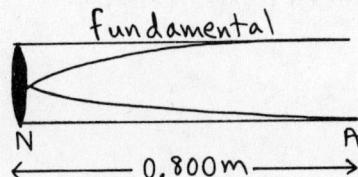
fundamental
N ←—— 0.800m ——→ A

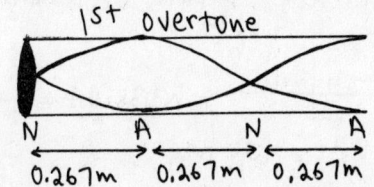
1st overtone
0.267m 0.267m 0.267m

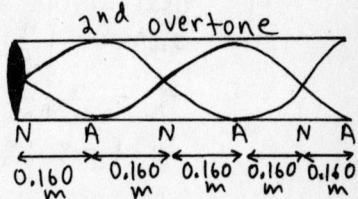
2nd overtone
0.160m 0.160m 0.160m 0.160m 0.160m

Location of the displacement antinodes (A), measured from the closed end:
fundamental   0.800 m
1st overtone   0.267 m, 0.800 m
2nd overtone   0.160 m, 0.480 m, 0.800 m

b) The pressure antinodes correspond to the displacement nodes shown in the sketch. Location of the pressure antinodes (displacement nodes (N)) measured from closed end:
fundamental   0
1st overtone   0, 0.534 m
2nd overtone   0, 0.320 m, 0.640 m

20-19

a) open at both ends ⇒ displacement antinode at each end

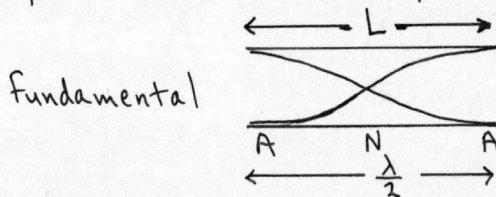
fundamental

$\frac{\lambda}{2} = L \Rightarrow \lambda = 2L$

$f = \frac{v}{\lambda} = \frac{v}{2L} = \frac{344 \text{ m/s}}{2(4.88 \text{ m})} = \underline{35.2 \text{ Hz}}$

b) closed at one end ⇒ displacement node at closed end

fundamental

$\frac{\lambda}{4} = L \Rightarrow \lambda = 4L$

$f = \frac{v}{\lambda} = \frac{v}{4L} = \frac{344 \text{ m/s}}{4(4.88 \text{ m})} = \underline{17.6 \text{ Hz}}$

20-25

A   B   Q
←— 2.00m —→←1.00m→

a) Path difference from points A and B to point Q is 3.00 m − 1.00 m = 2.00 m.
Reinforcement (constructive interference) ⇒ path difference = $n\lambda$, $n = 1, 2, 3, \ldots$

$2.00 \text{ m} = n\lambda \Rightarrow \lambda = \frac{2.00 \text{ m}}{n}$

20-25 (cont)

$$f = \frac{v}{\lambda} = \frac{nv}{2.00\,m} = \frac{n(344\,m/s)}{2.00\,m} = n(172\,Hz), \quad n=1,2,3,\ldots$$
$$\Rightarrow f = 172\,Hz,\ 344\,Hz,\ 516\,Hz,\ \ldots$$

b) Destructive interference $\Rightarrow$ path difference $= \left(\frac{n}{2}\right)\lambda, \quad n=1,3,5,\ldots$
$$2.00\,m = \left(\frac{n}{2}\right)\lambda \Rightarrow \lambda = \frac{4.00\,m}{n}$$
$$f = \frac{v}{\lambda} = \frac{nv}{4.00\,m} = \frac{n(344\,m/s)}{4.00\,m} = n(86.0\,Hz), \quad n=1,3,5,\ldots$$
$$\Rightarrow f = 86\,Hz,\ 258\,Hz,\ 430\,Hz,\ \ldots$$

20-27

a)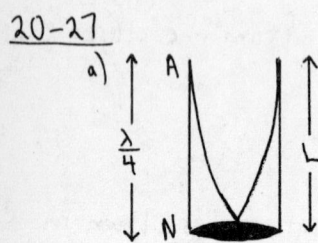

$$\frac{\lambda}{4} = L$$
$$\lambda = 4L = 4(0.12\,m) = 0.48\,m$$
$$f = \frac{v}{\lambda} = \frac{344\,m/s}{0.48\,m} = \underline{717\,Hz}$$

b) Now the length L of the air column becomes $\frac{1}{2}(0.12\,m) = 0.060\,m$ and $\lambda = 4L = 0.24\,m$. $f = \frac{v}{\lambda} = \frac{344\,m/s}{0.24\,m} = \underline{1430\,Hz}$

Problems

20-29

a) For reflection from a free end of a string the reflected wave is <u>not</u> inverted, so $y(x,t) = y_1(x,t) + y_2(x,t)$, where
$$y_1(x,t) = A\sin(\omega t + kx) \quad \text{(traveling to left)}$$
$$y_2(x,t) = A\sin(\omega t - kx) \quad \text{(traveling to right)}$$

$$y(x,t) = A[\sin(\omega t + kx) + \sin(\omega t - kx)]$$
Apply the trig identity $\sin(a \pm b) = \sin a \cos b \mp \cos a \sin b$ with $a = \omega t$ and $b = kx$
$$\Rightarrow \sin(\omega t + kx) = \sin\omega t \cos kx - \cos\omega t \sin kx$$
$$\sin(\omega t - kx) = \sin\omega t \cos kx + \cos\omega t \sin kx$$
and $y(x,t) = (2A\cos kx)\sin\omega t$ (the other two terms cancel)

b) $x = 0 \Rightarrow \cos kx = 1$ and $y(x,t) = 2A\sin\omega t$. The amplitude of the simple harmonic motion at $x=0$ is $2A$, which is the maximum for this standing wave, so $x=0$ is an antinode.

c) $y_{max} = \underline{2A}$ from part (b)
$$v_y = \frac{\partial y}{\partial t} = \frac{\partial}{\partial t}[(2A\cos kx)\sin\omega t] = 2A\cos kx \frac{\partial \sin\omega t}{\partial t} = 2A\omega \cos kx \cos\omega t$$
At $x=0$, $v_y = 2A\omega \sin\omega t$ and $(v_y)_{max} = \underline{2A\omega}$

20-29 (cont)

$$a_y = \frac{\partial^2 y}{\partial t^2} = \frac{\partial v_y}{\partial t} = 2A\omega \cos kx \frac{\partial \cos \omega t}{\partial t} = -2A\omega^2 \cos kx \sin \omega t$$

At $x=0$, $a_y = -2A\omega^2 \sin \omega t$ and $(a_y)_{max} = \underline{2A\omega^2}$

20-31

a) Plank oscillates with maximum amplitude at its center ⇒ it is oscillating in its fundamental mode.

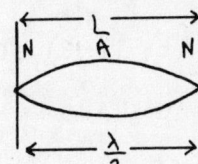

$\frac{\lambda}{2} = L \Rightarrow \lambda = 2L = 2(5.00\text{ m}) = 10.0\text{ m}$

Student jumps upward two times per second ⇒ $f = 2.00$ Hz

$v = f\lambda = (2.00\text{ Hz})(10.0\text{ m}) = \underline{20.0\text{ m/s}}$

b) There now must be an antinode 1.25 m (a distance of $\frac{L}{4}$) from one end. The nodal structure for the standing wave must be:

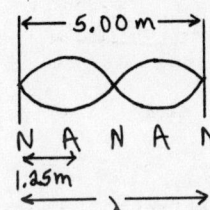

v depends on the properties of the plank, so is the same as in part (a). Now $\lambda = L = 5.00$ m.

$f = \frac{v}{\lambda} = \frac{20.0\text{ m/s}}{5.00\text{ m}} = \underline{4.00\text{ Hz}}$

The student now has to jump four times each second.

20-33

a) For an open pipe the successive harmonics are $f_n = nf_1$, $n = 1, 3, 5, ...$
For a stopped pipe the successive harmonics are $f_n = nf_1$, $n = 1, 2, 3, ...$

If the pipe is open and the harmonics are successive, then
$f_n = 400$ Hz and $f_{n+1} = 560$ Hz
$nf_1 = 400$ Hz $(n+1)f_1 = 560$ Hz

Subtract first equation from the second
$(n+1)f_1 - nf_1 = 560\text{ Hz} - 400\text{ Hz}$
$f_1 = 260$ Hz

Then $n = \frac{400\text{ Hz}}{f_1} = \frac{400\text{ Hz}}{260\text{ Hz}} = 1.54$.

But n must be an integer, so the pipe cannot be open.

If the pipe is stopped and the harmonics are successive, then
$f_n = 400$ Hz and $f_{n+2} = 560$ Hz (in this case successive harmonics differ in n by 2)
$nf_1 = 400$ Hz $(n+2)f_1 = 560$ Hz

Subtract ⇒ $2f_1 = 160$ Hz
$f_1 = 80$ Hz

Then $n = \frac{400\text{ Hz}}{f_1} = 5$ so $400\text{ Hz} = 5f_1$ and $560\text{ Hz} = 7f_1$.

This solution gives integer n as it should; the pipe is stopped.

20-33 (cont)

b) From part (a) these are the 5th and 7th harmonics.

c) From part (a) $f_1 = 80\,Hz$.

For a stopped pipe $f_1 = \frac{v}{4L} \Rightarrow L = \frac{v}{4f_1} = \frac{344\,m/s}{4(80\,Hz)} = \underline{1.08\,m}$.

20-35

a) The tension F is related to the wave speed by $v = \sqrt{\frac{F}{\mu}}$ (Eq. 19-13), so use the information given to calculate v.

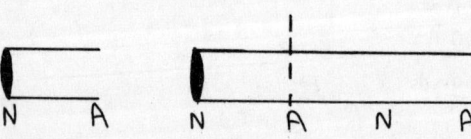

fundamental

$\frac{\lambda}{2} = L \Rightarrow \lambda = 2L = 2(0.600\,m) = 1.20\,m$

$v = f\lambda = (220\,Hz)(1.20\,m) = 264\,m/s$

$\mu = \frac{m}{L} = \frac{1.42 \times 10^{-3}\,kg}{0.600\,m} = 2.367 \times 10^{-3}\,kg/m$

Then $v = \sqrt{\frac{F}{\mu}} \Rightarrow F = \mu v^2 = (2.367 \times 10^{-3}\,kg/m)(264\,m/s)^2 = \underline{165\,N}$.

b) $F = \mu v^2$ and $v = f\lambda \Rightarrow F = \mu f^2 \lambda^2$

μ is a property of the string so is constant.
λ is determined by the length of the string so stays constant.

μ, λ constant $\Rightarrow \frac{F}{f^2} = \mu \lambda^2 =$ constant $\Rightarrow \frac{F_1}{f_1^2} = \frac{F_2}{f_2^2}$

$F_2 = F_1 \left(\frac{f_2}{f_1}\right)^2 = (165\,N)\left(\frac{233\,Hz}{220\,Hz}\right)^2 = 185\,N$

The percent change in F is $\frac{F_2 - F_1}{F_1} = \frac{185\,N - 165\,N}{165\,N} = 0.12 = \underline{12\%}$

20-39

a)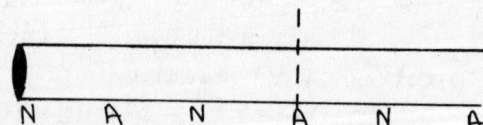

The successive lengths differ by an additional $\frac{\lambda}{2}$ length (the A to A distance)

$\Rightarrow \frac{\lambda}{2} = 55.5\,cm - 18.0\,cm = 93.0\,cm - 55.5\,cm = 37.5\,cm$

$\lambda = 2(37.5\,cm) = 75.0\,cm = 0.750\,m$

$v = f\lambda = (500\,Hz)(0.750\,m) = \underline{375\,m/s}$

b) $v = \sqrt{\frac{\gamma RT}{M}}$ (Eq. 19-27)

$\gamma = \frac{Mv^2}{RT} = \frac{(28.8 \times 10^{-3}\,kg/mol)(375\,m/s)^2}{(8.3145\,J/mol\cdot K)(350\,K)} = \underline{1.39}$

20-41

The node to node distance is $\lambda/2$, so $\lambda/2 = 0.0677$ m.
$$\lambda = 2(0.0677 \text{ m}) = 0.1354 \text{ m}.$$
$$v = f\lambda = (1000 \text{ Hz})(0.1354 \text{ m}) = 135.4 \text{ m/s}.$$

If we assume that the iodine vapor can be treated as an ideal gas, then $v = \sqrt{\frac{\gamma RT}{M}}$ (Eq. 19-27).

If the vapor is monatomic then $M = 0.127$ kg/mol. For a monatomic ideal gas $\gamma = 1.67$ (Table 17-1). Solve for γ and see if agrees with this value.
$$\gamma = \frac{Mv^2}{RT} = \frac{(0.127 \text{ kg/mol})(135.4 \text{ m/s})^2}{(8.3145 \text{ J/mol·K})(400 \text{ K})} = 0.700$$

This is not the $\gamma = 1.67$ required for a monatomic ideal gas; the vapor is not monatomic.

If the vapor is diatomic then $M = 2(0.127 \text{ kg/mol}) = 0.254$ kg/mol. For a diatomic gas $\gamma = 1.40$. See what the data gives:
$$\gamma = \frac{Mv^2}{RT} = \frac{(0.254 \text{ kg/mol})(135.4 \text{ m/s})^2}{(8.3145 \text{ J/mol·K})(400 \text{ K})} = 1.40$$

This is the correct value for a diatomic ideal gas; the iodine vapor is <u>diatomic</u>.

CHAPTER 21

Exercises 1, 7, 11, 13, 15, 17, 21

Problems 25, 27, 29, 31, 37

Exercises

21-1

$p_{max} = BkA$ (Eq. 21-5)

As computed in Example 21-1 the adiabatic bulk modulus for air is $B = 1.42 \times 10^5$ Pa.

a) $f = 200$ Hz

Need to calculate k: $\lambda = \frac{v}{f}$ and $k = \frac{2\pi}{\lambda} \Rightarrow k = \frac{2\pi f}{v} = \frac{(2\pi \text{ rad})(200 \text{ Hz})}{344 \text{ m/s}} = 3.653$ rad/m.

Then $p_{max} = BkA = (1.42 \times 10^5 \text{ Pa})(3.653 \text{ rad/m})(0.0100 \times 10^{-3} \text{ m}) = \underline{5.19 \text{ Pa}}$.

This is below the pain threshold of 30 Pa.

b) f is larger by a factor of 10 so $k = \frac{2\pi f}{v}$ is larger by a factor of 10, and $p_{max} = BkA$ is larger by a factor of 10. $p_{max} = 51.9$ Pa, above the pain threshold.

c) There is again an increase in f, k, and p_{max} of a factor of 10, so $p_{max} = 519$ Pa, far above the pain threshold.

21-7

a) $\omega = 2\pi f = (2\pi \text{ rad})(300 \text{ Hz}) = 1.885 \times 10^3$ rad/s

$k = \frac{2\pi}{\lambda} = \frac{2\pi f}{v} = \frac{\omega}{v} = \frac{1.885 \times 10^3 \text{ rad/s}}{344 \text{ m/s}} = 5.48$ rad/m

$B = 1.42 \times 10^5$ Pa (Example 21-1)

Then $p_{max} = BkA = (1.42 \times 10^5 \text{ Pa})(5.48 \text{ rad/m})(6.00 \times 10^{-6} \text{ m}) = \underline{4.67 \text{ Pa}}$.

b) Eq. (21-6): $I = \frac{1}{2}\omega BkA^2 = \frac{1}{2}(1.885 \times 10^3 \text{ rad/s})(1.42 \times 10^5 \text{ Pa})(5.48 \text{ rad/m})(6.00 \times 10^{-6} \text{ m})^2$

$I = \underline{0.0264 \text{ W/m}^2}$

c) Eq. (21-11): $\beta = (10 \text{ dB}) \log\left(\frac{I}{I_0}\right)$, with $I_0 = 1 \times 10^{-12}$ W/m².

$\beta = (10 \text{ dB}) \log\left(\frac{0.0264 \text{ W/m}^2}{1 \times 10^{-12} \text{ W/m}^2}\right) = \underline{104 \text{ dB}}$

21-11

Let 1 refer to the mother and 2 to the father.

From Example 21-7, $\beta_2 - \beta_1 = (10 \text{ dB}) \log\left(\frac{I_2}{I_1}\right)$

Eq. (21-10): $\frac{I_1}{I_2} = \frac{r_2^2}{r_1^2} \Rightarrow \frac{I_2}{I_1} = \frac{r_1^2}{r_2^2}$

$\Delta\beta = \beta_2 - \beta_1 = (10 \text{ dB}) \log\left(\frac{I_2}{I_1}\right) = (10 \text{ dB}) \log\left(\frac{r_1}{r_2}\right)^2 = (20 \text{ dB}) \log\left(\frac{r_1}{r_2}\right)$

21-21 (cont)
b)

$\tan\alpha = \dfrac{1200\,m}{v_s t}$

$t = \dfrac{1200\,m}{v_s \tan\alpha}$

From part (a) $\alpha = 23.6°$.

Mach $2.50 \Rightarrow v_s = 2.50\,v = 2.50(344\,m/s) = 860\,m/s$

$t = \dfrac{1200\,m}{(860\,m/s)(\tan 23.6°)} = \underline{3.19\,s}$

Problems

21-25

a) Use the intensity level β to calculate I at this distance.

$\beta = (10\,dB) \log\left(\dfrac{I}{I_0}\right)$

$55.0\,dB = (10\,dB) \log\left(\dfrac{I}{10^{-12}\,W/m^2}\right)$

$\log\left(\dfrac{I}{10^{-12}\,W/m^2}\right) = 5.50 \Rightarrow I = 3.16 \times 10^{-7}\,W/m^2$

Then use Eq. (21-9) to calculate p_{max}:

$I = \dfrac{p_{max}^2}{2\rho v} \Rightarrow p_{max} = \sqrt{2\rho v I}$

From Example 21-3, $\rho = 1.20\,kg/m^3$ for air at $20°C$.

$p_{max} = \sqrt{2\rho v I} = \sqrt{2(1.20\,kg/m^3)(344\,m/s)(3.16\times 10^{-7}\,W/m^2)} = \underline{1.62\times 10^{-2}\,Pa}$

b) Eq. (21-5): $p_{max} = BkA \Rightarrow A = \dfrac{p_{max}}{Bk}$

For air $B = 1.42\times 10^5\,Pa$ (Example 21-1).

$k = \dfrac{2\pi}{\lambda} = \dfrac{2\pi f}{v} = \dfrac{(2\pi\,rad)(440\,Hz)}{344\,m/s} = 8.037\,rad/m$

$A = \dfrac{p_{max}}{Bk} = \dfrac{1.62\times 10^{-2}\,Pa}{(1.42\times 10^5\,Pa)(8.037\,rad/m)} = \underline{1.42\times 10^{-8}\,m}$

c) $\beta_2 - \beta_1 = (10\,dB) \log\left(\dfrac{I_2}{I_1}\right)$ (Example 21-7)

Eq. (21-10): $\dfrac{I_1}{I_2} = \dfrac{r_2^2}{r_1^2} \Rightarrow \dfrac{I_2}{I_1} = \dfrac{r_1^2}{r_2^2}$

$\beta_2 - \beta_1 = (10\,dB) \log\left(\dfrac{r_1}{r_2}\right)^2 = (20\,dB) \log\left(\dfrac{r_1}{r_2}\right)$

Let $\beta_2 = 55.0\,dB$ and $r_2 = 5.00\,m$. Then $\beta_1 = 30.0\,dB \Rightarrow r_1 = ?$.

$55.0\,dB - 30.0\,dB = (20\,dB) \log\left(\dfrac{r_1}{r_2}\right)$

$25.0\,dB = (20\,dB) \log\left(\dfrac{r_1}{r_2}\right)$

$\log\left(\dfrac{r_1}{r_2}\right) = 1.25 \Rightarrow r_1 = 17.8\,r_2 = \underline{89.0\,m}$

21-11 (cont)

$$\Delta \beta = (20\,dB)\log\left(\frac{3.00\,m}{0.30\,m}\right) = \underline{20\,dB}$$

21-13

$f_{beat} = f_1 - f_2$ (Eq. 21-12)
2.6 beats per second $\Rightarrow f_1 - f_2 = 2.6\,Hz$
One frequency is 440.0 Hz $\Rightarrow$ the other is $440.0\,Hz + 2.6\,Hz = \underline{442.6\,Hz}$ or
$440.0\,Hz - 2.6\,Hz = \underline{437.4\,Hz}$.

21-15

Note: When the source is at rest $\lambda = \frac{v}{f_s} = \frac{344\,m/s}{500\,Hz} = 0.688\,m$.

a) Eq. (21-15): $\lambda = \frac{v - v_s}{f_s} = \frac{344\,m/s - 30.0\,m/s}{500\,Hz} = \underline{0.628\,m}$

b) Eq. (21-16): $\lambda = \frac{v + v_s}{f_s} = \frac{344\,m/s + 30.0\,m/s}{500\,Hz} = \underline{0.748\,m}$

c) $f_L = \frac{v}{\lambda}$ (since $v_L = 0$) $\Rightarrow f_L = \frac{344\,m/s}{0.628\,m} = \underline{548\,Hz}$

d) $f_L = \frac{v}{\lambda} = \frac{344\,m/s}{0.748\,m} = \underline{460\,Hz}$

21-17

a) $v_s = -35.0\,m/s$; $v_L = +15.0\,m/s$ (L to S positive)

$f_L = \left(\frac{v + v_L}{v + v_s}\right)f_s = \left(\frac{344\,m/s + 15.0\,m/s}{344\,m/s - 35.0\,m/s}\right)(300\,Hz) = \underline{349\,Hz}$
(Listener and source approaching $\Rightarrow f_L > f_s$.)

b) $v_L = -15.0\,m/s$; $v_s = +35.0\,m/s$ (L to S positive)

$f_L = \left(\frac{v + v_L}{v + v_s}\right)f_s = \left(\frac{344\,m/s - 15.0\,m/s}{344\,m/s + 35.0\,m/s}\right)(300\,Hz) = \underline{260\,Hz}$
(Listener and source moving away from each other $\Rightarrow f_L < f_s$.)

21-21

a) $\sin\alpha = \frac{v}{v_s}$ (Eq. 21-19)
Mach 2.50 $\Rightarrow \frac{v_s}{v} = 2.50$

$\sin\alpha = \frac{1}{2.50} = 0.400 \Rightarrow \underline{\alpha = 23.6°}$

21-27

a) $I = \frac{1}{2} B\omega k A^2$ (Eq. 21-6)

The distance between the minimum and maximum radius is $2A$, so
$A = \frac{12.2\,cm - 11.8\,cm}{2} = 0.2\,cm$.

For air, $B = 1.42 \times 10^5$ Pa (Example 21-1).
$\omega = 2\pi f = 2\pi (800\,Hz) = 5027\,rad/s$
$k = \frac{2\pi}{\lambda} = \frac{2\pi f}{v} = \frac{\omega}{v} = \frac{5027\,rad/s}{344\,m/s} = 14.6\,rad/m$
$I = \frac{1}{2} B\omega k A^2 = \frac{1}{2}(1.42 \times 10^5\,Pa)(5027\,rad/s)(14.6\,rad/m)(0.2 \times 10^{-2}\,m)^2 = \underline{2.08 \times 10^4\,W/m^2}$

b) The intensity I is the average power per unit area. The surface of the sphere has area $4\pi r^2$, so $I = \frac{P_{av}}{4\pi r^2}$. Use $r = \frac{1}{2}(11.8\,cm + 12.2\,cm) = 12.0\,cm = 0.120\,m$, the equilibrium radius of the sphere.

$P_{av} = 4\pi r^2 I = 4\pi (0.120\,m)^2 (2.08 \times 10^4\,W/m^2) = \underline{3.77 \times 10^3\,W}$

c) Eq. (21-10): $\frac{I_1}{I_2} = \frac{r_2^2}{r_1^2}$. Let r_1 be at the surface of the sphere and $r_2 = 10.0\,m$.
$I_2 = I_1 \left(\frac{r_1^2}{r_2^2}\right) = 2.08 \times 10^4\,W/m^2 \left(\frac{0.120\,m}{10.0\,m}\right)^2 = \underline{3.00\,W/m^2}$

Eq. (21-6): $I = \frac{1}{2} B\omega k A^2$
ω and k depend on the frequency and wave speed so don't change with distance from the source.

$\frac{r_2^2}{r_1^2} = \frac{I_1}{I_2} = \frac{\frac{1}{2} B\omega k A_1^2}{\frac{1}{2} B\omega k A_2^2} = \frac{A_1^2}{A_2^2} \Rightarrow \frac{A_1}{A_2} = \frac{r_2}{r_1}$

$A_2 = A_1 \left(\frac{r_1}{r_2}\right) = (0.2\,cm)\left(\frac{0.120\,m}{10.0\,m}\right) = 2.4 \times 10^{-3}\,cm = \underline{2.4 \times 10^{-5}\,m}$

$p_{max} = B k A = (1.42 \times 10^5\,Pa)(14.6\,rad/m)(2.4 \times 10^{-5}\,m) = \underline{49.8\,Pa}$

21-29

a) $\beta = (10\,dB) \log \frac{I}{I_0}$, where $I_0 = 10^{-12}\,W/m^2$.
$I = \frac{p_{max}^2}{2\rho v}$, $I_0 = \frac{p_{max-0}^2}{2\rho v}$ so $\frac{I}{I_0} = \left(\frac{p_{max}}{p_{max-0}}\right)^2$

$\beta = (10\,dB) \log \left(\frac{p_{max}}{p_{max-0}}\right)^2 = (20\,dB) \log \left(\frac{p_{max}}{p_{max-0}}\right)$
(We have used the property of the log that $\log x^n = n \log x$.)

b) $p_{max} = \sqrt{2\rho v I} = \sqrt{2(1.20\,kg/m^3)(344\,m/s)(1 \times 10^{-12}\,W/m^2)} = \underline{2.87 \times 10^{-5}\,Pa}$

c) From Example 21-7, $\beta_2 - \beta_1 = (20\,dB) \log \frac{I_2}{I_1}$
But $\frac{I_2}{I_1} = \left(\frac{p_{max-2}}{p_{max-1}}\right)^2$, so $\beta_2 - \beta_1 = (10\,dB) \log \left(\frac{p_{max-2}}{p_{max-1}}\right)^2 = (20\,dB) \log \left(\frac{p_{max-2}}{p_{max-1}}\right)$

21-29 (cont)

$\beta_2 - \beta_1 = 43\,dB \Rightarrow 43\,dB = (20\,dB)\log\frac{P_{max-2}}{P_{max-1}}$

$2.15 = \log\frac{P_{max-2}}{P_{max-1}}$

$\frac{P_{max-2}}{P_{max-1}} = \underline{141}$

21-31

a) $f = \frac{1}{1.65} = 0.625\,Hz$, $v = 0.40\,m/s$

The crest to crest distance is the wavelength λ, so $\lambda = 0.18\,m$.
Use Eq. (21-15) for the wavelength in front of the moving source:

$\lambda = \frac{v - v_s}{f_s} \Rightarrow v_s = v - \lambda f_s = 0.40\,m/s - (0.18\,m)(0.625\,Hz) = \underline{0.29\,m/s}$

b) Use Eq. (21-16) for the wavelength behind the moving source.

$\lambda = \frac{v + v_s}{f_s} = \frac{0.40\,m/s + 0.29\,m/s}{0.625\,Hz} = \underline{1.1\,m}$

21-37

a) $\lambda = \frac{v}{f} = \frac{1480\,m/s}{25.0 \times 10^3\,Hz} = \underline{0.0592\,m}$

b) The Problem-Solving Strategy 3. on p. 658 describes how to do this problem. The frequency of the directly radiated waves is $f_s = 25,000\,Hz$. The moving whale first plays the role of a moving listener, receiving waves with frequency f_L'. The whale then acts as a moving source, emitting waves with the same frequency $f_s' = f_L'$ with which they were received. Let the speed of the whales be v_w.

whale receives waves

$\leftarrow v_w$ $v_s = 0$ $v_L = -v_w$

f_L' $\xrightarrow{+}_{L\,to\,S}$ f_s $f_L' = f_s\left(\frac{v + v_L}{v + v_s}\right) = f_s\left(\frac{v - v_w}{v}\right)$

whale re-emits the waves

$\leftarrow v_w$ $v_L = 0$ $v_s = +v_w$

$f_s' = f_L'$ $\xrightarrow{+}_{L\,to\,S}$ f_L $f_L = f_s'\left(\frac{v + v_L}{v + v_s}\right) = f_s'\left(\frac{v}{v + v_w}\right)$

But $f_s' = f_L' \Rightarrow f_L = f_s\left(\frac{v - v_w}{v}\right)\left(\frac{v}{v + v_w}\right) = f_s\left(\frac{v - v_w}{v + v_w}\right)$

Then $\Delta f = f_s - f_L = f_s\left(1 - \frac{v - v_w}{v + v_w}\right) = f_s\left(\frac{v + v_w - v + v_w}{v + v_w}\right) = \frac{2 f_s v_w}{v + v_w}$

$\Delta f = \frac{2(2.50 \times 10^4\,Hz)(5.85\,m/s)}{1480\,m/s + 5.85\,m/s} = \underline{197\,Hz}$

(Listener and source moving away from each other $\Rightarrow$ frequency is lowered.)